T0321221

LINEAR ALGEBRAIC GROUPS AND FINITE GROUPS OF LIE TYPE

Originating from a summer school taught by the authors, this concise treatment includes many of the main results in the area. An introductory chapter describes the fundamental results on linear algebraic groups, culminating in the classification of semisimple groups. The second chapter introduces more specialized topics in the subgroup structure of semisimple groups, and describes the classification of the maximal subgroups of the simple algebraic groups. The authors then systematically develop the subgroup structure of finite groups of Lie type as a consequence of the structural results on algebraic groups. This approach will help students to understand the relationship between these two classes of groups.

The book covers many topics that are central to the subject, but missing from existing textbooks. The authors provide numerous instructive exercises and examples for those who are learning the subject as well as more advanced topics for research students working in related areas.

GUNTER MALLE is a Professor in the Department of Mathematics at the University of Kaiserslautern, Germany.

DONNA TESTERMAN is a Professor in the Mathematics Section at the École Polytechnique Fédérale de Lausanne, Switzerland.

CAMBRIDGE STUDIES IN ADVANCED MATHEMATICS

All the titles listed below can be obtained from good booksellers or from Cambridge University Press. For a complete series listing visit: http://www.cambridge.org/mathematics

Linear Algebraic Groups and Finite Groups of Lie Type

GUNTER MALLE
University of Kaiserslautern, Germany

DONNA TESTERMAN
École Polytechnique Fédérale de Lausanne, Switzerland

CAMBRIDGE
UNIVERSITY PRESS

University Printing House, Cambridge CB2 8BS, United Kingdom

Published in the United States of America by Cambridge University Press, New York

Cambridge University Press is part of the University of Cambridge.

It furthers the University's mission by disseminating knowledge in the pursuit of education, learning and research at the highest international levels of excellence.

www.cambridge.org
Information on this title: www.cambridge.org/9781107008540

© G. Malle and D. Testerman 2011

First published 2011

A catalogue record for this publication is available from the British Library

ISBN 978-1-107-00854-0 Hardback

Contents

Preface

These notes grew out of a summer school on "Finite Groups and Related Geometrical Structures" held in Venice from September 5th to September 15th 2007. The aim of the course was to introduce an audience consisting mainly of PhD students and postdoctoral researchers working in finite group theory and neighboring areas to results on the subgroup structure of linear algebraic groups and the related finite groups of Lie type.

As will be seen in Part I, a linear algebraic group is an affine variety which is equipped with a group structure in such a way that the binary group operation and inversion are continuous maps. A connected (irreducible) linear algebraic group has a maximal solvable connected normal subgroup such that the quotient group is a central product of simple algebraic groups, a so-called semisimple algebraic group. Thus, one is led to the study of semisimple groups and connected solvable groups. A connected solvable linear algebraic group is the semidirect product of the normal subgroup consisting of its unipotent elements with an abelian (diagonalizable) subgroup (for example, think of the group of invertible upper triangular matrices). While one cannot expect to classify unipotent groups, remarkably enough this is possible for the semisimple quotient.

The structure theory of semisimple groups was developed in the middle of the last century and culminated in the classification of the semisimple linear algebraic groups defined over an algebraically closed field, a result essentially due to Chevalley, first made available via the *Séminaire sur la classification des groupes de Lie algébriques* at the Ecole Normale Supérieure in Paris, during the period 1956–1958 ([15]). Analogously to the work of Cartan and Killing on the classification of the complex semisimple Lie algebras, Chevalley showed that the semisimple groups are determined up to isomorphism by a set of combinatorial data, based principally upon a root system (as for the semisimple Lie algebras) and a dual root system. Moreover, the set of possible combinatorial data does not depend on the characteristic of the underlying field. Part I of this text is devoted to developing the tools necessary for describing this classification. We have followed the development in two very good texts ([32] and [66]) on linear algebraic groups, and we often refer to these books for the proofs which we have omitted. Our aim is to give the reader a feel for the group-theoretic ingredients of this classification, without going into the details of the underlying algebraic geometric foundations, and then to move on to the material of Parts II and III, which should perhaps be seen as the distinguishing feature of this text.

The 20- to 30-year period following the classification was a productive time in "semisimple" theory, during which many actors, notably Borel, Bruhat, Springer, Steinberg and Tits, played a role in the further study of these groups. The conjugacy classes, endomorphisms, representations, and subgroup structure were among the topics of consideration, with the principal aim of reducing their classification and the description of their structure to combinatorial data related to the root system and the Weyl group of the ambient group. Part II of this book treats some of these subjects. In particular, we describe the Tits BN-pair for a semisimple linear algebraic group and obtain a Levi decomposition for the associated parabolic subgroups; we discuss conjugacy classes of semisimple elements and their centralizers; we describe the parametrization of the irreducible representations of these groups and their automorphism groups. We leave the discussion of the general endomorphisms of simple algebraic groups to Part III, where these are used to construct the finite groups of Lie type.

In the last chapters of Part II, we turn to more recent developments in the theory of semisimple algebraic groups, where we describe the classification of their maximal positive-dimensional subgroups. These results can be seen as an extension of the fundamental work of Dynkin on the maximal subalgebras of the semisimple complex Lie algebras. It is the subject of several very long research articles by Liebeck, Seitz and others and was completed in 2004.

In the course of the classification of finite simple groups, attention turned to the analogues of algebraic groups over finite fields. These so-called finite groups of Lie type were eventually shown to comprise, together with the alternating groups, almost all the finite simple groups. Steinberg found a unified approach to constructing not only the well-known finite classical groups, but also their twisted "Steinberg variations", as well as further seemingly sporadic examples, the Suzuki and Ree groups, as fixed point subgroups of certain endomorphisms of simple linear algebraic groups defined over $\overline{\mathbb{F}_p}$. There does not yet seem to be a generally accepted terminology for such endomorphisms, and we call them "Steinberg endomorphisms" in this text.

Part III is devoted to the definition and study of these finite groups. We first classify the endomorphisms whose fixed point subgroups are finite, following the work of Steinberg, and hence are able to describe the full set of finite groups "of Lie type" thus obtained. The theorem of Lang–Steinberg then provides the necessary machinery for applying the results of Parts I and II to the study of these fixed point subgroups. For example, we give the proof of the existence of a BN-pair for these groups, which allows one to deduce a formula for their order. Furthermore, we study their Sylow subgroups and touch on some other aspects of the subgroup structure. Finally, we return to the question of the maximal subgroups, this time sketching a proof of Aschbacher's reduction theorem for the maximal subgroups of the finite classical groups and indicating how it has been applied (by Kleidman, Liebeck and others) and how it must still be applied if one hopes to determine the maximal subgroups of the finite classical groups. We conclude with a discussion of what is known about the maximal subgroups of the exceptional finite groups of Lie type, including work of Liebeck, Saxl, Seitz and others. We then come full-circle and sketch the proof of a result which enables one to lift certain embeddings of finite groups of Lie type to embeddings of algebraic groups, where one can apply the more complete information of Part II.

The course is not self-contained in several aspects. First, in order to keep the size manageable we assume the reader to be familiar with some basic notions of affine and projective algebraic varieties. In the development of the general theory of algebraic groups we include those proofs of a more group theoretical nature, or which just use the basic notions of connectedness and dimension, and refer to the standard texts for the others which require deeper methods from algebraic geometry, like properties of morphisms, tangent spaces, etc. Secondly, we do not explain the Steinberg presentation of semisimple algebraic groups, although some of its consequences are mentioned and needed in the text. Also, while we have included an appendix with a self-contained development of the basic theory of root systems and Weyl groups, as far as it is relevant for the development in the main text, we haven't repeated the proof of the classification of indecomposable root systems, which has already been laid out in many texts. In any case we give references to the results we need, and some statements form part of the exercises.

We hope that this text will be useful to doctoral students and researchers who are working in areas which rely upon a general knowledge of the groups of Lie type, without needing to understand every detail of the proof of the classification of semisimple groups. In particular, Parts II and III should give a good overview of much of what is known about the subgroup structure of these groups and to a lesser extent their conjugacy classes and representation theory. The numerous exercises are intended to supplement and illustrate the theory, and should help the book fulfill its objective of serving as the basis for a first-year graduate level course.

We began working on this project at the Mathematisches Forschungsinstitut at Oberwolfach even before the start of the summer school, and then continued at various places, including the EPFL (Lausanne), the Isaac Newton Institute at Cambridge, and the Banff International Research Station. We thank all these institutions for providing an inspiring atmosphere and enough fresh air, and for their hospitality. The second author would also like to acknowledge the support of the Swiss National Science Foundation through grants numbers PP002-68710 and 200021-122267.

We are grateful to Clara Franchi, Maria Silvia Lucido, Enrico Jabara, Mario Mainardis and John van Bon for organizing and inviting us to teach at the Venice summer school. We also thank Stephen Clegg, Kivanc Ersoy, Andreas Glang, Daniele Toller and Pinar Urgurlu, for providing TeX-files of their notes taken during our classes, made available to us shortly afterwards, and which constituted the basis for this manuscript. We thank Meinolf Geck for many clarifying discussions on various topics, Olivier Brunat and Ulrich Thiel for a careful reading of a preliminary version which lead to various improvements, and Thomas Gobet, Claude Marion and Britta Späth for proofreading parts of the manuscript.

Finally, we would like to dedicate this volume to Maria Silvia Lucido, who died in an accident half a year after the summer school. We were both impressed by her enthusiasm, energy and joyfulness during our brief acquaintance.

Tables

Notation

We have tried to conform to standard notation whenever that exists. There are a few key notions for which different conventions exist in the literature. For us, a *reductive group* is not necessarily connected, while a *semisimple group* always is. A *simple algebraic group* is a non-trivial semisimple group with no proper positive dimensional normal subgroup. A root system which cannot be decomposed into an orthogonal union of subroot systems will be called *indecomposable* (sometimes the term irreducible is used in the literature). There does not seem to be an accepted standard notation for the various orthogonal groups. Here, the full isometry group of a non-degenerate quadratic form is denoted by GO, and its connected component of the identity by SO. In particular, in characteristic 2, SO is *not* the intersection of GO with the special linear group SL. We have chosen the name *Steinberg endomorphisms* for what some authors call (generalized) Frobenius maps, to acknowledge Steinberg's role in the study of these endomorphisms. Aschbacher gave a first subdivision of natural subgroups of classical groups over finite fields into classes which he called C_i. Later, many of these classes were redefined by various authors. We follow the notation of Liebeck and Seitz in their paper [50] for these classes of subgroups. Finally, we write Z_n for the cyclic group of order n since C_n is already used for one of the root systems. We write $\mathbb{N} = \{1, 2, \ldots\}$ for the set of positive natural numbers and set $\mathbb{N}_0 := \mathbb{N} \cup \{0\}$.

PART I

LINEAR ALGEBRAIC GROUPS

In this part we introduce the main objects of study, linear algebraic groups over algebraically closed fields.

We assume that the reader is familiar with basic concepts and results from commutative algebra and algebraic geometry. More specifically, the reader should know about affine and projective varieties, their associated coordinate ring, their dimension, the Zariski topology, and basic properties thereof.

In Chapter 1 we define our main objects of study. The examples which will guide us throughout the text are certain subgroups or quotient groups of the isometry group of a finite-dimensional vector space equipped with a bilinear or quadratic form. We state the important result which says that any linear algebraic group is a closed subgroup of some group of invertible matrices over our fixed field, which is nearly obvious for all of our examples (the proof will be given in Chapter 5). In Chapter 2, we show that the Jordan decomposition of a matrix results in a uniquely determined Jordan decomposition of elements in a linear algebraic group. This in turn gives us the notion of semisimple and unipotent elements in these groups. We establish the important result that any group consisting entirely of unipotent elements is conjugate to a subgroup of the upper unitriangular matrices.

Chapter 3 is devoted to the structure theory of commutative linear algebraic groups. In particular, Theorem 3.1 focuses attention on groups consisting entirely of unipotent elements or of semisimple elements. Theorem 3.2 classifies the connected one-dimensional linear algebraic groups. While one can say something about the structure of connected commutative groups consisting entirely of unipotent elements, these will not play a role in this text. Hence we turn at this point to commutative groups consisting entirely of semisimple elements and introduce the notion of a torus, its character group and its cocharacter group. These will play a crucial role in the classification of semisimple groups. We turn in Chapter 4 to the structure theory of connected solvable groups, for which the prototype is the group of upper triangular invertible matrices. Indeed, the Lie–Kolchin theorem (Theorem 4.1 and Corollary 4.2) shows that any such group is isomorphic to a closed subgroup of the group of upper triangular matrices. The importance of closed connected solvable subgroups will become apparent in Chapter 6.

But before defining these so-called Borel subgroups, we must extend our theory to cover group actions and in particular quotient groups; this is the content of Chapter 5. The results on homogeneous spaces prepare the terrain for establishing the main result of Chapter 6, the Borel fixed point theorem, Theorem 6.1, some of whose many applications we discuss. We can also finally define the radical of a linear algebraic group and establish its connection with Borel subgroups. In these two chapters, we omit some essential geometric arguments and notions. In particular, we do not prove results on complete varieties but restrict ourselves only to projective varieties.

The last three chapters of this part are devoted to introducing the combinatorial data which classifies semisimple algebraic groups and to establishing structural results and the classification theorem. The most important ingredient of the data is a root system, which is obtained via the adjoint representation of the group, acting on its tangent space; this theory is described in Chapter 7. Theorem 8.17 is the main structural result on reductive groups and we study in detail the case of the group SL_2 in order to sketch a proof of this result. The final chapter describes the classification of semisimple algebraic groups in terms of the data mentioned above. We conclude by explaining where our standard examples appear in this classification.

1
Basic concepts

Throughout, k denotes an algebraically closed field of arbitrary characteristic.

1.1 Linear algebraic groups and morphisms

Recall that a subset X of k^n of the form

$$X = X(I) = \{(x_1, \ldots, x_n) \in k^n \mid f(x_1, \ldots, x_n) = 0 \text{ for all } f \in I\}$$

for some ideal $I \lhd k[T_1, \ldots, T_n]$ is called an *algebraic set*. Taking complements of algebraic sets as open sets defines a topology on k^n, the *Zariski topology*.

An *affine algebraic variety* is an algebraic set together with the induced Zariski topology. (We will often omit the word "algebraic".) For $X \subseteq k^n$ an affine algebraic variety, let $I \lhd k[T_1, \ldots, T_n]$ denote the (radical) ideal of polynomials vanishing identically on X. The quotient ring $k[X] = k[T_1, \ldots, T_n]/I$ is called the *coordinate algebra* or *algebra of regular functions on X* since it can be naturally identified with the algebra of all polynomial functions on X with values in k.

If $X \subseteq k^n$, $Y \subseteq k^m$ are affine varieties, their cartesian product $X \times Y$ is naturally an algebraic set in k^{n+m}, hence possesses the structure of an affine variety. We will always consider the product $X \times Y$ equipped with the Zariski topology, not with the product topology, which in general is different. Note that $k[X \times Y] \cong k[X] \otimes_k k[Y]$.

A map $\varphi : X \to Y$ between two affine varieties X, Y, which can be defined by polynomial functions in the coordinates, is called a *morphism* of affine varieties. Note that morphisms are continuous in the Zariski topology. A morphism $\varphi : X \to Y$ induces functorially a k-algebra homomorphism $\varphi^* : k[Y] \to k[X]$ via $\varphi^*(f) := f \circ \varphi$ for $f \in k[Y]$.

$$X \xrightarrow{\varphi} Y$$

$$\varphi^*(f) \searrow \quad \downarrow f$$

$$k$$

In fact, the above defines a contravariant equivalence between the category of affine varieties with morphisms of varieties and the category of finitely generated reduced k-algebras with k-algebra homomorphisms, the so-called affine k-algebras, see [32, §1.5].

We can now define our main object of study.

Definition 1.1   A *linear algebraic group* is an affine algebraic variety equipped with a group structure such that the group operations (multiplication and inversion)

$$\mu : G \times G \longrightarrow G, \qquad\qquad i : G \longrightarrow G,$$
$$(g, h) \longmapsto gh, \qquad\qquad g \longmapsto g^{-1},$$

are morphisms of varieties. (Recall our convention on the topology on $G \times G$.)

Example 1.2   The base field k provides two natural examples of algebraic groups:

(1) The additive group $G = (k, +)$ of k is defined by the zero ideal $I = (0)$ in $k[T]$, and addition is given by a polynomial; hence G is an algebraic group, with coordinate ring $k[G] = k[T]$. The group G is called the *additive group*, noted $\mathbf{G_a}$.

(2) The multiplicative group $G = (k^\times, \cdot)$ of k can be identified with the set of pairs $\{(x, y) \in k^2 \mid xy = 1\}$ (where multiplication is componentwise, again given by polynomials), which is the algebraic set defined by the ideal $I = (XY - 1) \lhd k[X, Y]$. So here $k[G] = k[X, Y]/(XY - 1) \cong k[X, X^{-1}]$. The group G is called the *multiplicative group* and noted $\mathbf{G_m}$.

It is not immediately obvious from the above definition that the *general linear group*

$$\mathrm{GL}_n := \{A \in k^{n \times n} \mid \det A \neq 0\}$$

of invertible $n \times n$-matrices over k is an algebraic group, since the determinant condition is not a closed condition. But as for $\mathbf{G_m}$ above, GL_n can be identified with the closed subset

$$\{(A, y) \in k^{n \times n} \times k \mid \det A \cdot y = 1\},$$

with componentwise multiplication, via $A \mapsto (A, \det A^{-1})$. Clearly multiplication is a polynomial map, and by Cramer's rule, the same holds for inversion. Thus GL_n is a further (and very important) example of a linear algebraic group. Its coordinate ring is given by

$$k[\mathrm{GL}_n] = k[T_{ij}, Y \mid 1 \le i, j \le n]/(\det(T_{ij})Y - 1)$$
$$\cong k[T_{ij} \mid 1 \le i, j \le n]_{\det(T_{ij})},$$

the localization of $k[T_{ij} \mid 1 \le i, j \le n]$ at the multiplicative set generated by $\det(T_{ij})$. Note that $\mathrm{GL}_1 = \mathbf{G}_\mathrm{m}$.

Maps between linear algebraic groups should preserve not only the group structure, but also the structure as an affine variety:

Definition 1.3 A map $\varphi : G_1 \to G_2$ of linear algebraic groups is a *morphism of linear algebraic groups* if it is a group homomorphism and also a morphism of varieties, that is, the induced map $\varphi^* : k[G_2] \to k[G_1]$ is a k-algebra homomorphism.

Example 1.4 (1) If $G \le \mathrm{GL}_n$ is a closed subgroup then the natural embedding $G \hookrightarrow \mathrm{GL}_n$ is a morphism of linear algebraic groups.

(2) The determinant map $\det : \mathrm{GL}_n \to \mathbf{G}_\mathrm{m}$, $A \mapsto \det A$, is a group homomorphism and clearly also a morphism of varieties, so a morphism of algebraic groups.

Proposition 1.5 *Kernels and images of morphisms of algebraic groups are closed.*

For the proof of the above statement, we will make use of the following property of morphisms of varieties, which will also be used in subsequent chapters (see [66, Thm. 1.9.5] or [26, Cor. 2.2.8]):

Proposition 1.6 *Let $\varphi : X \to Y$ be a morphism of varieties. Then $\varphi(X)$ contains a non-empty open subset of $\overline{\varphi(X)}$.*

Proof of Proposition 1.5 Let $\varphi : G \to H$ be a morphism of algebraic groups. Since $\ker(\varphi) = \varphi^{-1}(1)$ and φ is a continuous map, $\ker(\varphi)$ is closed. By Proposition 1.6, $\varphi(G)$ contains a non-empty open subset of $\overline{\varphi(G)}$; then $\varphi(G)$ is closed, by Exercise 10.3(d). $\qquad\square$

It is clear that any closed subgroup of GL_n inherits the structure of a linear algebraic group. In fact, the converse is also true:

Theorem 1.7 *Let G be a linear algebraic group. Then G can be embedded as a closed subgroup into GL_n for some n.*

The proof of this crucial characterization will be given as a corollary to Theorem 5.5. For example, the map

$$\mathbf{G}_a \longrightarrow \mathrm{GL}_2, \qquad c \mapsto \begin{pmatrix} 1 & c \\ 0 & 1 \end{pmatrix},$$

defines an embedding of the additive group $\mathbf{G}_a$ as a closed subgroup in GL_2; note that this map is in fact an isomorphism of algebraic groups onto its image.

1.2 Examples of algebraic groups

We introduce some further important examples of linear algebraic groups which will show up throughout the text. We start with three natural subgroups of GL_n. Clearly the group of invertible upper triangular matrices

$$\mathrm{T}_n := \left\{ \begin{pmatrix} * & & * \\ & \ddots & \\ 0 & & * \end{pmatrix} \text{ invertible} \right\} = \{(a_{ij}) \in \mathrm{GL}_n \mid a_{ij} = 0 \text{ for } i > j\},$$

its subgroup of upper triangular matrices with 1's on the diagonal

$$\mathrm{U}_n := \left\{ \begin{pmatrix} 1 & & * \\ & \ddots & \\ 0 & & 1 \end{pmatrix} \right\} = \{(a_{ij}) \in \mathrm{T}_n \mid a_{ii} = 1 \text{ for } 1 \leq i \leq n\},$$

and the group of diagonal invertible matrices

$$\mathrm{D}_n := \left\{ \begin{pmatrix} * & & 0 \\ & \ddots & \\ 0 & & * \end{pmatrix} \text{ invertible} \right\} = \{\mathrm{diag}(a_1, \ldots, a_n) \mid a_i \neq 0 \text{ for } 1 \leq i \leq n\},$$

are closed subgroups of GL_n, hence linear algebraic groups.

Recall that a group is called *nilpotent* if its *descending central series* defined by

$$\mathcal{C}^0 G := G, \qquad \mathcal{C}^i G := [\mathcal{C}^{i-1} G, G] \text{ for } i \geq 1,$$

eventually reaches 1. It is not hard to see that U_n is a nilpotent group, with $\mathcal{C}^{n-1}(\mathrm{U}_n) = 1$. (One uses the filtration of U_n by normal subgroups $V_m = \{(a_{ij}) \in \mathrm{U}_n \mid a_{ij} = 0 \text{ for } 1 \leq j - i \leq m\}$, for $1 \leq m \leq n - 1$.)

Furthermore, the *derived series* of a group G is defined by

$$G^{(0)} := G, \qquad G^{(i)} := [G^{(i-1)}, G^{(i-1)}] \text{ for } i \geq 1.$$

If there exists some d with $G^{(d)} = 1$, then G is called *solvable*, and the minimal such d is the *derived length* of G. Clearly $G^{(i)} \leq \mathcal{C}^i G$, so any nilpotent group is solvable.

In our example, T_n is solvable, with $T_n^{(1)} = U_n$. (U_n is generated by elementary matrices, all of which can be written as commutators.) We will see later on (Corollary 4.2) that T_n is in some sense the prototype of a connected solvable linear algebraic group.

We now define the various families of *classical groups* as groups of isometries of non-degenerate bilinear or quadratic forms on finite-dimensional vector spaces. Recall that k is assumed to be algebraically closed.

The special linear groups

The *special linear group*

$$\mathrm{SL}_n := \left\{ (a_{ij}) \in k^{n \times n} \mid \det(a_{ij}) = 1 \right\}$$

of $n \times n$-matrices of determinant 1 is a closed subgroup of GL_n, with coordinate ring

$$k[\mathrm{SL}_n] = k[T_{ij} \mid 1 \le i, j \le n]/(\det(T_{ij}) - 1).$$

As k is algebraically closed, we clearly have $\mathrm{GL}_n = Z(\mathrm{GL}_n) \cdot \mathrm{SL}_n$.

The symplectic groups

For $n \ge 1$ let $J_{2n} := \begin{pmatrix} 0 & K_n \\ -K_n & 0 \end{pmatrix}$ where $K_n := \begin{pmatrix} 0 & & 1 \\ & \cdot\cdot\cdot & \\ 1 & & 0 \end{pmatrix}$. The *symplectic group* in dimension $2n$ is the closed subgroup

$$\mathrm{Sp}_{2n} = \left\{ A \in \mathrm{GL}_{2n} \mid A^{\mathrm{tr}} J_{2n} A = J_{2n} \right\}$$

of GL_{2n}; so it is the group of invertible linear transformations of the even-dimensional vector space k^{2n} leaving invariant the non-degenerate skew-symmetric bilinear form with Gram matrix J_{2n} (a so-called *symplectic form*). Here, it is no longer so easy to explicitly write down the coordinate ring.

One can show that Sp_{2n} is generated by transvections (see [79, 8.5]), and hence $\mathrm{Sp}_{2n} \le \mathrm{SL}_{2n}$, and that for $n = 1$, any matrix of determinant 1 is symplectic. So $\mathrm{Sp}_2 = \mathrm{SL}_2$, while for all $n \ge 2$, Sp_{2n} is a proper subgroup of SL_{2n}.

The *conformal symplectic group* is the closed subgroup of GL_{2n} defined as

$$\mathrm{CSp}_{2n} := \left\{ A \in \mathrm{GL}_{2n} \mid A^{\mathrm{tr}} J_{2n} A = c J_{2n} \text{ for some } c \in k^{\times} \right\},$$

the group of transformations leaving J_{2n} invariant up to a non-zero scalar. It contains Sp_{2n} as a closed normal subgroup.

The odd-dimensional orthogonal groups

First assume that $\mathrm{char}(k) \neq 2$. For $n \geq 1$ the *orthogonal group* in (odd) dimension $2n + 1$ is defined by

$$\mathrm{GO}_{2n+1} = \left\{ A \in \mathrm{GL}_{2n+1} \mid A^{\mathrm{tr}} K_{2n+1} A = K_{2n+1} \right\}$$

with K_{2n+1} as above. Thus, this is the group of invertible linear transformations leaving invariant the non-degenerate symmetric bilinear form with Gram matrix K_{2n+1}.

If $\mathrm{char}(k) = 2$, skew-symmetric and symmetric bilinear forms coincide, and the previous construction just yields the symplectic group in dimension $2n$. For arbitrary k the orthogonal groups have to be defined using the quadratic form

$$f : k^{2n+1} \longrightarrow k, \quad f(x_1, \ldots, x_{2n+1}) := x_1 x_{2n+1} + x_2 x_{2n} + \cdots + x_n x_{n+2} + x_{n+1}^2,$$

on k^{2n+1} associated to K_{2n+1}. The group of isometries

$$\mathrm{GO}_{2n+1} = \{ A \in \mathrm{GL}_{2n+1} \mid f(Ax) = f(x) \text{ for all } x \in k^{2n+1} \}$$

of f is the *odd-dimensional orthogonal group* over k. (For $\mathrm{char}(k) \neq 2$ this defines the same group as before.)

Again there is a conformal version

$$\mathrm{CO}_{2n+1} := \left\{ A \in \mathrm{GL}_{2n+1} \mid \exists c \in k^{\times} : f(Ax) = cf(x) \text{ for all } x \in k^{2n+1} \right\},$$

the odd-dimensional *conformal orthogonal group*, containing GO_{2n+1} as a closed normal subgroup.

The even-dimensional orthogonal groups

For even dimension $2n \geq 2$ the orthogonal group is defined using the quadratic form

$$f : k^{2n} \longrightarrow k, \quad f(x_1, \ldots, x_{2n}) := x_1 x_{2n} + x_2 x_{2n-1} + \cdots + x_n x_{n+1},$$

on k^{2n} associated to K_{2n}. The group of isometries

$$\mathrm{GO}_{2n} = \{ A \in \mathrm{GL}_{2n} \mid f(Ax) = f(x) \text{ for all } x \in k^{2n} \}$$

of f is the *even-dimensional orthogonal group* over k. For $\mathrm{char}(k) \neq 2$ we can also obtain this as the group of invertible linear transformations leaving invariant the non-degenerate symmetric bilinear form with Gram matrix K_{2n}:

$$\mathrm{GO}_{2n} = \left\{ A \in \mathrm{GL}_{2n} \mid A^{\mathrm{tr}} K_{2n} A = K_{2n} \right\}.$$

The even-dimensional *conformal orthogonal group* is defined as before as

$$\mathrm{CO}_{2n} := \left\{ A \in \mathrm{GL}_{2n} \mid \exists c \in k^{\times} : f(Ax) = cf(x) \text{ for all } x \in k^{2n} \right\}.$$

Our choice of symmetric, skew-symmetric and quadratic forms above may seem a bit arbitrary. In fact, any non-degenerate symmetric bilinear form leads to the same group up to conjugacy, and similarly for non-degenerate skew-symmetric bilinear forms, respectively quadratic forms (see for example [2, §7]), but for the choices made above, certain natural subgroups have a particularly nice shape, as will become apparent later.

As a final example, let G be a finite group. Then G has a faithful permutation representation, that is, there is an embedding $G \hookrightarrow \mathfrak{S}_n$ into a *symmetric group* $\mathfrak{S}_n$ for some n. Moreover, $\mathfrak{S}_n \hookrightarrow \mathrm{GL}_n$ via the natural permutation representation. Combining these two homomorphisms we get an embedding $G \hookrightarrow \mathrm{GL}_n$ whose image is a closed subgroup (i.e., the set of zeros of a finite set of polynomial functions). Therefore, any finite group can be considered as a linear algebraic group, with the discrete topology.

1.3 Connectedness

We now recall a topological notion which will play a crucial role in the study of linear algebraic groups.

Definition 1.8 A topological space X is called *irreducible* if it cannot be decomposed as $X = X_1 \cup X_2$ where X_i is a non-empty proper closed subset for $i = 1, 2$.

In view of the importance of this concept, we present some further elementary characterizations of irreducibility:

Proposition 1.9 *The following are equivalent for an affine algebraic variety X:*

(i) *X is irreducible.*

(ii) *Every non-empty open subset of X is dense.*

(iii) *Any two non-empty open subsets of X intersect non-trivially.*

(iv) *The vanishing ideal I of X is a prime ideal.*

(v) *$k[X]$ is an integral domain.*

Proof (i)⇔(ii): Indeed, if $X_1 \subseteq X$ is open then $X = \overline{X}_1 \cup (X \setminus X_1)$. Next, (ii)⇔(iii) is obvious. The equivalence (i)⇔(iv) is shown in [32, Prop. 1.3C]. Finally, the equivalence of (iv) and (v) is well known. □

Furthermore, we need the following basic properties (see [32, Prop. 1.3A, 1.3B and 1.4] and also Exercise 10.1):

Proposition 1.10 *Let X, Y be affine varieties. Then we have:*

(a) *A subset Z of X is irreducible if and only if its closure $\overline{Z}$ is irreducible.*
(b) *If X is irreducible and $\varphi : X \to Y$ is a morphism then $\varphi(X)$ is irreducible.*
(c) *If X, Y are irreducible then $X \times Y$ is irreducible.*
(d) *X has only finitely many maximal irreducible subsets X_i, and $X = \bigcup X_i$. In other words, every variety is a finite union of its maximal irreducible subsets.*

The maximal irreducible subsets in the preceding statement are called the *irreducible components* of X. Note that by (a) irreducible components are necessarily closed.

Definition 1.11 A topological space X is said to be *connected* if it cannot be decomposed as a disjoint union $X = X_1 \sqcup X_2$, where the X_i's are non-empty closed subsets.

Note that any irreducible set is connected; the converse is not true in general. See Exercise 10.2.

Example 1.12 Let's next look at some linear algebraic groups.

(1) $\mathbf{G}_a$ and $\mathbf{G}_m$ are connected by Proposition 1.9(v) since $k[\mathbf{G}_a] = k[T]$ and $k[\mathbf{G}_m] = k[T, T^{-1}]$ are integral domains.
(2) GL_n is connected since $k[\mathrm{GL}_n] = k[T_{ij}]_{\det(T_{ij})}$ is an integral domain, being a localization of the polynomial ring $k[T_{ij}]$.

The next result gives a first example of how the Zariski topology on a linear algebraic group allows one to deduce group theoretic structural results.

Proposition 1.13 *Let G be a linear algebraic group.*

(a) *The irreducible components of G are pairwise disjoint, so they are the connected components of G.*
(b) *The irreducible component G° containing $1 \in G$ is a closed normal subgroup of finite index in G.*
(c) *Any closed subgroup of G of finite index contains G°.*

Proof (a) Let X, Y be two irreducible components of G. Assume that $g \in X \cap Y$. Since multiplication by g^{-1} is a morphism of G onto itself, $g^{-1}X, g^{-1}Y$ are irreducible by Proposition 1.10(b) and $1 \in g^{-1}X \cap g^{-1}Y$. Therefore, without loss of generality we may assume that $1 \in X \cap Y$. Now, $\mu(X \times Y) = XY$ is irreducible by Proposition 1.10(b), (c). As $X = X \cdot 1 \subseteq XY$ and

$Y = 1 \cdot Y \subseteq XY$, by the maximality of X, Y we get $X = XY = Y$. Therefore, distinct components are disjoint.

(b) As G° is an irreducible component, so is $(G^\circ)^{-1}$, and $1 \in G^\circ \cap (G^\circ)^{-1}$. Thus, by part (a), $G^\circ = (G^\circ)^{-1}$ and similarly $G^\circ \cdot G^\circ \subseteq G^\circ$, therefore G° is a subgroup of G. For $g \in G$, $g^{-1}G^\circ g$ is again an irreducible component, as an isomorphic image of G° and $1 \in G^\circ \cap g^{-1}G^\circ g$. Again by part (a) we get $G^\circ = g^{-1}G^\circ g$, so G° is normal.

Let X be any irreducible component of G. If $g \in X$ then $1 \in g^{-1}X$ and so $g^{-1}X = G^\circ$. It follows that $X = gG^\circ$, that is, the components are cosets of G°. Since by Proposition 1.10(d) there are finitely many components, $|G : G^\circ|$ is finite.

(c) Let $H \leq G$ be a closed subgroup of finite index. Then $H^\circ \leq G^\circ \leq G$. Now, $|G : H^\circ| = |G : H| \cdot |H : H^\circ|$ is finite by part (b). Thus $G^\circ = \bigsqcup gH^\circ$ is a finite disjoint union of the closed cosets of H° and, being connected, is equal to H°. So $G^\circ \leq H$. □

Thus for linear algebraic groups, the concepts of irreducibility and connectedness coincide. We will henceforth refer to the connected (irreducible) components of a linear algebraic group G as simply the *components* of G.

Example 1.14 (1) Let G be a linear algebraic group, H a proper closed subgroup of finite index. Then G is not connected by Proposition 1.13(c). In particular, for a finite algebraic group G we always have $G^\circ = 1$.

(2) Recall the orthogonal group GO_{2n+1} from Section 1.2 and consider the morphism $\det : \mathrm{GO}_{2n+1} \to \mathbf{G}_{\mathrm{m}}$. Clearly, $\mathrm{im}(\det) \subseteq \{\pm 1\}$. Since $-I_{2n+1} \in \mathrm{GO}_{2n+1}$ we have $\mathrm{im}(\det) = \{\pm 1\}$. Thus, if $\mathrm{char}(k) \neq 2$, then GO_{2n+1} is not connected by Proposition 1.10(b) since the image is not connected. Our arguments show that in fact $\mathrm{GO}_{2n+1} \cong \ker(\det) \times \langle -I_{2n+1} \rangle$.

Similarly, one can show that GO_{2n} is not connected in any characteristic by exhibiting a closed subgroup of index 2. (See Exercise 10.8 for $\mathrm{char}(k) \neq 2$. In characteristic 2, the determinant has to be replaced by the so-called pseudodeterminant, which we will not define here, see for example [30, p. 124–131].)

Definition 1.15 For $n \geq 2$, the *special orthogonal group* $\mathrm{SO}_n := \mathrm{GO}_n^\circ$ is the connected component of the identity in GO_n.

It can be shown that when n is odd or $\mathrm{char}(k) \neq 2$, SO_n is the kernel of the determinant, so $\mathrm{SO}_n = \mathrm{GO}_n \cap \mathrm{SL}_n$; when n is even and $\mathrm{char}(k) = 2$, SO_n is the kernel of the pseudodeterminant. Thus SO_n is of index 2 in GO_n unless n is odd and $\mathrm{char}(k) = 2$ see for example [26, §1.7] or [30, §14]. Analogously, the even-dimensional conformal orthogonal group CO_{2n} is not connected, and we have $|CO_{2n} : CO^\circ_{2n}| = 2$.

The following fact from algebraic geometry allows one to establish the connectedness of some algebraic groups (see [26, Thm. 2.4.6]):

Proposition 1.16 *Let G be a linear algebraic group and $\varphi_i : Y_i \to G$, $i \in I$, a family of morphisms from irreducible affine varieties Y_i such that $1 \in G_i := \varphi_i(Y_i)$ for all $i \in I$. Then $H := \langle G_i \mid i \in I \rangle$ is a closed and connected subgroup. Moreover, there exist $n \in \mathbb{N}$ and $(i_1, \ldots, i_n) \in I^n$ such that $H = G_{i_1}^{\pm 1} \cdots G_{i_n}^{\pm 1}$.*

Example 1.17 (1) Let's see an application of the above criterion. Clearly,
$$SL_2 = \langle U_2, U_2^- \rangle \text{ with } U_2 = \left\{ \begin{pmatrix} 1 & * \\ 0 & 1 \end{pmatrix} \right\} \cong \mathbf{G}_a, \ U_2^- = \left\{ \begin{pmatrix} 1 & 0 \\ * & 1 \end{pmatrix} \right\} \cong \mathbf{G}_a.$$
Both U_2 and U_2^- are closed and connected by Example 1.12(1); therefore SL_2 is connected by Proposition 1.16.

This argument can be extended to show that SL_n is connected for all $n \geq 2$. Similarly, one sees that T_n, U_n and D_n are connected for all $n \geq 1$ (see Exercise 10.7).

(2) This example shows that in general centralizers of elements need not be connected, even in a connected group. Let $G = SL_2$ over a field of characteristic $\mathrm{char}(k) \neq 2$, and $g = \begin{pmatrix} 1 & 1 \\ 0 & 1 \end{pmatrix} \in G$. Then
$$C_G(g) = \left\{ \begin{pmatrix} a & b \\ 0 & a \end{pmatrix} \ \middle| \ a, b \in k, \ a^2 = 1 \right\} = H \sqcup \begin{pmatrix} -1 & 0 \\ 0 & -1 \end{pmatrix} H,$$
with
$$H := \left\{ \begin{pmatrix} 1 & b \\ 0 & 1 \end{pmatrix} \ \middle| \ b \in k \right\}.$$
Since H is a closed subgroup of index 2 in $C_G(g)$, this centralizer is disconnected by Proposition 1.13(c).

As another application of our connectedness criterion we prove the following result on commutators which will be needed later on:

Proposition 1.18 *Let H, K be subgroups of a linear algebraic group G where K is closed and connected. Then $[H, K]$ is closed and connected.*

Proof For $h \in H$, define $\varphi_h : K \to G$ by $g \mapsto [h, g]$. Since φ_h is a composition of multiplication and inversion, it is a morphism. Moreover, $1 = \varphi_h(1) \in \varphi_h(K)$ for all h. Now, $[H, K] = \langle \varphi_h(K) \mid h \in H \rangle$, therefore $[H, K]$ is closed and connected by Proposition 1.16. $\qquad \square$

In particular, for G a connected group the derived subgroup $[G, G]$ and more generally all terms in the derived series and in the descending central series are closed and connected.

1.4 Dimension

Another fundamental invariant of algebraic varieties is their dimension.

For an irreducible variety X, the coordinate ring $k[X]$ is an integral domain by Proposition 1.9(v). Let $k(X)$ be the field of fractions of $k[X]$. We define the *dimension* of X by $\dim(X) := \operatorname{trdeg}_k(k(X))$, the transcendence degree of $k(X)$ over k. Equivalently, the dimension of X equals the maximal length of descending chains of prime ideals in $k[X]$. If X is a reducible affine algebraic variety, then according to Proposition 1.10(d) it can be decomposed as a finite union $X = X_1 \cup \cdots \cup X_t$ of its irreducible components X_i, and we set $\dim(X) := \max\{\dim(X_i) \mid 1 \le i \le t\}$.

For G a linear algebraic group, $\dim(G) = \dim(G^\circ)$ since G is the union of the finitely many cosets gG° of the irreducible subgroup G°, by Proposition 1.13, and $\dim(G) = \dim(gG^\circ) = \dim(G^\circ)$ by the above definition. In particular, $\dim(G) = 0$ if and only if G is a finite algebraic group.

Dimension behaves well with respect to morphisms in the following sense (see [26, Cor. 2.2.9] or [32, Thm. 4.3]):

Proposition 1.19 *Let $\varphi : X \to Y$ be a morphism of irreducible varieties with $\varphi(X)$ dense in Y. Then there exists a non-empty open subset $U \subseteq Y$ with $U \subseteq \varphi(X)$ such that*

$$\dim \varphi^{-1}(y) = \dim(X) - \dim(Y) \qquad \text{for all } y \in U.$$

In particular, for short exact sequences of linear algebraic groups this gives:

Corollary 1.20 *Let $\varphi : G_1 \to G_2$ be a morphism of linear algebraic groups. Then*

$$\dim\left(\operatorname{im}(\varphi)\right) + \dim\left(\ker(\varphi)\right) = \dim(G_1).$$

Proof Every fiber of φ is a coset of $\ker(\varphi)$, hence of the same dimension. Now apply Proposition 1.19 to $X = G_1^\circ$, $Y = \operatorname{im}(\varphi)^\circ$. $\quad\square$

Example 1.21 We compute the dimension of some linear algebraic groups.

(1) $\dim(\mathbf{G_a}) = 1$, as $k[\mathbf{G_a}] = k[T]$ with field of fractions $k(\mathbf{G_a}) = k(T)$.
(2) $\dim(\mathbf{G_m}) = 1$, since again $k(\mathbf{G_m}) = k(T)$.

(3) $\dim(\mathrm{GL}_n) = n^2$ since the field of fractions of $k[\mathrm{GL}_n] = k[T_{ij}]_{\det(T_{ij})}$ is just the rational function field $k(T_{ij} \mid 1 \leq i, j \leq n)$.

(4) $\dim(\mathrm{SL}_n) = n^2 - 1$, by applying Corollary 1.20 to the (surjective) determinant map $\det : \mathrm{GL}_n \to \mathbf{G}_\mathrm{m}$, with kernel SL_n.

For inductive purposes, the following result is very helpful:

Proposition 1.22 *If Y is a proper closed subset of an irreducible variety X, then $\dim(Y) < \dim(X)$.*

Proof Let $Y_1 \subseteq Y$ be an irreducible component. The embedding $\varphi : Y_1 \hookrightarrow Y \hookrightarrow X$ induces a surjective homomorphism $\varphi^* : k[X] \to k[Y_1]$. Since Y_1 is irreducible, $k[Y_1]$ is an integral domain by Proposition 1.9(v), so $\ker(\varphi^*)$ is a prime ideal of $k[X]$, non-zero since $Y_1 \subset X$ is proper. Now any chain of prime ideals in $k[Y_1]$ can be lifted to a chain in $k[X]$ through $\ker(\varphi^*)$, hence of greater length since X is irreducible. □

2
Jordan decomposition

As a first step towards a structure theory of linear algebraic groups, we investigate the properties of single elements.

2.1 Decomposition of endomorphisms

Recall the additive Jordan decomposition for endomorphisms: If V is a finite-dimensional vector space over k and $a \in \text{End}(V)$ an endomorphism of V, then there exist unique $s, n \in \text{End}(V)$ such that s is *semisimple*, i.e., diagonalizable, n is nilpotent, $a = s + n$ and $sn = ns$. Moreover, $s = P(a)$ and $n = Q(a)$ for polynomials $P, Q \in T \cdot k[T]$. We now derive a multiplicative version of this.

Definition 2.1 An endomorphism $u \in \text{End}(V)$ of a finite-dimensional vector space V is called *unipotent* if $u - 1$ is nilpotent.

Equivalently, u is unipotent if and only if all its eigenvalues are equal to 1. Note that over a field of characteristic $\text{char}(k) = p > 0$, u is unipotent if and only if it has p-power order, since then $(u - 1)^{p^i} = u^{p^i} - 1$ for all $i \geq 0$.

The additive Jordan decomposition can be used to obtain a multiplicative version:

Proposition 2.2 *For $g \in \text{GL}(V)$, there exist unique $s, u \in \text{GL}(V)$ such that $g = su = us$, where s is semisimple and u is unipotent.*

Proof Let s, n be the semisimple and nilpotent parts of g, as in the additive Jordan decomposition above. Since g is invertible, so is s and we may set $u = 1 + s^{-1}n$. As n is nilpotent and $sn = ns$, $s^{-1}n = u - 1$ is also nilpotent. Therefore u is unipotent and $su = s(1 + s^{-1}n) = s + n = g$. If $g = su$ is any such decomposition, where $u = 1 + n$ with n nilpotent and commuting with

s, then $g = s + sn$ is the unique additive Jordan decomposition of g, so s and u are uniquely determined. $\qquad\square$

Definition 2.3 Let $a \in \mathrm{GL}(V)$. We call s, respectively n, u as above the *semisimple*, respectively *nilpotent*, *unipotent part* of a.

In order to transfer this definition to an arbitrary linear algebraic group G we embed it as a closed subgroup in some $\mathrm{GL}(V)$ (see Theorem 1.7). We will first need a result on the "local finiteness" of certain automorphisms of $k[G]$. Each $x \in G$ defines a morphism $G \to G$ given by $g \mapsto gx$. We denote the corresponding $k[G]$-algebra homomorphism by $\rho_x : k[G] \to k[G]$, so $\rho_x(f)(g) = f(gx)$ for $f \in k[G]$, $g \in G$. This defines an action of G as abstract group on $k[G]$. We then have the following:

Proposition 2.4 *Let G be a linear algebraic group and V a finite-dimensional subspace of $k[G]$. Then there exists a finite-dimensional G-invariant subspace X of $k[G]$ containing V. In particular, $k[G]$ is a union of finite-dimensional G-invariant subspaces. Moreover, the restriction to any such finite-dimensional subspace X affords a morphism of algebraic groups $\rho :$ $G \to \mathrm{GL}(X)$.*

Proof It suffices to prove the statement for the case $V = \langle f \rangle$, a one-dimensional subspace of $k[G]$. Let $\mu^*(f) = \sum_{i \in I} f_i \otimes g_i$, so that $\rho_x(f) = \sum_i g_i(x) f_i$. Hence the finite-dimensional subspace generated by the $\{f_i \mid i \in I\}$ contains $\rho_x(f)$, for all $x \in G$. The subspace X generated by $\{\rho_x(f) \mid x \in G\}$ is therefore a finite-dimensional G-invariant subspace of $k[G]$. It is clear from the above construction that the coordinates of ρ_x are polynomial functions in x. Hence the map $x \mapsto (\rho_x)|_X$ affords a morphism of algebraic groups $G \to \mathrm{GL}(X)$. $\qquad\square$

We can now return to the Jordan decomposition in G. Given an arbitrary vector space V over k, we say that an element $x \in \mathrm{End}(V)$ is *locally finite* if V is a union of finite-dimensional x-stable subspaces and we say that x is *locally semisimple*, respectively *locally nilpotent*, if its restriction to any finite-dimensional x-stable subspace is semisimple, respectively nilpotent. One can then show that any locally finite endomorphism of V has a Jordan decomposition as above. Moreover, by Proposition 2.4, we see that ρ_x is a locally finite endomorphism of $k[G]$, for all $x \in G$. We are now ready to prove:

Theorem 2.5 (Jordan decomposition) *Let G be a linear algebraic group.*

(a) *For any embedding ρ of G into some $\mathrm{GL}(V)$ and for any $g \in G$, there* *exist unique* $g_s, g_u \in G$ *such that* $g = g_s g_u = g_u g_s$, *where* $\rho(g_s)$ *is semisimple and* $\rho(g_u)$ *is unipotent.*

(b) *The decomposition* $g = g_s g_u = g_u g_s$ *is independent of the chosen embedding.*

(c) *Let* $\varphi : G_1 \to G_2$ *be a morphism of algebraic groups. Then* $\varphi(g_s) = \varphi(g)_s$ *and* $\varphi(g_u) = \varphi(g)_u$.

Sketch of proof For $g \in G$, ρ_g is an invertible locally finite linear transformation of $k[G]$. Since ρ_g is an algebra homomorphism, one can deduce that $(\rho_g)_s$ and $(\rho_g)_u$ are also algebra homomorphisms. In particular, $f \mapsto ((\rho_g)_s(f))(1)$ defines a homomorphism $k[G] \to k$, that is, a point g_s in G, and similarly for $(\rho_g)_u$. It remains to show that $(\rho_g)_s = \rho_{g_s}$ and $(\rho_g)_u = \rho_{g_u}$. See [66, 2.4.8] for the details.

We give a proof of (b) and (c) only for the case $k = \overline{\mathbb{F}_p}$: We may assume that $G \leq \mathrm{GL}_n(\overline{\mathbb{F}_p})$. As $\overline{\mathbb{F}_p}$ is the union $\cup_{i \geq 1} \mathbb{F}_{p^i}$ of finite fields, any $g \in G$ lies in $\mathrm{GL}_n(\mathbb{F}_q)$ for some power q of p, so has finite order. Then g_s is diagonalizable, hence of order prime to p, while g_u has only eigenvalues 1, so is of p-power order. Since $g = g_s g_u$ it follows that g_s is the p'-part of g and g_u is the p-part of g. By the uniqueness of the p- and p'-parts of elements of finite groups, the claim follows. $\qquad\square$

See [32, Thm. 15.3] for a proof in the general case.

Definition 2.6 Let G be a linear algebraic group. The decomposition $g = g_s g_u = g_u g_s$ in Theorem 2.5 is called the *Jordan decomposition* of $g \in G$, and g is called *semisimple*, respectively *unipotent*, if $g = g_s$, respectively $g = g_u$. We write

$$G_u := \{g \in G \mid g \text{ is unipotent}\},$$
$$G_s := \{g \in G \mid g \text{ is semisimple}\}.$$

for the subsets of unipotent, respectively semisimple elements of G. If G consists entirely of unipotent elements then we say that G is a *unipotent group*.

We note that the term "semisimple group" is reserved for a different concept; see Definition 6.14.

Example 2.7 We have already encountered groups with $G = G_u$ and with $G = G_s$:

(1) $\mathbf{G_a} \cong \left\{ \begin{pmatrix} 1 & a \\ 0 & 1 \end{pmatrix} \Big| a \in k \right\}$ is a unipotent group and, more generally, so is U_n for all $n \geq 2$.

(2) $\mathbf{G_m}$, and more generally $\mathbf{D}_n = \{\mathrm{diag}(t_1, \ldots, t_n) \mid \prod t_i \neq 0\}$ for $n \geq 1$, consist of semisimple elements.

Let G be a linear algebraic group. Then the set G_u of unipotent elements is closed (see Exercise 10.12), while G_s is, in general, neither closed nor open. For example, in GL_n the set of semisimple elements is dense:

Example 2.8 We will observe that G_s is "badly behaved". Let $G = \mathrm{GL}_2$ and $g \in G \setminus G_s$. Then g is conjugate to $\begin{pmatrix} a & 1 \\ 0 & a \end{pmatrix}$ for some $a \in k^\times$. Define

$\varphi : \mathbf{G_m} \to \mathrm{GL}_2$ by $\varphi(t) = \begin{pmatrix} at & 1 \\ 0 & at^{-1} \end{pmatrix}$. Then φ is a morphism and $\varphi(t)$ is semisimple for $at \neq at^{-1}$, hence for $t \neq \pm 1$. So, $\varphi(\mathbf{G_m} \setminus \{\pm 1\}) \subseteq G_s$. But $\mathbf{G_m} \setminus \{\pm 1\}$ is open in $\mathbf{G_m}$, and dense by Proposition 1.9(ii) since $\mathbf{G_m}$ is irreducible. Hence $\overline{\mathbf{G_m} \setminus \{\pm 1\}} = \mathbf{G_m}$ and

$$\varphi(\mathbf{G_m}) = \varphi(\overline{\mathbf{G_m} \setminus \{\pm 1\}}) \subseteq \overline{\varphi(\mathbf{G_m} \setminus \{\pm 1\})} \subseteq \overline{G_s},$$

so all non-semisimple elements of G lie in $\overline{G_s}$. Hence $\overline{G_s} = G$.

2.2 Unipotent groups

We now prove the basic structure result for unipotent groups. Recall from Theorem 1.7 that any linear algebraic group has an embedding into some GL_n, so we may and will restrict attention to unipotent matrix groups.

Proposition 2.9 *Let $G \leq \mathrm{GL}_n$ be a unipotent group. Then there exists $g \in \mathrm{GL}_n$ such that $g^{-1}Gg \leq \mathrm{U}_n$.*

Proof Write $V = k^n$. We will argue by induction on n.

The only unipotent element of GL_1 is the identity, so the claim is clear for $n = 1$. Now suppose that $n > 1$. If there exists a G-invariant proper subspace $0 \neq W < V$, then by choosing an appropriate basis we may assume that $G \leq \left\{ \left(\begin{array}{c|c} * & * \\ \hline 0 & * \end{array} \right) \right\}$.

The G-invariance of W induces natural homomorphisms $\varphi : G \to \mathrm{GL}(W)$ and $\Phi : G \to \mathrm{GL}(V/W)$. Since $\dim(W)$ and $\dim(V/W)$ are both smaller than

$\dim(V)$, by induction we get (up to a change of basis, so up to conjugation)

$$G \leq \left\{ \left(\begin{array}{c|c} \begin{matrix} 1 & \ast \\ & \ddots \\ 0 & 1 \end{matrix} & \ast \\ \hline 0 & \begin{matrix} 1 & \ast \\ & \ddots \\ 0 & 1 \end{matrix} \end{array} \right) \right\} = U_n$$

as claimed.

If, on the other hand, G acts irreducibly on V, then the elements of G generate the full endomorphism algebra $\mathrm{End}(V)$ by Burnside's double centralizer theorem [39, Thm. 1.16]. Let $g \in G$. Since any element of G, being unipotent, has trace n we find $\mathrm{tr}((g-1)h) = \mathrm{tr}(gh) - \mathrm{tr}(h) = 0$ for all $h \in G$. Therefore, $\mathrm{tr}((g-1)x) = 0$ for all $x \in \mathrm{End}(V)$. Choosing for x matrices with only one non-zero entry one easily sees that this is only possible if $g - 1 = 0$, that is $g = 1$ and so $G = 1$, contradicting the irreducibility of G on V. $\quad\square$

As U_n is a nilpotent group (see Section 1.2) Proposition 2.9 immediately implies the following:

Corollary 2.10   *A unipotent linear algebraic group is nilpotent, hence solvable.*

Over $k = \overline{\mathbb{F}_p}$, unipotent elements are p-elements, thus the above statement can be seen as an algebraic group analogue of the fact that finite p-groups are nilpotent.

3

Commutative linear algebraic groups

The Jordan decomposition of elements leads to a similar decomposition of abelian algebraic groups. The case when all elements are semisimple will be of particular importance to us later.

3.1 Jordan decomposition of commutative groups

The structure of commutative algebraic groups is described by the following result:

Theorem 3.1 *Let G be a commutative linear algebraic group.*

(a) *The sets G_s and G_u are closed subgroups of G.*
(b) *The product map $\pi : G_u \times G_s \to G$ is an isomorphism of algebraic groups.*
(c) *If G is connected then so are G_s and G_u.*

Proof (a) Choose an embedding of G into some $\mathrm{GL}(V)$ according to Theorem 1.7. Let g, $g' \in G$ be two (commuting) elements. Commuting endomorphisms can be simultaneously trigonalized, so if g, g' are semisimple (unipotent) so are gg' and g^{-1}. Therefore G_s, G_u are subgroups of G.

Note that G_u is closed by Exercise 10.12. Also, G_s consists of commuting semisimple endomorphisms, so can be simultaneously diagonalized. Let V_i, $1 \le i \le t$, denote the various common eigenspaces of G_s on V, that is, $V_i = \{v \in V \mid gv = \varphi_i(g)v \text{ for all } g \in G_s\}$ for suitable morphisms $\varphi_i : G_s \to \mathbf{G_m}$, and $V = \bigoplus_{i=1}^{t} V_i$. Since G_u commutes with G_s, it preserves all V_i. So G stabilizes each V_i, therefore by Proposition 2.9 there exists a basis such that G_u acts by upper unitriangular matrices on each V_i, while G_s acts by scalars.

So with respect to this basis, $G \leq T_n$ and $G_s = D_n \cap G$, where $n := \dim(V)$. Hence G_s is closed.

(b) The product map $\pi : G_s \times G_u \to G$, $\pi(s, u) = su$, is a group homomorphism since G is commutative. It is surjective by Jordan decomposition (Theorem 2.5) and injective by the uniqueness in Theorem 2.5. Moreover, π is a morphism of varieties since multiplication $\mu : G \times G \to G$ is a morphism.

It remains to show that the inverse map

$$\pi^{-1} : G \longrightarrow G_s \times G_u, \qquad \pi^{-1}(g) = (g_s, g_s^{-1}g),$$

is also a morphism. For this it suffices to notice that the map

$$\psi : g \mapsto g_s, \quad \text{that is} \quad \begin{pmatrix} d_1 & & * \\ & \ddots & \\ 0 & & d_n \end{pmatrix} \longmapsto \begin{pmatrix} d_1 & & 0 \\ & \ddots & \\ 0 & & d_n \end{pmatrix},$$

is a morphism.

(c) As G is assumed connected and $G \to G_s$, $g \mapsto g_s$, is a morphism, its image G_s is connected by Proposition 1.10(b). Similarly $G \to G_u$, $g \mapsto g_u = gg_s^{-1}$, is a morphism so has a connected image. □

Compare this again to the situation for finite groups: a finite abelian group is the direct product of its Sylow subgroups.

The structure of one-dimensional connected groups is very restricted (see [66, Thm. 3.4.9]):

Theorem 3.2 *If G is a connected linear algebraic group of dimension 1 then $G \cong \mathbf{G_a}$ or $G \cong \mathbf{G_m}$.*

About the proof The proof of this seemingly innocent result is rather difficult, in particular the case where G consists of unipotent elements. We only prove here that G is commutative and either $G = G_s$ or $G = G_u$.

First, we claim that G is commutative. For $g \in G$ define a morphism $\varphi_g : G \to G$ by $\varphi_g(x) := x^{-1}gx$. As $\varphi_g(G)$ is irreducible, $\overline{\varphi_g(G)}$ is also an irreducible closed subvariety of G. So, either, $\overline{\varphi_g(G)}$ is of dimension 1 and then $\overline{\varphi_g(G)} = G$ by Proposition 1.22, or $\overline{\varphi_g(G)}$ is of dimension 0.

If $\overline{\varphi_g(G)}$ is of dimension 0 for all $g \in G$ then $\varphi_g(G) = \{g\}$ by connectedness and hence G is commutative.

So we may assume that $\varphi_g(G)$ is dense in G for some $g \in G$. We fix an embedding of G in some GL_n. Now $\varphi_g(G)$ contains a non-empty open subset of its closure by Proposition 1.6. Thus $G \setminus \varphi_g(G)$ is contained in a closed proper subset, so of dimension 0 by Proposition 1.22, and hence finite.

As $h \in \varphi_g(G)$ has characteristic polynomial equal to the characteristic polynomial of g and there is only a finite number of elements in $G \setminus \varphi_g(G)$, there exist finitely many characteristic polynomials $p_1, \ldots, p_t$ for elements of G. Now

$$G = \bigcup_{i=1}^{t} ((\text{zeros of } p_i) \cap G)$$

is a finite union of closed sets. As G is connected, we get $t = 1$ and hence $p_1 = (T - 1)^n$ (for $1 \in G$). Therefore G is unipotent, and, by Corollary 2.10, solvable. It follows from Proposition 1.18 that $[G, G]$ is a closed connected proper subgroup of G, so $[G, G] = 1$ by Proposition 1.22, and G is commutative, a contradiction.

By Theorem 3.1, G_s and G_u are closed connected subgroups of G. If either is proper, then it is of dimension 0 by Proposition 1.22, hence trivial by connectedness. Thus either $G = G_u$ or $G = G_s$ as claimed. □

See Definition 3.3 and Proposition 3.9 for the remaining step in the case that $G = G_s$.

3.2 Tori, characters and cocharacters

We now turn to the study of connected groups containing only semisimple elements. By Theorem 3.2, any such group of dimension 1 is isomorphic to $\mathbf{G}_\mathrm{m}$. This motivates the following definition:

Definition 3.3 A linear algebraic group is called a *torus* if it is isomorphic to a direct product $\mathbf{G}_\mathrm{m} \times \cdots \times \mathbf{G}_\mathrm{m}$ of copies of $\mathbf{G}_\mathrm{m}$, that is, to D_n for some $n \geq 0$.

Tori are best studied via their characters:

Definition 3.4 Let G be a linear algebraic group. A *character* of G is a morphism of algebraic groups $\chi : G \to \mathbf{G}_\mathrm{m}$. The set of characters of G is denoted by $X(G)$. Note that it can naturally be considered as a subset of $k[G]$.

A *cocharacter* of G is a morphism of algebraic groups $\gamma : \mathbf{G}_\mathrm{m} \to G$. The set of cocharacters of G is denoted by $Y(G)$.

Clearly, $X(G)$ is an abelian group (usually written additively) with respect to

$$(\chi_1 + \chi_2)(g) := \chi_1(g)\chi_2(g) \qquad \text{for } \chi_1, \chi_2 \in X(G), \ g \in G.$$

Similarly, if G is commutative then $Y(G)$ is an abelian group with respect to

$$(\gamma_1 + \gamma_2)(x) := \gamma_1(x)\gamma_2(x) \qquad \text{for } \gamma_1, \gamma_2 \in Y(G), \ x \in \mathbf{G}_\mathrm{m}.$$

Example 3.5 (Characters and cocharacters of D_n)

Let $G = \mathrm{D}_n$, an n-dimensional torus. If $n = 1$ then $G = \mathbf{G}_\mathrm{m}$, so $X(G) = \mathrm{End}(\mathbf{G}_\mathrm{m}) = \{t \mapsto t^j \mid j \in \mathbb{Z}\} \cong \mathbb{Z}$ by Exercise 10.11.

For arbitrary n, define $\chi_i \in X(G)$ by $\chi_i(g) := t_i$ if $g = \mathrm{diag}(t_1, \ldots, t_n) \in G$, $1 \le i \le n$. Then any $\chi \in X(G)$ can be uniquely decomposed as $\chi = \chi_1^{a_1} \cdots \chi_n^{a_n}$ for some $a_i \in \mathbb{Z}$, so $X(\mathrm{D}_n) \cong \mathbb{Z}^n$. Note that here the characters are just the monomials in $k[\mathrm{D}_n] = k[T_1^{\pm 1}, \ldots, T_n^{\pm 1}]$, and hence form a basis of the coordinate ring.

Similarly, for $(a_1, \ldots, a_n) \in \mathbb{Z}^n$ the map $\mathbf{G}_\mathrm{m} \to \mathrm{D}_n$, $t \mapsto \mathrm{diag}(t^{a_1}, \ldots, t^{a_n})$, is a morphism, hence a cocharacter. Conversely, by restricting a cocharacter $\gamma : \mathbf{G}_\mathrm{m} \to \mathrm{D}_n$ to the ith diagonal position one sees that any cocharacter has the above form. Therefore $Y(\mathrm{D}_n) \cong \mathbb{Z}^n$.

As a result, for an n-dimensional torus T we have $X(T) \cong Y(T) \cong \mathbb{Z}^n$. Now, for $\chi \in X := X(T)$, $\gamma \in Y := Y(T)$, the composite

$$\chi \circ \gamma : \mathbf{G}_\mathrm{m} \to \mathbf{G}_\mathrm{m}, \quad t \mapsto \chi(\gamma(t)),$$

is an endomorphism of $\mathbf{G}_\mathrm{m}$. So by Exercise 10.11 it acts by raising to some integer power $\langle \chi, \gamma \rangle \in \mathbb{Z}$, that is, $\chi(\gamma(t)) = t^{\langle \chi, \gamma \rangle}$.

Proposition 3.6 *Let T be a torus with character group X, cocharacter group Y. The map $\langle \, , \, \rangle : X \times Y \to \mathbb{Z}$ is a perfect pairing between X and Y, that is, any homomorphism $X \to \mathbb{Z}$ is of the form $\chi \mapsto \langle \chi, \gamma \rangle$ for some $\gamma \in Y$, and any homomorphism $Y \to \mathbb{Z}$ is of the form $\gamma \mapsto \langle \chi, \gamma \rangle$ for some $\chi \in X$. This defines group isomorphisms $Y \cong \mathrm{Hom}(X, \mathbb{Z})$ and $X \cong \mathrm{Hom}(Y, \mathbb{Z})$.*

Proof We may assume that $T = \mathbf{G}_\mathrm{m} \times \cdots \times \mathbf{G}_\mathrm{m}$ (n copies) and $X = \{\chi_1^{a_1} \cdots \chi_n^{a_n} \mid (a_1, \ldots, a_n) \in \mathbb{Z}^n\}$ with χ_i as defined above. Any homomorphism $\varphi : X \to \mathbb{Z}$ is uniquely determined by $\varphi(\chi_i) =: d_i \in \mathbb{Z}$ since

$$\varphi\left(\prod_{i=1}^n \chi_i^{a_i}\right) = \sum_{i=1}^n a_i \varphi(\chi_i) = \sum_{i=1}^n a_i d_i.$$

Let $\gamma \in Y$ with $\gamma(t) := \begin{pmatrix} t^{d_1} & & 0 \\ & \ddots & \\ 0 & & t^{d_n} \end{pmatrix}$ for $t \in \mathbf{G}_\mathrm{m}$. Then

$$t^{\langle \chi, \gamma \rangle} = \chi(\gamma(t)) = \prod_{i=1}^n \chi_i^{a_i}(\gamma(t)) = t^{\sum_{i=1}^n a_i d_i} = t^{\varphi(\chi)}.$$

Therefore, $\varphi(\chi) = \langle \chi, \gamma \rangle$ for all $\chi \in X$.

Thus the map $Y \to \operatorname{Hom}(X, \mathbb{Z})$, $\gamma \mapsto \langle -, \gamma \rangle$, is surjective. For injectivity, assume $\langle \chi, \gamma \rangle = 0$ for all $\chi \in X$, then

$$t^{d_i} = \chi_i(\gamma(t)) = t^{\langle \chi_i, \gamma \rangle} = t^0 = 1$$

for all $t \in \mathbf{G}_{\mathrm{m}}$ and all i. Thus $d_i = 0$ for all i, that is, $\gamma = 0$. Hence $Y \cong \operatorname{Hom}(X, \mathbb{Z})$. Similarly one argues that $X \cong \operatorname{Hom}(Y, \mathbb{Z})$. $\qquad \square$

Definition 3.7 Let T be a torus. For a closed subgroup $H \leq T$ we define

$$H^\perp := \{\chi \in X(T) \mid \chi(h) = 1 \text{ for all } h \in H\},$$

a subgroup of $X(T)$. Similarly, for a subgroup $X_1 \leq X(T)$ we let

$$X_1^\perp := \{t \in T \mid \chi(t) = 1 \text{ for all } \chi \in X_1\} = \bigcap_{\chi \in X_1} \ker(\chi),$$

a closed subgroup of T.

The following properties of these constructions will be needed later (see Exercise 10.15 for a proof):

Proposition 3.8 *Let T be a torus.*

(a) *For any closed subgroup $H \leq T$, restriction defines an isomorphism*

$$X(T)/H^\perp \cong X(H).$$

(b) *For any subgroup $X_1 \leq X(T)$, $X_1^{\perp\perp}/X_1$ is a finite p-group, where $p = \operatorname{char}(k)$; in particular, $X_1^{\perp\perp} = X_1$ if $X(T)/X_1$ has no p-torsion.*

The following can be shown using properties of characters (see [32, §16] or [66, 3.2.7]):

Proposition 3.9 *Any closed connected subgroup of D_n is a torus.*

This yields the missing part in the proof of Theorem 3.2 in case $G = G_s$: embed G into some GL_n and diagonalize.

The following crucial result shows the "rigidity" of tori: their normalizers are only a finite bit larger than their centralizers. It will later allow us to define the Weyl group, a finite group which plays an important role in controlling the structure of semisimple groups (see [66, Thm. 3.2.9] for a proof):

Theorem 3.10 *Let G be a linear algebraic group and $T \leq G$ be a torus. Then, $N_G(T)^\circ = C_G(T)^\circ$, and $N_G(T)/C_G(T)$ is finite.*

Example 3.11 (To illustrate Theorem 3.10) Let $G = \mathrm{GL}_n$, $m \leq n$, and

$$T = \left\{ \begin{pmatrix} * & & & & \\ & \ddots & & & \\ & & *1 & & \\ & & & \ddots & \\ & & & & 1 \end{pmatrix} \right\} = \{\mathrm{diag}(t_1, \ldots, t_n) \mid t_i = 1 \text{ for } m+1 \leq i \leq n\},$$

an m-dimensional torus in G, which is isomorphic to $\mathbf{G}_\mathrm{m} \times \cdots \times \mathbf{G}_\mathrm{m}$ (m copies). Then

$$C_G(T) = \left\{ \left(\begin{array}{ccc|c} * & & & \\ & \ddots & & 0 \\ & & * & \\ \hline & 0 & & A \end{array} \right) \;\middle|\; A \in \mathrm{GL}_{n-m} \right\} \cong \mathrm{D}_m \times \mathrm{GL}_{n-m},$$

and $N_G(T) \cong N_{\mathrm{GL}_m}(\mathrm{D}_m) \times \mathrm{GL}_{n-m}$ where $N_{\mathrm{GL}_m}(\mathrm{D}_m) = M$ is the set of all monomial $m \times m$-matrices. As D_m has finite index in M, $M^\circ = \mathrm{D}_m$. Also, GL_{n-m} is connected, so we get

$$N_G(T)^\circ \cong \mathrm{D}_m \times \mathrm{GL}_{n-m} \cong C_G(T) = C_G(T)^\circ,$$

and $N_G(T)/C_G(T) \cong M/\mathrm{D}_m \cong \mathfrak{S}_m$ is a symmetric group, hence indeed finite.

4

Connected solvable groups

We extend our structure results from commutative to solvable groups G. For this, however, connectedness of G is crucial.

4.1 The Lie–Kolchin theorem

If $G \leq \mathrm{GL}(V)$ is unipotent, then by Proposition 2.9 it always stabilizes a one-dimensional subspace of V. This is true for a larger class of groups.

Recall from Proposition 1.18 that commutator subgroups are closed and connected if the original group is.

Theorem 4.1 (Lie–Kolchin) *Let G be a connected solvable subgroup of $\mathrm{GL}(V)$, $V \neq 0$. Then G has a common eigenvector on V, that is, there exists a one-dimensional subspace of V stabilized by G.*

Proof We may assume that G is closed since if G is solvable then so is its closure $\overline{G}$ ([4, §2.4]).

We will use induction on $n = \dim V$ and on the derived length d of G (see Section 1.2).

For $n = 1$ the result is trivially true. Now assume that $n > 1$. If $d = 1$ then G is commutative and the result follows by the proof of Theorem 3.1(a).

Now assume that $d \geq 2$. If there exists a G-invariant proper subspace $0 \neq W < V$ of V, choose an appropriate basis such that $G \subseteq \left\{ \left(\begin{array}{c|c} * & * \\ \hline 0 & * \end{array} \right) \right\}$.

Then more precisely, any $g \in G$ can be represented by $\left(\begin{array}{c|c} \varphi(g) & * \\ \hline 0 & \psi(g) \end{array} \right)$ where $\varphi : G \to \mathrm{GL}(W)$ is the canonical restriction morphism, and $\psi : G \to \mathrm{GL}(V/W)$. Now, $\varphi(G)$ is solvable, connected, acting on a vector space of smaller dimension $\dim W < \dim V$, so by the inductive hypothesis there exists a common eigenvector $w \in W < V$.

Therefore, we may assume that G acts irreducibly on V. As $G' := [G, G]$ is a closed, connected solvable subgroup of derived length $d - 1$, by induction there exists a common eigenvector $v \in V$ of G'. Since $G' \lhd G$, gv is again a common eigenvector of G' for all $g \in G$.

Let W denote the span of all common eigenvectors of G'. By the observation above, W is G-invariant and $W \neq 0$. As G acts irreducibly, necessarily $W = V$. Therefore, V has a basis consisting of common eigenvectors of G' and so G' acts diagonally, implying that G' is commutative.

Now for fixed $y \in G'$ all conjugates $x^{-1}yx$ by elements $x \in G$ lie in G', hence act diagonally with the same eigenvalues as y. Therefore, there are only finitely many possibilities for $x^{-1}yx$. It follows that the morphism $\varphi_y : G \to G'$, $x \mapsto x^{-1}yx$, has finite image. So, by the connectedness of G, $\varphi_y(G) = \{y\}$, implying $G' \leq Z(G)$. Since G acts irreducibly on V, $Z(G)$ acts by scalars by Schur's lemma. As all elements of G' have determinant 1, it follows that G' is finite. But since it is connected, $G' = 1$ and hence G is commutative, in contradiction to our assumption that $d \geq 2$. $\square$

Corollary 4.2 *Let G be a connected, solvable subgroup of GL_n. Then, G is conjugate to a subgroup of T_n.*

Proof Apply Theorem 4.1 to inductively construct a basis with respect to which G is triangular. $\square$

Remark 4.3 The assertion of the above corollary becomes false if G is no longer assumed to be connected. Indeed, consider any finite non-trivial solvable subgroup of GL_n, $n > 1$, which acts irreducibly.

4.2 Structure of connected solvable groups

We have seen that connected solvable groups are isomorphic to subgroups of T_n. This has strong implications for the structure of such groups. We have a natural split exact sequence

$$1 \longrightarrow \mathrm{U}_n \longrightarrow \mathrm{T}_n \overset{\pi}{\longrightarrow} \mathrm{D}_n \longrightarrow 1$$

where π is the natural morphism with

$$\pi \begin{pmatrix} t_1 & & * \\ & \ddots & \\ 0 & & t_n \end{pmatrix} = \begin{pmatrix} t_1 & & 0 \\ & \ddots & \\ 0 & & t_n \end{pmatrix}.$$

This sequence is also split as an exact sequence of algebraic groups; the inclusion $D_n \to T_n$ is a morphism and a section to π.

Now let $G \le T_n$ be a closed connected subgroup. The restriction of π to G has kernel $G_u = G \cap U_n$, the closed normal subgroup of G consisting of all unipotent elements. The image $T := \pi(G)$ is then a closed connected subgroup of D_n, hence a torus by Proposition 3.9. So restriction yields the exact sequence

$$1 \longrightarrow G_u \longrightarrow G \overset{\pi}{\longrightarrow} T \longrightarrow 1.$$

Since the torus T is abelian, it follows that $[G, G] \le G_u$.

This almost proves the first part of the following structure result for connected solvable groups:

Theorem 4.4 *Let G be a connected, solvable linear algebraic group. Then:*

(a) *G_u is a closed, connected, normal subgroup of G and $[G, G] \le G_u$.*
(b) *All maximal tori of G are conjugate and if T is any such maximal torus, then $G = G_u \rtimes T$ and $N_G(T) = C_G(T)$.*

Here a *maximal torus* of G is a subtorus of G which is maximal with respect to inclusion. Note that by the theorem, maximal tori are all of the same dimension, hence of maximal possible dimension. The semidirect product of two algebraic groups X and N is defined as for abstract groups, where we require that X act as a group of algebraic group automorphisms of N. An algebraic group H is isomorphic to the semidirect product of two closed subgroups X and N if $N \lhd H$, $N \cap X = 1$ and the product map $N \rtimes X \to H$ is an isomorphism of algebraic groups.

Proof of part (a) It remains to show that G_u is connected. Set $\tilde{G} := G/G'$, where $G' = [G, G]$. We will see later (in Proposition 5.7) that this factor group by a closed normal subgroup has a natural structure of linear algebraic group in such a way that the natural surjection $\varphi : G \to \tilde{G}$ is a morphism of linear algebraic groups.

Since $\tilde{G}$ is commutative and connected, $\tilde{G} = \tilde{G}_u \times \tilde{G}_s$ with $\tilde{G}_u$ connected by Theorem 3.1. Then, $\varphi(G_u) \le \tilde{G}_u$ by Theorem 2.5, and therefore, $G_u \le \varphi^{-1}(\tilde{G}_u)$. On the other hand, by our above considerations before Theorem 4.4, $\ker(\varphi) = G' \le G_u$. If $g \in \varphi^{-1}(\tilde{G}_u)$, with Jordan decomposition $g = g_s g_u$, then $\varphi(g_s) = \varphi(g)_s = 1$. Therefore, $g_s \in \ker(\varphi) \le G_u$, so $g_s = 1, g = g_u$ and it follows that $\varphi^{-1}(\tilde{G}_u) = G_u$. Now $G' = \ker(\varphi) \le G_u$ is connected by Proposition 1.18, so contained in G_u°. Thus

$$|G_u : G_u^\circ| = |G_u/G' : G_u^\circ/G'| = |\varphi(G_u) : \varphi(G_u^\circ)| = |\tilde{G}_u : \varphi(G_u^\circ)|.$$

Hence $\varphi(G_u^{\circ})$ is a closed subgroup of finite index in the connected group $\tilde{G}_u$, so of index 1. Hence, G_u is also connected.

The proof of part (b) is more difficult and requires results about actions of diagonalizable groups (see [32, Thm. 19.3]).

Finally, the assertion on the normalizer follows from the Frattini argument: if $g = ut \in N_G(T)$, with $u \in G_u$ and $t \in T$, then $uTu^{-1} = T$. But, for $s \in T$, $usu^{-1} = (u\,su^{-1}s^{-1})s$ lies in T if and only if $u\,su^{-1}s^{-1} = 1$, that is, u centralizes s. $\qquad\square$

Note the similarity of the second part with the Schur–Zassenhaus Theorem for finite groups ([2, §18.1]). Indeed, G is already known to be an extension of the unipotent group G_u by a torus $\pi(G)$, all of whose elements are semisimple. The theorem asserts that a complement to G_u exists and that furthermore all complements are conjugate. If k is the algebraic closure of a finite field $\mathbb{F}_p$, then unipotent elements have p-power order, while semisimple elements have order prime to p. Thus G_u is the analogue of a normal Sylow p-subgroup of G, and any maximal torus is a p'-complement. In that situation, the assertion in (b) is just as in the Schur–Zassenhaus Theorem.

If G is not connected, anything may happen. For example, take $G = N_{\mathrm{GL}_2}(D_2)$, an extension of D_2 by an element of order 2, and assume that $\mathrm{char}(k) = 2$. Then, although G is solvable, G_u is not a subgroup, in particular not normal, and the maximal torus D_2 is not self-normalizing.

Corollary 4.5 *Let G be a connected, solvable linear algebraic group. Then any semisimple element of G lies in a maximal torus and any unipotent element of G lies in a connected unipotent subgroup of G.*

The result for unipotent elements follows from Theorem 4.4(a); for semisimple elements it follows from the omitted proof of Theorem 4.4(b).

Finally, we need a result on centralizers of tori in connected solvable groups which will later on be generalized to arbitrary connected groups (see Theorem 6.19) (for the proof see [66, Cor. 6.3.6]):

Proposition 4.6 *Let G be connected solvable, $S \le G$ a torus. Then $C_G(S)$ is connected.*

5
G-spaces and quotients

One aspect of the theory of linear algebraic groups which has been missing up to now is that of a quotient group. We need to first see how to give the structure of variety to a quotient and it will become clear that we cannot limit ourselves to affine varieties. Thus, we begin by recalling some basic aspects of the general theory of varieties and morphisms.

5.1 Actions of algebraic groups

In group theory, it is often helpful to consider actions of groups, for example the action of a group on itself by conjugation. We will find it necessary to consider actions of linear algebraic groups on affine and projective varieties.

For this recall that projective n-space $\mathbb{P}^n$ may be defined as the set of equivalence classes of $k^{n+1} \setminus \{(0,0,\ldots,0)\}$ modulo the diagonal action of $k^\times$ by multiplication. Taking common zeros of a collection of homogeneous polynomials in $k[T_0, T_1, \ldots, T_n]$ as closed sets defines a topology on $\mathbb{P}^n$. A *projective variety* is then a closed subset of $\mathbb{P}^n$ equipped with the induced topology.

The k-algebra of regular functions on an affine variety here needs to be replaced by a *sheaf of functions*, as follows. First, for X an irreducible affine variety and $x \in X$, let $I(x) \lhd k[X]$ be the ideal of functions vanishing at x and let $\mathcal{O}_x$ be the localization of $k[X]$ with respect to the prime ideal $I(x)$. Then setting $\mathcal{O}_X(U) = \bigcap_{x \in U} \mathcal{O}_x$, for $U \subseteq X$ open, defines a sheaf of functions $\mathcal{O}_X$ on X. Clearly the fields of fractions of $\mathcal{O}_X(U)$ and $k[X]$ agree. More generally, if X is reducible, say $X = X_1 \cup \ldots \cup X_t$ with irreducible components X_i, then setting

$$\mathcal{O}_X(U) := \{f : U \to k \mid f|_{U \cap X_i} \in \mathcal{O}_{X_i}(U \cap X_i)\},$$

for $U \subseteq X$ open, defines a sheaf on X. One can show that $\mathcal{O}_X(X) = k[X]$, the usual k-algebra of regular functions on X.

Now consider the covering $\mathbb{P}^n = \bigcup_{i=0}^n U_i$, where U_i consists of the points in $\mathbb{P}^n$ with non-zero ith homogeneous coordinate. Then U_i can be identified with affine n-space k^n via

$$(x_0, x_1, \ldots, x_n) \mapsto \left(\frac{x_0}{x_i}, \ldots, \frac{x_{i-1}}{x_i}, \frac{x_{i+1}}{x_i} \ldots, \frac{x_n}{x_i} \right).$$

Moreover, the induced topology on the sets U_i is precisely the topology of the affine space k^n. For each $x \in U_i$, we have the ring of functions $\mathcal{O}_x$ as above, using the identification of U_i with k^n. Then for $U \subseteq \mathbb{P}^n$, set $\mathcal{O}(U) := \bigcap_{x \in U} \mathcal{O}_x$. The sheaf of functions for an arbitrary projective variety is defined by restriction of the given sheaf on $\mathbb{P}^n$. It is then clear that the only globally defined functions on $\mathbb{P}^n$ are the constant functions. This is more generally true for an arbitrary projective variety.

The *dimension* of an irreducible projective variety is the dimension of any affine open subset. For a general projective variety, one simply takes the maximum dimension of the irreducible components of the variety. (See [66, Prop. 1.2.4, §1.8.1].)

Now a *morphism of varieties* is a continuous map $\varphi : X \to Y$ between two (affine or projective) varieties such that for all $V \subseteq Y$ open and $U = \varphi^{-1}(V)$ and all $f \in \mathcal{O}_Y(V)$, we have $f \circ \varphi \in \mathcal{O}_X(U)$.

Henceforth, we will include projective varieties when we consider a general variety, though we continue to limit ourselves to linear algebraic groups, which are always affine varieties. We point out that Propositions 1.6, 1.19, and 1.22 remain valid in the setting of projective varieties.

Important examples of projective varieties are provided by partial flag varieties of an n-dimensional vector space V over k. Recall that a *flag* in V is a chain of subspaces $0 \subsetneq V_1 \subsetneq \cdots \subsetneq V_r = V$. One can equip the set of flags having a fixed sequence of dimensions $(\dim V_1, \ldots, \dim V_r)$ with the structure of a projective variety, the partial flag variety. The case where $\dim V_i = i$ for all i is said to be the *flag variety* of V. (See for example [26, §3.3] for details.)

Definition 5.1 Let G be a linear algebraic group. A variety X is a *G-space* if there exists a group action $G \times X \to X$, $(g, x) \mapsto g.x$, of G on X which is also a morphism of varieties. Let X and Y be G-spaces. A *morphism of G-spaces* is a morphism of varieties $\varphi : X \to Y$ such that $\varphi(g.x) = g.\varphi(x)$ for all $g \in G$, $x \in X$. A G-space X is said to be *homogeneous* if the action of G is transitive on X.

Proposition 5.2 *Let X be a G-space.*

(a) *For each $x \in X$, the stabilizer $G_x := \{g \in G \mid g.x = x\}$ is a closed subgroup.*

(b) *The fixed point set $X^G := \{x \in X \mid g.x = x \text{ for all } g \in G\}$ is closed.*

Proof By the definition of G-spaces, for any $x \in X$ the *orbit map*

$$\varphi : G \longrightarrow X, \qquad g \longmapsto g.x,$$

is a morphism of varieties. Hence $\varphi^{-1}(x) = G_x$ is closed, so assertion (a) follows. Part (b) is Exercise 10.18. $\square$

Example 5.3 We describe two important sources for G-spaces.

(1) Equip G with the structure of a G-space by considering the action of G on itself by conjugation. Then Proposition 5.2 shows that centralizers of elements and the center $Z(G)$ are closed.

(2) Suppose that V is a finite-dimensional vector space over k. A morphism $\varphi : G \to \mathrm{GL}(V)$ (of algebraic groups) is called a *rational representation* of G, and the space V is called a *(rational) kG-module*. Then, V is a G-space via the action $(g, v) \mapsto \varphi(g)v$. Furthermore, G also acts on the associated projective space $\mathbb{P}(V)$ via $(g, \langle v \rangle) \mapsto \langle \varphi(g)v \rangle$. With this action $\mathbb{P}(V)$ becomes a G-space.

In contrast to the situation for stabilizers considered above, it is not true, in general, that orbits are always closed. But at least we have the following weaker statement:

Proposition 5.4 *Let $X \neq \emptyset$ be a G-space.*

(a) *Every orbit $G.x$ is open in its closure.*

(b) *Orbits of minimal dimension are closed.*

Proof By Proposition 1.6 (which continues to hold in this more general setting) applied to the orbit map $G \to X$, $g \mapsto g.x$, the orbit $G.x$ contains an open subset Y of its closure. As $G.x$ is the union of G-translates of Y, (a) follows.

For (b), note that for $x \in X$, $g.\overline{G.x}$ is closed and contains $G.x$. Therefore, $\overline{G.x} \subseteq g.\overline{G.x}$. Since the same is true for g^{-1}, we get $g.\overline{G.x} = \overline{G.x}$ for all $g \in G$. It follows that $\overline{G.x}$ is a union of G-orbits.

Choose $x \in X$ with $\dim G.x$ minimal. If $G.x$ is not closed, we claim that the union of G-orbits $\overline{G.x} \setminus G.x$ is of strictly smaller dimension, contradicting minimality. To see this, let $Y \subseteq \overline{G.x}$ be an irreducible component intersecting $G.x$. Then $G.x \cap Y$ is open in Y; hence $(\overline{G.x} \setminus G.x) \cap Y$ is closed in Y and so by Proposition 1.22 has smaller dimension. (As pointed out above, the proof of the latter result goes through in that case as well.) $\square$

5.2 Existence of rational representations

One expects that a transitive group action of an algebraic group G will correspond to the action of G on a coset space G/H, for a subgroup $H \leq G$. In order to make this identification, one must define a structure of variety on the coset space G/H.

Now, if $H \leq G$ is a closed subgroup, and $I \lhd k[G]$ denotes the ideal of functions vanishing on H, then it is easily seen that

$$H = \{x \in G \mid \rho_x(I) \subseteq I\},$$

with ρ_x the right translation by x defined in Section 2.1.

The next theorem (see [32, §11.2]) is the first main ingredient in the construction of quotients:

Theorem 5.5 (Chevalley) *Let $H \leq G$ be a closed subgroup of a linear algebraic group G. Then there exists a rational representation $\varphi : G \to \mathrm{GL}(V)$ and a one-dimensional subspace $W \leq V$ such that $H = \{g \in G \mid \varphi(g)W = W\}$.*

Proof The ideal $I \lhd k[G]$ of functions vanishing on H is finitely generated, say $I = (F_j \mid j \in J)$ for some finite set J. Proposition 2.4 then implies that there exists a finite-dimensional G-invariant subspace X of $k[G]$ containing the F_j, $j \in J$, and a corresponding rational representation $\rho : G \to \mathrm{GL}(X)$. Then $M := X \cap I$ is H-invariant. Conversely, if $x \in G$ with $\rho_x(M) = M$, then, since M generates the ideal I we have

$$\rho_x(I) = \rho_x(M)\rho_x(k[G]) = Mk[G] = I,$$

so $H = \{x \in G \mid \rho_x(M) = M\}$.

If $d = \dim M$, we then set $V = \wedge^d X$, the dth exterior power of X, with the representation $\varphi : G \to \mathrm{GL}(V)$ induced by the natural (rational) G-action. Then the one-dimensional subspace $W = \wedge^d M$ of V is clearly $\varphi(H)$-invariant.

Now assume that $\varphi(g)(W) = W$ for some $g \in G$. Let $w_1, \ldots, w_d$ be a basis of M and $w_{l+1}, \ldots, w_{l+d}$ be a basis of $g(M)$. Then $\varphi(g)(w_1 \wedge \cdots \wedge w_d) \in W$ by assumption. But by the choice of the second basis, it is also a multiple of $w_{l+1} \wedge \cdots \wedge w_{l+d}$. But this then implies that $\rho_g(M) = M$, so $g \in H$. Hence $W \leq V$ satisfy the assertions of the theorem. $\square$

Let's point out two important consequences. The first is the characterization of linear algebraic groups already stated in Theorem 1.7:

Corollary. *Any linear algebraic group can be embedded as a closed subgroup into* GL_n *for some* n.

Proof Choosing $H = 1$ in Theorem 5.5, we obtain a faithful rational representation $\rho : G \to \mathrm{GL}(V) \cong \mathrm{GL}_n$, $n = \dim(V)$, of our linear algebraic group G in GL_n, hence an embedding as a closed subgroup. □

Secondly, for arbitrary $H \leq G$, and V as in Theorem 5.5, let $\overline{v} \in \mathbb{P}(V)$ be the point in $\mathbb{P}(V)$ corresponding to the line $\langle v \rangle$ stabilized by H. Set $X = G.\overline{v} \subseteq \mathbb{P}(V)$. Then X is a homogeneous G-space and the action provides a surjective map $\varphi : G \to X$, $g \mapsto g.\overline{v}$, with fibers the cosets of H, which induces a bijection $\overline{\varphi} : G/H \to X$.

Using this bijection $\overline{\varphi}$, we can endow G/H with the structure of a variety. Indeed, the orbit closure $\overline{G.\overline{v}} \subseteq \mathbb{P}(V)$ is a closed subset of $\mathbb{P}(V)$, so it is a projective variety. By Proposition 5.4(a), $G.\overline{v}$ is open in its closure. Therefore, $G.\overline{v}$ is what is called a *quasi-projective variety*.

We call G/H, endowed with this structure of a variety, the *quotient space of G by H*. Then, by construction, the natural map $\pi : G \to G/H$ is a morphism of varieties. It can be shown that the variety structure on G/H obtained in this way is independent of the chosen rational representation [32, 12.4]. Moreover by Proposition 1.19 we have:

Proposition 5.6 *Let $H \leq G$ be a closed subgroup of a linear algebraic group G. Then* $\dim(G/H) = \dim(G) - \dim(H)$.

In the case of a normal subgroup, even more is true (see [4, Thm. 6.8]):

Proposition 5.7 *Let H be a closed normal subgroup of a linear algebraic group G. Then:*

(a) *G/H is an affine variety.*
(b) *G/H with the usual group structure is a linear algebraic group.*

About the proof. We apply Theorem 5.5 to obtain a morphism $\varphi : G \to \mathrm{GL}(V)$ and a one-dimensional subspace $L \subseteq V$ whose stabilizer is H. Now H acts on L by scalar multiplications, so there is an associated character $H \to k^{\times}$ which defines this action. Given a character $\chi \in X(H)$, set

$$V_\chi = \{v \in V \mid \varphi(h)v = \chi(h)v \text{ for all } h \in H\}.$$

Then as H is normal in G, G permutes the V_χ, one of which contains L. As the sum of the V_χ is direct there are only finitely many χ for which $V_\chi \neq 0$. We may then replace V by this sum and assume henceforth that V is the direct sum of V_χ, with H acting as scalar multiplications on each V_χ.

Let $W = \{x \in \text{End}(V) \mid x(V_\chi) \subseteq V_\chi \text{ for all } \chi \in X(H)\}$, which is naturally isomorphic to $\bigoplus_\chi \text{End}(V_\chi)$. Now $\text{GL}(V)$ acts via conjugation on $\text{End}(V)$ and $\varphi(G) \leq \text{GL}(V)$ stabilizes W. Thus we have a group homomorphism $\psi : G \to \text{GL}(W)$, which is a rational representation. We claim that $\ker \psi = H$. Indeed, if $g \in H$ then $\varphi(g)$ acts as a scalar on each V_χ and so conjugating by $\varphi(g)$ is the identity on W. Conversely, if $\psi(g) = 1$, then $\varphi(g)$ stabilizes each V_χ and commutes with $\text{End}(V_\chi)$, which implies that $\varphi(g)$ acts as a scalar on each V_χ. In particular $\varphi(g)$ stabilizes the line L and by Theorem 5.5, $g \in H$. We now have a group isomorphism $G/H \cong \text{im}(\psi)$. To see that this is indeed an isomorphism of algebraic varieties, refer to [4, Thm. 6.8]. $\qquad\square$

Remarks 5.8 (a) In the situation of Proposition 5.7 let $\pi : G \to G/H$ be the canonical surjection. One checks that $\pi^* : k[G/H] \to k[G]$ is injective. Hence $k[G/H]$ may be viewed as a subalgebra of $k[G]$. One can show that

$$k[G/H] \cong \{f \in k[G] \mid \rho_x(f) = f \text{ for all } x \in H\},$$

that is, $k[G/H]$ is precisely the algebra of H-invariant functions on G.

(b) It follows from Proposition 5.7 that the quotient G/G' considered in the proof of Theorem 4.4 is a linear algebraic group.

Example 5.9 Let $G = \text{GL}_n$, $H = Z(G) = \{tI_n \mid t \in k^\times\}$. Then the *projective general linear group* $\text{PGL}_n := \text{GL}_n/Z(\text{GL}_n)$ is a linear algebraic group of dimension

$$\dim(\text{PGL}_n) = \dim(\text{GL}_n) - \dim(Z(\text{GL}_n)) = n^2 - 1$$

(see Exercise 10.17 for a direct proof).

Similarly one obtains the *projective conformal symplectic group*

$$\text{PCSp}_{2n} := \text{CSp}_{2n}/Z(\text{CSp}_{2n})$$

as well as the *projective conformal orthogonal group* and its connected component of the identity

$$\text{PCO}_{2n} := \text{CO}_{2n}/Z(\text{CO}_{2n}), \qquad \text{PCO}_{2n}^\circ = \text{CO}_{2n}^\circ/Z(\text{CO}_{2n}^\circ)$$

as linear algebraic groups.

6

Borel subgroups

As seen in Section 4.2, the structure of connected solvable linear algebraic groups is well-understood. We intend to exploit this by studying a particular family of connected solvable subgroups of an arbitrary linear algebraic group.

By the Lie–Kolchin Theorem any connected solvable subgroup $G \leq \mathrm{GL}_n$ can be embedded in T_n. In particular G stabilizes a flag $\mathcal{F} : 0 = V_0 \subset V_1 \subset \cdots \subset V_n = k^n$ of subspaces. Moreover, for G an arbitrary closed subgroup of GL_n, it is clear that the stabilizer $G_\mathcal{F} \leq G$ of any such flag is a solvable group, and the quotient variety $G/G_\mathcal{F}$ is a quasi-projective variety, i.e., an open subset of a projective space. (See the remarks at the beginning of Chapter 5 and Proposition 5.4.) Now if we choose $\mathcal{F}$ such that its G-orbit is of minimal dimension, then this orbit is in fact closed (loc. cit.) and so a projective space. We will obtain such minimal-dimensional orbits by choosing $G_\mathcal{F}$ of maximal possible dimension among flag stabilizers. This leads us to our definition of *Borel subgroups* (see Definition 6.3 below).

6.1 The Borel fixed point theorem

The principal ingredient for the study of Borel subgroups is the following fixed point theorem:

Theorem 6.1 (Borel fixed point theorem)   *Let G be a connected, solvable linear algebraic group acting on a non-empty projective G-space X. Then there exists $x \in X$ such that $g.x = x$ for all $g \in G$.*

Sketch of proof   We argue by induction on $\dim(G)$. If $\dim(G) = 0$ then $G = 1$, so all $x \in X$ are fixed points. Now assume that $\dim(G) > 0$. By Proposition 1.18, $G' := [G, G]$ is connected, of smaller dimension by Proposition 1.22 (since G is solvable). Thus, the set Y of fixed points of G' on X

is non-empty. By Proposition 5.2(b), Y is closed in X, hence also projective, and it is stabilized by G since G' is normal. Thus, we may replace X by Y.

Then, all stabilizers G_x, for $x \in X$, contain G', hence are normal in G. So G/G_x is an affine variety, by Proposition 5.7. By Proposition 5.4(b) there exists $x \in X$ whose orbit is closed, hence projective. Then the canonical morphism $G/G_x \to G.x$ is bijective, with G/G_x affine irreducible and $G.x$ projective. One must now appeal to a geometric argument based upon [32, Lemma 21.1], to see that the above conditions force G/G_x to be reduced to one point, that is, x is a fixed point for G. $\square$

Remarks 6.2 (a) The conclusion of the theorem in fact holds more generally when X is just assumed to be a complete variety (see [32, Thm. 21.2]).

(b) The Lie–Kolchin Theorem (Theorem 4.1) asserts that for G a connected, solvable subgroup of $\mathrm{GL}(V)$, there exists $0 \neq v \in V$ such that $g.v \in \langle v \rangle$ for all $g \in G$, that is, G, acting on $\mathbb{P}(V)$, has a fixed point. Borel's fixed point theorem can hence be used to give a short proof of this assertion.

Definition 6.3 Let G be a linear algebraic group. A *Borel subgroup* of G is a closed, connected, solvable subgroup B of G which is maximal with respect to all these properties.

Clearly, Borel subgroups exist: just take a closed, connected, solvable subgroup of maximal possible dimension. In fact, there is not much choice (see [32, §21.3]):

Theorem 6.4 *Let G be a linear algebraic group. Then:*

(a) *All Borel subgroups of G are conjugate.*

(b) *If G is connected, G/B is projective for any Borel subgroup B of G.*

Proof of (a) assuming (b) Firstly, we consider the case $G = \mathrm{GL}_n$. The group T_n of upper triangular matrices is a closed, connected, solvable subgroup of G. If H is any closed, connected, solvable subgroup of G, then Corollary 4.2 gives $H \leq g^{-1}\mathrm{T}_n g$ for some $g \in G$. Therefore, T_n is maximal among closed, connected, solvable subgroups of G, hence a Borel subgroup, and any Borel subgroup is conjugate to T_n.

For the general case we can replace G by G° as Borel subgroups of G° are Borel subgroups of G. Let H be a Borel subgroup of G. Assuming G/B is projective for a fixed Borel subgroup B, H acts on the projective space G/B via $gB \mapsto hgB$. By Theorem 6.1, there exists a fixed point, that is, there exists $g \in G$ such that $HgB = gB$. Therefore, $g^{-1}Hg \leq B$. By maximality of H, we conclude that $H = gBg^{-1}$. $\square$

As in the case of solvable groups we say that a subtorus $T \leq G$ is a *maximal torus* of G if it is maximal among subtori with respect to inclusion.

Corollary 6.5 *Let G be a linear algebraic group. Then all maximal tori of G are conjugate.*

Proof Let T_1, T_2 be maximal tori of G. Since T_i is connected solvable, there exists a Borel subgroup B_i of G with $T_i \leq B_i$, for $i = 1, 2$. By Theorem 6.4, there exists $g \in G$ such that $B_2 = B_1^g$. Therefore, $T_1^g \leq B_2$, hence T_2 and T_1^g are two maximal tori of the connected solvable group B_2. By Theorem 4.4, there exists $b \in B_2$ such that $T_2 = T_1^{gb}$. $\qquad\square$

Definition 6.6 The *rank* of a linear algebraic group G is the dimension of a maximal torus of G, denoted by $\mathrm{rk}(G)$.

Note that by Corollary 6.5, as for connected solvable groups, maximal tori are the tori of maximal dimension so the rank is well-defined.

Example 6.7 Let's find Borel subgroups, maximal tori and the rank of classical groups.

(1) For $G = \mathrm{GL}_n$, T_n is a Borel subgroup (see the proof of Theorem 6.4), and D_n is a maximal torus of T_n, hence of G. Therefore, $\mathrm{rk}(\mathrm{GL}_n) = \dim(D_n) = n$.

(2) For $G = \mathrm{SL}_n$, if we repeat the argument given for GL_n, we observe that $T_n \cap \mathrm{SL}_n$ is a Borel subgroup and $D_n \cap \mathrm{SL}_n$ is a maximal torus of SL_n. The exact sequence

$$1 \longrightarrow D_n \cap \mathrm{SL}_n \longrightarrow D_n \xrightarrow{\det} \mathbf{G}_{\mathrm{m}} \longrightarrow 1$$

then shows with Corollary 1.20 that $\mathrm{rk}(\mathrm{SL}_n) = n - 1$.

(3) For the special orthogonal group SO_{2n},

$$T := \left\{ \left(\begin{matrix} t_1 & & & & & 0 \\ & \ddots & & & & \\ & & t_n & & & \\ & & & t_n^{-1} & & \\ & & & & \ddots & \\ 0 & & & & & t_1^{-1} \end{matrix} \right) \;\middle|\; t_i \in k^\times \right\} \leq \mathrm{SO}_{2n}$$

is a connected subgroup isomorphic to $\mathbf{G}_{\mathrm{m}}^n$, hence a torus of dimension $\dim(T) = n$. Therefore $\mathrm{rk}(\mathrm{SO}_{2n}) \geq n$. Now let $T_1 \geq T$ be a maximal torus of SO_{2n}. Choose $s \in T$ with distinct eigenvalues (which is clearly possible since k is infinite). Then

$$T_1 \leq C_{\mathrm{SO}_{2n}}(T) \leq C_{\mathrm{GL}_{2n}}(T) \leq C_{\mathrm{GL}_{2n}}(s) = D_{2n}.$$

But it is easily checked from the definition of GO_{2n} in Section 1.2 that $D_{2n} \cap GO_{2n} = T$, so $T = T_1$ is a maximal torus and $rk(SO_{2n}) = n$.

Analogously one shows that $rk(Sp_{2n}) = rk(SO_{2n+1}) = n$ (see Exercise 10.19).

(4) Let $B := SO_{2n} \cap T_{2n}$. A short calculation shows that $\begin{pmatrix} A_1 & A_2 \\ 0 & A_3 \end{pmatrix} \in SO_{2n}$
if and only if $A_1^{tr} K_n A_3 = K_n$, and $A_3^{tr} K_n A_2 = -A_2^{tr} K_n A_3$, that is, $A_3 = K_n A_1^{-tr} K_n$, and $A_3^{tr} K_n A_2 = K_n A_1^{-1} A_2$ is skew-symmetric. (Recall that, by our definition, $SO_{2n} = GO_{2n}^\circ$ in any characteristic.) In particular, B is the product of

$$\left\{ \begin{pmatrix} I_n & K_n S \\ 0 & I_n \end{pmatrix} \;\middle|\; S \in k^{n \times n} \text{ skew-symmetric} \right\}$$

with $\{\mathrm{diag}(A, K_n A^{-tr} K_n) \mid A \in T_n\} \cong T_n$, of dimension

$$\dim(B) = \begin{cases} \binom{n}{2} + \binom{n+1}{2} = n^2, & \text{if } n \geq 2 \\ \binom{n+1}{2} = n^2, & \text{if } n = 1. \end{cases}$$

We claim that B is a Borel subgroup of SO_{2n}. Let $B_1 \geq B$ be a Borel subgroup of SO_{2n}. Then B_1 lies in a Borel subgroup of GL_{2n}, so it stabilizes a complete flag on $V = k^{2n}$. The only one-dimensional subspace invariant under B is $\langle v_1 \rangle$, generated by the first standard basis vector. Thus, this must also be fixed by B_1. But then B_1 also fixes $\langle v_1 \rangle^{\perp}$, a subspace of codimension 1. Now applying induction to the action of B_1, B on $\langle v_1 \rangle^{\perp} / \langle v_1 \rangle$ one sees that B_1 can only fix the standard flag, hence is contained in T_{2n}. It follows that $B_1 = B$, as claimed.

Analogously one shows that $Sp_{2n} \cap T_{2n}$, respectively $SO_{2n+1} \cap T_{2n+1}$ is a Borel subgroup of Sp_{2n}, respectively SO_{2n+1} (see Exercise 10.19).

6.2 Properties of Borel subgroups

Here we collect some further results which illustrate the importance of Borel subgroups and which will be needed later on.

Proposition 6.8 *Let G be connected, B a Borel subgroup of G. Then:*

(a) *An automorphism of G which fixes B pointwise is the identity.*
(b) $Z(G)^\circ \subseteq Z(B) \subseteq C_G(B) = Z(G)$.

Proof For (a) let σ be an automorphism of G which centralizes B. Then the morphism $\varphi : G \to G$, $x \mapsto \sigma(x)x^{-1}$, sends B to 1, so factors through

G/B. By Theorem 6.4(b), G/B is projective. Now the image of a projective variety under a morphism is closed (see for example [32, Prop. 6.1(c) and Thm. 6.2]), so $\varphi(G)$ is closed in G, hence affine. Now for any two distinct points in this affine variety there exists a function f separating them, whence $f \circ \varphi$ would be a globally defined non-constant function on the projective variety G/B, which does not exist. Thus, $\varphi(G) = 1$ as claimed.

For (b), $Z(G)^\circ$ is a closed connected solvable subgroup and hence lies in some Borel subgroup of G, and so by Theorem 6.4(a) in all of them. The second inclusion is obvious, as is $Z(G) \subseteq C_G(B)$. Finally, for $x \in C_G(B)$, conjugation by x satisfies the assumptions of (a), whence $x \in Z(G)$. $\square$

Corollary 6.9 *Let G be a connected linear algebraic group with nilpotent Borel subgroups. Then G is nilpotent. In particular, any connected two-dimensional group is solvable.*

This is an easy consequence of Proposition 6.8, see Exercise 10.20.

Theorem 6.10 *Let G be a connected linear algebraic group with Borel subgroup B. Then $G = \bigcup_{g \in G} g^{-1}Bg$.*

For GL_n we may take $B = \mathrm{T}_n$, and then the theorem follows from the existence of Jordan normal forms. As $\mathrm{GL}_n = \mathrm{SL}_n \cdot Z(\mathrm{GL}_n)$, this also applies to SL_n. For the general case see [32, §22.2].

Note that, in contrast, for a finite group G we can never have $G = \bigcup_{g \in G} g^{-1}Hg$ for a proper subgroup $H < G$.

As an immediate consequence we obtain:

Corollary 6.11 *Let G be a connected linear algebraic group. Then:*

(a) *Every semisimple element of G lies in a maximal torus.*
(b) *Every unipotent element of G lies in a closed connected unipotent subgroup.*
(c) *The maximal, closed, connected unipotent subgroups of G are all conjugate and they are of the form B_u where B is a Borel subgroup of G.*

Proof Every $g \in G$ lies in a Borel subgroup B of G by Theorem 6.10. If g is semisimple then it lies in a maximal torus of B, hence of G, by Corollary 4.5. If g is unipotent in B, then $u \in B_u$ which is closed by Theorem 4.4(a). If $U \le G$ is closed connected unipotent, it is solvable by Corollary 2.10, so contained in a Borel subgroup B, hence in its unipotent normal subgroup B_u again by Theorem 4.4(a), which shows (c). $\square$

A further important result asserts that Borel subgroups of connected groups are self-normalizing. This is easy to see for GL_n, taking the Borel subgroup T_n. In the general case let $N_G(B)$ be the normalizer of a Borel subgroup B of the connected group G. Then clearly B is also a Borel subgroup of $N_G(B)^\circ$, and it is normal, so $N_G(B)^\circ = B$ by Theorem 6.10. For the proof of the following stronger assertion one must use properties of complete varieties, see [32, Thm. 23.1].

Theorem 6.12 *Let G be a connected linear algebraic group and B a Borel subgroup of G. Then $N_G(B) = B$.*

If $N, N' \trianglelefteq G$ are closed connected (and solvable) normal subgroups, then NN' is closed connected normal (and solvable) by Proposition 1.16. This gives meaning to the following definition:

Definition 6.13 The maximal closed connected solvable normal subgroup of a linear algebraic group G is called the *radical* $R(G)$ of G.

By Theorem 4.4, $R(G)_u$ is a normal connected unipotent subgroup of the connected solvable group $R(G)$. Any closed connected normal unipotent subgroup of G is solvable by Corollary 2.10, hence contained in $R(G)$, hence in $R(G)_u$. Thus $R(G)_u$ is the maximal closed connected normal unipotent subgroup of G, the so-called *unipotent radical* $R_u(G)$ of G. Clearly $R_u(G) \leq R(G) \leq G^\circ$.

In the case where $k = \overline{\mathbb{F}_p}$, the unipotent radical is the largest connected normal subgroup consisting entirely of p-elements, so the analogue of the maximal normal p-subgroup $\mathcal{O}_p(G)$ for a finite group G.

Definition 6.14 A linear algebraic group G is called *reductive* if $R_u(G) = 1$. It is called *semisimple* if it is connected and $R(G) = 1$.

Unipotent groups can have a very complicated structure, see for example Exercise 10.16. On the other hand, connected reductive and semisimple groups can be classified; this will be discussed in Section 9.2.

It is easy to see that in general, for a connected group G, $G/R(G)$ is semisimple, and $G/R_u(G)$ is reductive (see Exercise 10.21).

Definition 6.15 The *semisimple rank* of a reductive group G is defined to be $\mathrm{rk}_{\mathrm{ss}}(G) := \mathrm{rk}(G/R(G))$. We will see later (Corollary 8.22) that for connected reductive groups $\mathrm{rk}_{\mathrm{ss}}(G) = \mathrm{rk}([G,G])$.

The radical of G is controlled by its Borel subgroups (see Exercise 10.22):

Proposition 6.16 *Let G be connected. Then $R(G) = (\bigcap_B B)^\circ$, where B runs over all Borel subgroups of G.*

Example 6.17 (1) If G is semisimple, T a torus, then $G \times T$ is reductive.

(2) If G is solvable connected, then by Theorem 4.4 $R_u(G) = G_u$ and $R(G) = G$. So a connected reductive solvable group is a torus, a semisimple solvable group is trivial.

(3) The subgroup $\left\{ \left(\begin{array}{c|c} A & * \\ \hline 0 & * \end{array} \right) \;\middle|\; A \in \mathrm{GL}_2 \right\}$ of GL_3 is not reductive, since
$\left\{ \left(\begin{array}{c|c} I_2 & * \\ \hline 0 & 1 \end{array} \right) \right\}$ is a non-trivial connected normal unipotent subgroup.

(4) $G = \mathrm{GL}_n$ is reductive, since by Proposition 6.16 and Example 6.7(1)

$$R(G) \le \mathrm{T}_n \cap \mathrm{T}_n^- := \left\{ \left(\begin{array}{ccc} * & & * \\ & \ddots & \\ 0 & & * \end{array} \right) \right\} \cap \left\{ \left(\begin{array}{ccc} * & & 0 \\ & \ddots & \\ * & & * \end{array} \right) \right\} = \mathrm{D}_n,$$

hence $R_u(G) = R(G)_u = 1$. But GL_n is **not** semisimple, as $Z(\mathrm{GL}_n) = \{tI_n \mid t \in k^\times\} \cong \mathbf{G}_m$ is a connected solvable normal subgroup. In fact, $R(\mathrm{GL}_n) = Z(\mathrm{GL}_n)$ by Proposition 6.20, so PGL_n is semisimple.

(5) $G = \mathrm{SL}_n$ is semisimple, since as in (4), SL_n is reductive and

$$R(G) = Z(G)^\circ = (\{tI_n \mid t \in k^\times\} \cap \mathrm{SL}_n)^\circ = \{tI_n \mid t^n = 1\}^\circ = 1$$

by Proposition 6.20(a) below.

Using properties of Borel subgroups, the result of Theorem 3.10 on centralizers of tori can be strengthened:

Proposition 6.18 *Let G be a connected linear algebraic group and $T \le G$ a maximal torus. Then $C_G(T)^\circ$ is nilpotent and T is its maximal torus.*

Proof Set $C = C_G(T)^\circ$. The central torus T of C is contained in any Borel subgroup B of C and is a maximal torus thereof. Hence $B/T \cong B_u$ is a nilpotent group, and since T is central in B, B is also nilpotent. By Corollary 6.9 this implies that $C = B$ is nilpotent. $\qquad\square$

We can say something as well about centralizers of arbitrary subtori, see [66, Thm 6.4.7]:

Theorem 6.19 *Let S be a torus of a connected linear algebraic group G.*

(a) *The centralizer $C_G(S)$ is connected.*

(b) *If B is a Borel subgroup of G containing S, then $C_G(S) \cap B$ is a Borel subgroup of $C_G(S)$ and all Borel subgroups of $C_G(S)$ are obtained in this way.*

About the proof of (a). For $c \in C_G(S)$ let B be a Borel subgroup of G containing c (which exists by Theorem 6.10). Set $X = \{xB \in G/B \mid x^{-1}cx \in B\}$. Then one can show that X is a closed subvariety of G/B, and hence is a projective variety. Now S acts on X via left multiplication and so by Theorem 6.1, there exists a fixed point $xB \in X$ with $x^{-1}Sx \leq B$. That is, there exists a Borel subgroup containing both S and c. Thus, we are reduced to the case of a connected solvable group, for which the result is true by Proposition 4.6. □

It is quite difficult to exploit the fact that a group is reductive, that is, has no non-trivial connected unipotent normal subgroup. At least, we can show:

Proposition 6.20 *Let G be connected reductive. Then:*

(a) $R(G) = Z(G)^{\circ}$ *is a torus.*
(b) $R(G) \cap [G, G]$ *is finite.*
(c) $[G, G]$ *is semisimple.*

Proof (a) $R(G)$ is connected solvable, so $R(G) = R(G)_u \rtimes T$, where T is a torus, by Theorem 4.4(b). Since G is reductive, $R(G)_u = R_u(G) = 1$, hence $R(G) = T$ is a torus. Clearly $Z(G)^{\circ} \subseteq R(G)$. By Theorem 3.10, the factor group $G/C_G(R(G)) = N_G(R(G))/C_G(R(G))$ is finite. By Proposition 1.13(c), $C_G(R(G))$ contains G°, so we have equality $C_G(R(G)) = G$ since G is connected, and hence $R(G) \subseteq Z(G)^{\circ}$.

(b) Choose an embedding $G \hookrightarrow \mathrm{GL}(V)$ according to Theorem 1.7. The torus $R(G) = Z(G)^{\circ}$ acts diagonally on V. The common eigenspaces give a decomposition $V = \bigoplus_{\chi} V_{\chi}$, where χ runs over $X(R(G))$ (so $t.v = \chi(t)v$ for $v \in V_{\chi}$, $t \in R(G)$). Now $G \leq C_{\mathrm{GL}(V)}(R(G))$ by (a), so G stabilizes each of the V_{χ}. Thus it consists of block diagonal matrices: $G \leq \prod_{\chi} \mathrm{GL}(V_{\chi})$ and so $[G, G] \leq \prod_{\chi} \mathrm{SL}(V_{\chi})$. Thus $[G, G] \cap R(G)$ consists of matrices

$$\begin{pmatrix} \chi_1(t)I_{d_1} & & \\ & \chi_2(t)I_{d_2} & \\ & & \ddots \end{pmatrix}$$

with $d_i := \dim V_{\chi_i}$ and $\chi_i(t)^{d_i} = 1$. But there exist only finitely many such matrices.

(c) By Proposition 1.18, $[G, G]$ is connected. As $R([G, G])$ is a characteristic normal subgroup of $[G, G]$, $R([G, G]) \leq R(G)$. Thus $R([G, G]) \leq (R(G) \cap [G, G])^{\circ} = 1$ by part (b), hence $[G, G]$ is semisimple. □

We also have that $G = R(G)[G, G]$, but the proof of this statement requires considerably more work (see Corollary 8.22 later).

7

The Lie algebra of a linear algebraic group

To go further in our investigation of reductive groups we require the concept of the Lie algebra of a linear algebraic group. This allows one to "linearize" many questions. Here, we assume that the reader is familiar with the notion of a Lie algebra over a field. We will write $\mathfrak{gl}_n$ for the Lie algebra of $n \times n$-matrices over k with the Lie bracket $[X, Y] = XY - YX$.

7.1 Derivations and differentials

Let A be a k-algebra.

Definition 7.1   A k-linear map $D : A \to A$ with $D(fg) = fD(g) + D(f)g$ for all $f, g \in A$ is called a *derivation* of A.

An easy calculation shows that if D_1, D_2 are derivations of A, then so is

$$[D_1, D_2] := D_1 \circ D_2 - D_2 \circ D_1.$$

This endows $\mathrm{Der}_k(A)$, the k-vector space of derivations of A, with the structure of a Lie algebra.

We now take $A = k[G]$, the coordinate ring of a linear algebraic group G. Then, for $x \in G$ we have an action $\lambda_x : k[G] \to k[G]$ on $k[G]$ via $(\lambda_x.f)(g) := f(x^{-1}g)$ for $f \in k[G]$, $g \in G$.

Definition 7.2   The *Lie algebra of* G is the subspace

$$\mathrm{Lie}(G) := \{D \in \mathrm{Der}_k(k[G]) \mid D\lambda_x = \lambda_x D \text{ for all } x \in G\}$$

of *left invariant derivations* of $k[G]$, a Lie subalgebra of $\mathrm{Der}_k(k[G])$.

Example 7.3   Let's compute the Lie algebras of one-dimensional connected groups.

(1) Let $G = \mathbf{G}_a$, with $k[G] = k[T]$. Any derivation $D \in \mathrm{Der}_k(k[G])$ is determined by its value $D(T) =: p(T) \in k[T]$ on T. For $x \in G$ we have $\lambda_x T = T - x$. Therefore $\lambda_x f(T) = f(T - x)$ for $f \in k[T]$. So D is left invariant if and only if

$$(\lambda_x D)(T) = \lambda_x p(T) = p(T - x) \quad \text{equals} \quad (D\lambda_x)(T) = D(T - x) = p(T)$$

for all $x \in G$. Thus, $p(T)$ is constant and $D = a\frac{d}{dT}$ for $a = p(1) \in k$. Therefore $\mathrm{Lie}(G) = \langle \frac{d}{dT} \rangle_k \cong k$ is the unique one-dimensional (commutative) Lie algebra over k.

(2) Let $G = \mathbf{G}_m$, with $k[G] = k[T, T^{-1}]$. Again, $D \in \mathrm{Der}_k(k[G])$ is determined by $D(T) =: p(T)$. Here, for $x \in G$ we have $\lambda_x T = x^{-1}T$, so $\lambda_x f(T) = f(x^{-1}T)$ for $f \in k[G]$; hence left invariance gives $p(T) = aT$ for some $a \in k$, so $D = aT\frac{d}{dT}$. Thus in this case as well we have $\mathrm{Lie}(G) = \langle T\frac{d}{dT} \rangle_k \cong k$.

We now give a second, more geometric construction for the Lie algebra of an algebraic group G.

For an affine variety X we define the *tangent space of X at $x \in X$* by

$$T_x(X) := \{\delta : k[X] \to k \text{ linear} \mid \delta(fg) = f(x)\delta(g) + \delta(f)g(x) \text{ for } f, g \in k[X]\}$$

(the k-vector space of *point derivations at x*). In case X is a linear algebraic group G, then as G acts homogeneously on itself by left translation, the tangent space at any group element $g \in G$ is naturally isomorphic to $T_1(G)$. Thus, we will consider the tangent space $T_1(G)$. Now

$$\Theta : \mathrm{Lie}(G) \to T_1(G), \qquad \Theta(D)(f) := D(f)(1),$$

is a k-linear map. In fact, a straightforward calculation shows this to be an isomorphism of vector spaces, see [32, Thm. 9.1]. By this, the Lie algebra structure on $\mathrm{Lie}(G)$ can be transported to $T_1(G)$. With this alternative interpretation one can establish the following (see [66, Thm. 4.3.7, Cor. 4.4.6, §4.1.7] for (a) and (b), [32, §5.1] for (c)):

Theorem 7.4 *Let G, G_1, G_2 be linear algebraic groups. Then:*

(a) $\mathrm{Lie}(G) = \mathrm{Lie}(G^\circ)$.
(b) $\dim G = \dim G^\circ = \dim(\mathrm{Lie}(G))$.
(c) $\mathrm{Lie}(G_1 \times G_2) \cong \mathrm{Lie}(G_1) \oplus \mathrm{Lie}(G_2)$ *as Lie algebras.*

Example 7.5 For a more elaborate example, let $G = \mathrm{GL}_n$. We claim that $\mathrm{Lie}(\mathrm{GL}_n) \cong \mathfrak{gl}_n$. For this, define

$$\mathfrak{gl}_n \longrightarrow \mathrm{Der}_k(k[G]), \qquad X \longmapsto D_X,$$

by $D_X(T_{ij}) := \sum_{l=1}^n T_{il}X_{lj}$. (Recall that $(k[G] = k[T_{ij}, \frac{1}{\det(T_{ij})}]$.) It is straightforward to check that D_X is a derivation. Now for $g \in G$ write g_{ij}^{-1} for the (i,j)th entry of g^{-1}. Then

$$D_X(\lambda_g T_{ij}) = D_X(\sum_m g_{im}^{-1} T_{mj})$$

$$= \sum_m g_{im}^{-1} D_X(T_{mj})$$

$$= \sum_m g_{im}^{-1} \sum_l T_{ml}X_{lj} = \sum_m \sum_l g_{im}^{-1} T_{ml}X_{lj},$$

while

$$\lambda_g(D_X T_{ij}) = \lambda_g(\sum_l T_{il}X_{lj})$$

$$= \sum_l X_{lj}\lambda_g(T_{il})$$

$$= \sum_l X_{lj} \sum_m g_{im}^{-1} T_{ml} = \sum_l \sum_m X_{lj}g_{im}^{-1} T_{ml}.$$

Hence D_X is a left invariant derivation of $k[G]$. Moreover, one can check that the map $X \mapsto D_X$ defines an injective Lie algebra homomorphism. Since $\dim(\mathfrak{gl}_n) = n^2 = \dim(\mathrm{GL}_n)$, the given map is surjective by Theorem 7.4(b) and we have the desired isomorphism.

Furthermore, the tangent space approach allows one to complete the functorial connection between linear algebraic groups and their Lie algebras.

Definition 7.6 Let $\varphi: X \to Y$ be a morphism of affine varieties. The *differential* $d_x\varphi$ *of* φ *at* $x \in X$ is the map $d_x\varphi: T_x(X) \to T_{\varphi(x)}(Y)$ defined by $d_x\varphi(\delta) := \delta \circ \varphi^*$ for $\delta \in T_x(X)$.

$$k[Y] \xrightarrow{\varphi^*} k[X]$$
$$d_x\varphi(\delta) \searrow \quad \downarrow \delta$$
$$k$$

If $\varphi : G \to H$ is a morphism of algebraic groups we set $d\varphi := d_1\varphi$, the differential at 1. Taking differentials behaves functorially:

Proposition 7.7 Let $X \xrightarrow{\varphi} Y \xrightarrow{\psi} Z$ be morphisms of affine varieties, and $x \in X$. Then:

(a) $d_x(\psi \circ \varphi) = d_{\varphi(x)}\psi \circ d_x\varphi$.

(b) *If $X = G_1$ and $Y = G_2$ are linear algebraic groups and φ is a morphism of algebraic groups, then $d\varphi : \mathrm{Lie}(G_1) \to \mathrm{Lie}(G_2)$ is a Lie algebra homomorphism.*

(c) *If $X = G_1$ and $Y = G_2$ are linear algebraic groups and φ is a morphism of algebraic groups, then φ is an isomorphism of algebraic groups if and only if φ and $d\varphi$ are both bijective.*

The proof is given in [32, §5.4, Thm. 9.1] and [66, Cor. 5.3.3].

Example 7.8 (Computation of some differentials)

(1) Consider the morphism $\mu : G \times G \to G$, $(x, y) \mapsto xy$. For $f \in k[G]$, write $\mu^*(f) = \sum f_i \otimes g_i \in k[G] \otimes k[G]$, so that $f(xy) = \sum f_i(x) g_i(y)$, for all $x, y \in G$. In particular, we have $f = \sum f_i(1) g_i = \sum g_i(1) f_i$. Now let $(X, Y) \in T_1(G \times G) \cong T_1(G) \oplus T_1(G)$ and set $d\mu(X, Y) = Z$. Then

$$Z(f) = d\mu(X, Y)(f) = (X, Y)(\mu^* f) = (X, Y)\left(\sum f_i \otimes g_i\right),$$

which by the product rule for derivatives is equal to $\sum X(f_i) g_i(1) + \sum f_i(1) Y(g_i)$. But this is precisely the action of $X + Y$ on f. So we have $d\mu(X, Y) = X + Y$.

(2) The differential of the inversion morphism $i : G \to G$, $g \mapsto g^{-1}$, is given by $di(X) = -X$ for $X \in T_1(G)$. (See Exercise 10.25.) This fact and the previous example somewhat justify the statement that consideration of the Lie algebra "linearizes" problems.

(3) Let $\mathrm{char}(k) = p > 0$. Consider the bijective morphism

$$\varphi \colon \mathbf{G_a} \longrightarrow U_2, \qquad a \mapsto \begin{pmatrix} 1 & a^p \\ 0 & 1 \end{pmatrix}.$$

Let us identify the map $d\varphi \colon \mathrm{Lie}(\mathbf{G_a}) \to \mathrm{Lie}(U_2)$. We have $k[U_2] = k[T_{12}] \cong k[\mathbf{G_a}] = k[T]$. Now note that

$$\varphi^*(T_{12})(a) = T_{12}(\varphi(a)) = T_{12}\left(\begin{pmatrix} 1 & a^p \\ 0 & 1 \end{pmatrix}\right) = a^p \qquad \text{for } a \in \mathbf{G_a},$$

so $\varphi^*(T_{12}) = T^p$. Thus for $D \in \mathrm{Lie}(\mathbf{G_a})$ we obtain

$$d\varphi(D)(T_{12}) = D(\varphi^*(T_{12})) = D(T^p) = pT^{p-1}D(T) = 0.$$

Hence $d\varphi$ is not injective, and so φ is not an isomorphism of algebraic groups by Proposition 7.7(c). Indeed, on the coordinate rings, $\varphi^* : k[T_{12}] \to k[T]$, $T_{12} \mapsto T^p$, is not surjective.

We now consider Lie algebras of closed subgroups $H \leq G$. If H is defined by an ideal $I \unlhd k[G]$, then by definition $k[H] = k[G]/I$. Any $D \in \mathrm{Lie}(G)$ with $DI \subseteq I$ naturally defines a derivation of $k[H] = k[G]/I$, and similarly $\delta \in T_1(G)$ with $\delta I = 0$ naturally defines an element of $T_1(H)$. With these identifications we have:

Theorem 7.9 *Let $H \leq G$ with vanishing ideal $I \unlhd k[G]$. Then:*

(a) $\mathrm{Lie}(H) = \{D \in \mathrm{Lie}(G) \mid DI \subseteq I\}$, *and* $T_1(H) = \{\delta \in T_1(G) \mid \delta I = 0\}$.

(b) *If $H \unlhd G$ is normal, then $\mathrm{Lie}(H)$ is an ideal in $\mathrm{Lie}(G)$. Moreover, the differential of the canonical projection $G \to G/H$ is the canonical projection $\mathrm{Lie}(G) \to \mathrm{Lie}(G)/\mathrm{Lie}(H)$ and thus induces an isomorphism $\mathrm{Lie}(G/H) \cong \mathrm{Lie}(G)/\mathrm{Lie}(H)$.*

The proof of the first part is a direct calculation, see [32, Lemma 9.4], and uses the identification of $\mathrm{Lie}(H)$ with $T_1(H)$. For the second part see [32, Cor. 10.4A and Sec. 11.5].

Example 7.10   (1) $\mathrm{Lie}(\mathrm{SL}_n) = \mathfrak{sl}_n := \{A \in \mathfrak{gl}_n \mid \mathrm{tr}A = 0\}$ (see Exercise 10.26(b)).

(2) $\mathrm{Lie}(\mathrm{PGL}_n) \cong \mathrm{Lie}(\mathrm{GL}_n)/\mathrm{Lie}(Z(\mathrm{GL}_n)) = \mathfrak{gl}_n/\{cI_n \mid c \in k\}$ by Theorem 7.9(b).

We have the additive Jordan decomposition in $\mathfrak{gl}_n$. For $X \in \mathfrak{gl}_n$ write X_s, X_n for the semisimple, respectively nilpotent part of X. For any closed subgroup $G \leq \mathrm{GL}_n$ we may embed $\mathrm{Lie}(G)$ into $\mathfrak{gl}_n$ by Theorem 7.9(a) and so define semisimple and nilpotent parts of elements of $\mathrm{Lie}(G)$. These satisfy the expected compatibility properties (see [32, Thm. 15.3]):

Proposition 7.11 *Let $G \leq \mathrm{GL}_n$ be a linear algebraic group, $X \in \mathrm{Lie}(G) \subseteq \mathfrak{gl}_n$. Then we have:*

(a) $X_s, X_n \in \mathrm{Lie}(G)$ *and* $[X_s, X_n] = 0$.

(b) X_s, X_n *are independent of the chosen embedding $G \hookrightarrow \mathrm{GL}_n$.*

(c) *If $\varphi \colon G_1 \to G_2$ is a morphism of linear algebraic groups, with differential $d\varphi : \mathrm{Lie}(G_1) \to \mathrm{Lie}(G_2)$, then $(d\varphi)(X)_s = d\varphi(X_s)$ and $(d\varphi)(X)_n = d\varphi(X_n)$ for $X \in \mathrm{Lie}(G_1)$.*

Compare this result with the Jordan decomposition in G (Theorem 2.5).

7.2 The adjoint representation

An important use of the Lie algebra is that it defines for each linear algebraic group G a natural rational representation $G \to \mathrm{GL}(\mathrm{Lie}(G))$ on its Lie algebra, as follows: For $x \in G$ define $\mathrm{Int}_x : G \to G$ by $\mathrm{Int}_x(y) = xyx^{-1}$. Then $d\,\mathrm{Int}_x : \mathrm{Lie}(G) \to \mathrm{Lie}(G)$ is a Lie algebra automorphism. We write $\mathrm{Ad}\,x := d\,\mathrm{Int}_x$ for $x \in G$.

This defines a representation

$$\mathrm{Ad} : G \longrightarrow \mathrm{GL}(\mathrm{Lie}(G)), \qquad x \mapsto \mathrm{Ad}\,x,$$

the *adjoint representation* of G:

Theorem 7.12 *Let G be a linear algebraic group. Then:*

(a) $\mathrm{Ad} : G \to \mathrm{GL}(\mathrm{Lie}(G))$ *is a rational representation of G.*

(b) $d\mathrm{Ad} = \mathrm{ad}$, *where* $(\mathrm{ad}(X))(Y) = [X,Y]$ *for* $X, Y \in \mathrm{Lie}(G)$.

(c) *If G is connected reductive, then* $\ker(\mathrm{Ad}) = Z(G)$.

See [32, Prop. 10.3, Thm. 10.4] for (a) and (b). The statement of (c) will follow from Theorem 8.17, see Exercise 10.32. Thus, if G is connected reductive, Ad is almost faithful: only the center of G is lost.

The following is easily verified: if $H \leq G$ is closed, then $\mathrm{Lie}(H)$ (as a Lie-subalgebra of $\mathrm{Lie}(G)$ according to Theorem 7.9) is $\mathrm{Ad}_G(H)$-invariant and the restriction of the adjoint representation of G to H, acting on $\mathrm{Lie}(H)$, is just the adjoint representation of H, i.e., $\mathrm{Ad}_G|_H = \mathrm{Ad}_H$ on $\mathrm{Lie}(H)$.

Example 7.13 Let $G = \mathrm{GL}_n$, so $\mathrm{Lie}(G) \cong \mathfrak{gl}_n$ by Example 7.5 and for $X \in \mathfrak{gl}_n$ we have the derivation $D_X \in \mathrm{Lie}(G)$ given by $D_X(T_{ij}) = \sum_l T_{il} X_{lj}$. Via the vector space isomorphism Θ defined above, we can associate a point derivation δ_X to X, with $\delta_X(T_{ij}) = (\sum_l T_{il} X_{lj})(I_n) = X_{ij}$.

For $g = (g_{ij}) \in G$, we evaluate $(\mathrm{Int}_g)^*(T_{ij})$: For $x = (x_{ij}) \in G$,

$$(\mathrm{Int}_g)^*(T_{ij})(x) = T_{ij}(gxg^{-1}) = \sum_{l,m} g_{il} g_{mj}^{-1} x_{lm},$$

so $(\mathrm{Int}_g)^*(T_{ij}) = \sum_{l,m} g_{il} g_{mj}^{-1} T_{lm}$. Thus $\mathrm{Ad}\,g = d\mathrm{Int}_g : \mathrm{Lie}(G) \to \mathrm{Lie}(G)$

satisfies

$$(\operatorname{Ad} g)(\delta_X)(T_{ij}) = \delta_X((\operatorname{Int}_g)^* T_{ij})$$
$$= \delta_X(\sum_{l,m} g_{il} g_{mj}^{-1} T_{lm})$$
$$= \sum_{l,m} g_{il} g_{mj}^{-1} \delta_X(T_{lm})$$
$$= \sum_{l,m} g_{il} g_{mj}^{-1} X_{lm} = (gXg^{-1})_{ij}.$$

So $(\operatorname{Ad} g)(\delta_X) = \delta_{gXg^{-1}}$ and $\operatorname{Ad} g$ for GL_n is just conjugation by g on $\mathfrak{gl}_n$, and by the remarks preceding this example, the same holds for any closed subgroup $G \leq \operatorname{GL}_n$.

We conclude this discussion of the adjoint representation of G by stating a result ([66, Cor. 5.4.7]) which will be crucial in the investigation of the structure of reductive groups. This concerns the action of a torus on $\operatorname{Lie}(G)$.

Proposition 7.14 *Let G be a connected linear algebraic group and $S \leq G$ a subtorus of G. Then (under the identification in Theorem 7.9)*

$$\operatorname{Lie}(C_G(S)) = C_{\operatorname{Lie}(G)}(S) := \{X \in \operatorname{Lie}(G) \mid \operatorname{Ad}(s)X = X \text{ for all } s \in S\}.$$

Example 7.15 We verify the previous statement for $G = \operatorname{GL}_n$. Let $m \geq 1$ and let S be the $(n - m + 1)$-dimensional subtorus $\{\operatorname{diag}(t_1, \ldots, t_n) \mid t_i = t_j \text{ for } 1 \leq i, j \leq m\}$. Then

$$C_G(S) = \left\{ \begin{pmatrix} A & 0 & \cdots & 0 \\ 0 & t_{m+1} & & 0 \\ \vdots & & \ddots & \vdots \\ 0 & 0 & \cdots & t_n \end{pmatrix} \;\middle|\; A \in \operatorname{GL}_m, \, t_{m+1}, \ldots, t_n \in k^{\times} \right\}.$$

The ideal of $k[G] = k[T_{ij}, \det(T_{ij})^{-1}]$ defining $C_G(S)$ is

$$(T_{ij} \mid i \neq j, \, i > m \text{ or } j > m).$$

So $\operatorname{Lie}(C_G(S))$ can be identified with the set of matrices

$$\left\{ \begin{pmatrix} X & 0 & \cdots & 0 \\ 0 & a_{m+1} & & 0 \\ \vdots & & \ddots & \vdots \\ 0 & 0 & \cdots & a_n \end{pmatrix} \;\middle|\; X \in \mathfrak{gl}_m, \, a_{m+1}, \ldots, a_n \in k \right\}$$

which is precisely $C_{\mathfrak{gl}_n}(S)$.

The proposition cannot be generalized by replacing the torus by an arbitrary closed subgroup H of G. See Exercise 10.27(d).

8

Structure of reductive groups

It turns out that the best way to investigate reductive algebraic groups is via their adjoint action on the Lie algebra. We start by decomposing the Lie algebra according to the action of a maximal torus. Then, as a preparation for the case of an arbitrary semisimple group we consider the smallest non-solvable groups in some detail.

8.1 Root space decomposition

Let G be a linear algebraic group. Let $T \le G$ be a maximal torus and assume that $\dim(T) \ge 1$, which is the case for example if $G \neq 1$ is reductive. Write $\mathfrak{g} := \mathrm{Lie}(G)$ for the Lie algebra of G. The image of T under the adjoint representation $\mathrm{Ad}\, T \le \mathrm{GL}(\mathfrak{g})$ is a set of commuting semisimple elements, so can be simultaneously diagonalized. For $\chi \in X(T)$ write

$$\mathfrak{g}_\chi = \{v \in \mathfrak{g} \mid (\mathrm{Ad}\, t)(v) = \chi(t)v \text{ for all } t \in T\}$$

for the common T-eigenspaces in $\mathfrak{g}$. Then $\mathfrak{g} = \bigoplus_{\chi \in X(T)} \mathfrak{g}_\chi$.

Definition 8.1 The set of non-zero characters with non-zero eigenspace

$$\Phi(G) := \{\chi \in X(T) \mid \chi \neq 0,\ \mathfrak{g}_\chi \neq 0\}$$

occurring in the above decomposition is called the *set of roots* of G with respect to T, and $W := N_G(T)/C_G(T)$ is the *Weyl group* of G with respect to T (also denoted $W_G(T)$ when necessary).

Note that the Weyl group is finite by Theorem 3.10.

Example 8.2 We determine $\Phi(G)$ for some connected reductive groups G.

(1) Let $G = \mathrm{GL}_n$, with Lie algebra $\mathfrak{gl}_n$ (see Example 7.5). By Example 7.13 the adjoint action of G on $\mathfrak{gl}_n$ is by conjugation. Let $T = \mathrm{D}_n$. Then $\mathrm{Lie}(T)$ is the set of diagonal matrices in $\mathfrak{gl}_n$ by Exercise 10.26. Let $E_{ij} \in \mathfrak{gl}_n$ be the $n \times n$-matrix whose (l, m) entry is $\delta_{il}\delta_{jm}$. A straightforward calculation gives

$$\mathrm{Ad}\,(\mathrm{diag}(t_1, \ldots, t_n))(E_{ij}) = t_i t_j^{-1} E_{ij}.$$

So the character $\chi_{ij} : T \to \mathbf{G}_\mathrm{m}$, $\chi_{ij}(\mathrm{diag}(t_1, \ldots, t_n)) = t_i t_j^{-1}$, is a root of G whenever $i \neq j$. Moreover, $\mathfrak{gl}_n = \bigoplus_{i \neq j}\langle E_{ij}\rangle \oplus \mathrm{Lie}(T)$ and T acts trivially on $\mathrm{Lie}(T)$. So G has the set of roots

$$\Phi(G) = \{\chi_{ij} \mid 1 \leq i, j \leq n,\ i \neq j\},$$

of order $|\Phi(G)| = n(n-1)$. Note that all $\mathfrak{g}_\alpha$, $\alpha \in \Phi(G)$, are one-dimensional. The Weyl group is

$$W = N_G(T)/T = (\text{monomial matrices})/T$$
$$\cong \text{permutation matrices} \cong \mathfrak{S}_n,$$

the symmetric group on n letters.

(2) Let $G = \mathrm{SL}_n$. Recall from Example 7.10(a) that $\mathrm{Lie}(\mathrm{SL}_n) \cong \mathfrak{sl}_n$. By Example 7.13, the action of G on $\mathrm{Lie}(G)$ is given by conjugation. Then, with respect to the maximal torus $\mathrm{D}_n \cap \mathrm{SL}_n$, as above we obtain $\Phi(G) = \{\chi_{ij} \mid 1 \leq i, j \leq n,\ i \neq j\}$. The details are left to the reader.

By transport of structure, the Weyl group W acts naturally on the character group $X(T)$ as well as on the cocharacter group $Y(T)$ via

$$(w.\chi)(t) := \chi(t^w) \qquad\qquad \text{for all } w \in W,\ \chi \in X(T),\ t \in T,$$
$$(w.\gamma)(c) := {}^w(\gamma(c)) := \gamma(c)^{w^{-1}} \qquad \text{for all } w \in W,\ \gamma \in Y(T),\ c \in \mathbf{G}_\mathrm{m}.$$

These actions can easily be seen to be faithful, and moreover they are compatible with the pairing $\langle\,,\,\rangle : X(T) \times Y(T) \to \mathbb{Z}$ defined in Example 3.5 (see Exercise 10.30):

Lemma 8.3 *For all $w \in W$, $\chi \in X(T)$, $\gamma \in Y(T)$ we have*

$$\langle w.\chi, \gamma\rangle = \langle\chi, w^{-1}.\gamma\rangle.$$

We now claim that W stabilizes the set of roots $\Phi(G) \subseteq X(T)$. Indeed, for $n \in N_G(T)$, $\alpha \in \Phi(G)$ and $v \in \mathfrak{g}_\alpha$, $(\mathrm{Ad}\,n)(v)$ is again a common eigenvector for all elements of T, since

$$(\operatorname{Ad} t \operatorname{Ad} n)(v) = (\operatorname{Ad} n \operatorname{Ad} (n^{-1}tn))(v)$$
$$= \operatorname{Ad} n(\alpha(n^{-1}tn)v)$$
$$= \alpha(n^{-1}tn)(\operatorname{Ad} n)(v) \qquad \text{for all } t \in T.$$

As there exists $t \in T$ such that $\alpha(n^{-1}tn) \neq 1$, we have established the following:

Proposition 8.4 *Let G be reductive with maximal torus T, set of roots Φ and Lie algebra $\mathfrak{g}$. Then for all $\alpha \in \Phi$, $n \in N_G(T)$ and $w = nC_G(T) \in N_G(T)/C_G(T)$ we have $(\operatorname{Ad} n)(\mathfrak{g}_\alpha) \subseteq \mathfrak{g}_{w.\alpha}$. In particular, Φ is W-stable.*

8.2 Semisimple groups of rank 1

Let G be a connected non-solvable group. Then $\operatorname{rk}(G) \geq 1$ and $\dim(G) \geq 3$ by Corollaries 2.10 and 6.9. Assume throughout this section that $\operatorname{rk}(G) = 1$.

Let B be a Borel subgroup of G and $T \leq B$ a maximal torus, so $\dim(T) = 1$ and $T \cong \mathbf{G}_m$. Now the Weyl group $W = N_G(T)/C_G(T)$ acts faithfully as a group of algebraic group automorphisms on $T \cong \mathbf{G}_m$, so $|W| \leq 2$ by Exercise 10.11. We now require [66, Prop. 7.1.5] whose proof is geometric and will be omitted here:

Proposition 8.5 *Let G be a rank 1 connected non-solvable linear algebraic group with Borel subgroup B. Then W has order 2 and $\dim G/B = 1$.*

This result has an immediate application to the study of semisimple groups of rank 1 ([32, Thm. 25.3]):

Proposition 8.6 *Let G be a semisimple linear algebraic group of rank 1. Then there exists a surjective morphism $\varphi \colon G \to \operatorname{PGL}_2$, with finite kernel; in particular $\dim(G) = 3$.*

About the proof. Being semisimple, G is non-solvable by Example 6.17(2). By Theorem 6.4 and the previous proposition, G/B is a projective variety of dimension 1. In fact, one can show that $G/B \cong \mathbb{P}^1$. Moreover, G acts as automorphisms of G/B (via left translation) which induces a morphism of algebraic groups $\varphi \colon G \to \operatorname{Aut}_a(\mathbb{P}^1) \cong \operatorname{PGL}_2$.

Let $g \in \ker(\varphi)$. Then $ghB = hB$ for all $h \in G$ which implies that $g \in B^h$ for all $h \in G$. So $(\ker \varphi)^\circ \subseteq (\bigcap_h B^h)^\circ = R(G) = 1$, by Proposition 6.16. In particular $\ker \varphi$ is finite. We have already established that $\dim G \geq 3$, so

$$3 \leq \dim(G) = \dim(\varphi(G)) \leq \dim(\operatorname{PGL}_2) = 3,$$

whence $\dim(G) = 3$, and φ is surjective by Corollary 1.20 and Proposition 1.22. $\qquad\qquad\qquad\qquad\qquad\qquad\qquad\qquad\qquad\qquad\qquad\qquad\qquad\qquad\square$

We are thus led to consider carefully the structure of the rank 1 group $\mathrm{PGL}_2 = \mathrm{GL}_2/\{aI_2 \mid a \in k^\times\}$. For this, we investigate the action of PGL_2 on its Lie algebra $\mathrm{Lie}(\mathrm{PGL}_2)$. Write $\bar{X}$ for the image of $X \subseteq \mathrm{GL}_2$ under the natural projection $\pi : \mathrm{GL}_2 \to \mathrm{PGL}_2$. By dimension considerations one sees that $\overline{T_2}$ is a Borel subgroup of PGL_2 and $T := \overline{D_2}$ is a maximal torus contained in $\overline{T_2}$. A second Borel subgroup containing T is the image of the lower triangular matrices. By Example 7.10 we have

$$\mathrm{Lie}(\mathrm{PGL}_2) \cong \mathfrak{gl}_2/\{aI_2 \mid a \in k\}.$$

Assume in what follows that $\mathrm{char}(k) \neq 2$. Then we have $\mathfrak{gl}_2/\{aI_2 \mid a \in k\} \cong \mathfrak{sl}_2$ (the 2×2-matrices of trace 0, see Exercise 10.26). As in Example 7.13 one checks that the adjoint representation of PGL_2 acting on $\mathrm{Lie}(\mathrm{PGL}_2)$ is simply the natural action of PGL_2 on $\mathfrak{sl}_2$ by conjugation. We determine the root space decomposition. Fix an isomorphism $\varphi \colon \mathbf{G}_m \to T$, $c \mapsto \overline{\begin{pmatrix} c & \\ & 1 \end{pmatrix}}$. Then

$$\mathrm{Ad}\,\varphi(c)\left(\begin{pmatrix} 0 & 1 \\ 0 & 0 \end{pmatrix}\right) = \begin{pmatrix} 0 & c \\ 0 & 0 \end{pmatrix}, \qquad \mathrm{Ad}\,\varphi(c)\left(\begin{pmatrix} 0 & 0 \\ 1 & 0 \end{pmatrix}\right) = \begin{pmatrix} 0 & 0 \\ c^{-1} & 0 \end{pmatrix},$$

and $\begin{pmatrix} 1 & 0 \\ 0 & -1 \end{pmatrix}$ is fixed by $\mathrm{Ad}\,\varphi(c)$. Define $\beta, \gamma \in X(T)$ by

$$\beta\left(\overline{\begin{pmatrix} c & \\ & 1 \end{pmatrix}}\right) = c, \quad \gamma\left(\overline{\begin{pmatrix} c & \\ & 1 \end{pmatrix}}\right) = c^{-1}.$$

As $X(T)$ is written additively we have $\gamma = -\beta$, so $\mathfrak{sl}_2$ decomposes as the direct sum

$$\mathfrak{sl}_2 = (\mathfrak{sl}_2)_0 \oplus (\mathfrak{sl}_2)_\beta \oplus (\mathfrak{sl}_2)_{-\beta}$$

of common eigenspaces for $\mathrm{Ad}\,T$, with

$$(\mathfrak{sl}_2)_0 = \left\langle \begin{pmatrix} 1 & 0 \\ 0 & -1 \end{pmatrix} \right\rangle, \quad (\mathfrak{sl}_2)_\beta = \left\langle \begin{pmatrix} 0 & 1 \\ 0 & 0 \end{pmatrix} \right\rangle, \quad (\mathfrak{sl}_2)_{-\beta} = \left\langle \begin{pmatrix} 0 & 0 \\ 1 & 0 \end{pmatrix} \right\rangle.$$

Note that $X(T) = \langle \beta \rangle$.

In fact, the character β "occurs intrinsically" in PGL_2. That is, there exists a subgroup of PGL_2 whose Lie algebra corresponds to the second subspace in the above decomposition: Consider the image of U_2 in PGL_2.

This is the unipotent radical of our chosen Borel subgroup $\overline{T_2}$ and is therefore normalized by T. Fix an isomorphism

$$u\colon \mathbf{G}_a \longrightarrow \overline{U_2}, \qquad u(c) = \begin{pmatrix} 1 & c \\ 0 & 1 \end{pmatrix}.$$

Then $tu(c)t^{-1} = u(\beta(t)c)$ for all $c \in k$, $t \in T$. Finally, $W = N_G(T)/T$ is of order two, with representatives $\begin{pmatrix} 1 & 0 \\ 0 & 1 \end{pmatrix}, \begin{pmatrix} 0 & 1 \\ 1 & 0 \end{pmatrix}$.

A similar calculation goes through when $\mathrm{char}(k) = 2$.

We now consider G an arbitrary semisimple group of rank 1; for this, we will refine the statement of Proposition 8.6 and obtain structural information which resembles that which we found above for PGL_2. Let T be a maximal torus contained in a Borel subgroup B of G. Let $U = B_u$ and choose $n \in N_G(T) \setminus C_G(T)$, whence $n^2 \in C_G(T)$ and $ntn^{-1} = t^{-1}$ for $t \in T$ (see Proposition 8.5). Write $\mathfrak{g} := \mathrm{Lie}(G)$.

Proposition 8.7 *In the notation introduced above, we have:*

(a) $\dim(G) = 3$, $\dim(B) = 2$, $\dim(U) = 1$, $C_G(T) = T$ and $U \cap nUn^{-1} = 1$.
(b) *There is a unique* $\alpha \in X(T) \setminus \{0\}$ *such that* $\mathfrak{g}_\alpha = \mathrm{Lie}(U)$ *and* $\mathfrak{g}_{-\alpha} = \mathrm{Lie}(nUn^{-1})$, *and* $\mathfrak{g} = \mathrm{Lie}(T) \oplus \mathfrak{g}_\alpha \oplus \mathfrak{g}_{-\alpha}$.

Proof By Propositions 8.5 and 8.6, we have $\dim(G) = 3$ and $\dim(G/B) = 1$. Since G is non-solvable, B is not a torus (see Corollary 6.9) and so $\dim(U) = 1$. By Proposition 6.18 and Theorem 6.19(a), $C_G(T)$ is nilpotent and connected and so lies in some Borel subgroup of G. As $\dim(B) = 2$, either $C_G(T) = T$ or $C_G(T)$ is a Borel subgroup of G. But in the latter case, G would be solvable by Corollary 6.9, hence $C_G(T) = T$. This shows all statements in (a) but the last.

By Theorem 3.2, U is isomorphic to $\mathbf{G}_a$. Let $u : \mathbf{G}_a \to U$ be an isomorphism. As T normalizes U, it acts by automorphisms. By Exercise 10.10 the automorphisms of $\mathbf{G}_a$ are simply multiplications by non-zero scalars, so there exists $\alpha \in X(T)$ such that $tu(c)t^{-1} = u(\alpha(t)c)$ for all $t \in T$ and $c \in k$. Moreover, α is non-trivial as $C_G(T) = T$. Now we differentiate the action of T on U to obtain

$$\mathrm{Ad}\,(t)Y = \alpha(t)Y, \qquad \text{for all } t \in T,\ Y \in \mathrm{Lie}(U).$$

In particular, $\mathrm{Lie}(U)$ lies in the α-eigenspace $\mathfrak{g}_\alpha$ for T acting on $\mathfrak{g}$. As conjugation by n sends U to nUn^{-1}, $\mathrm{Lie}(nUn^{-1})$ then lies in $\mathfrak{g}_{-\alpha}$ by Proposition 8.4

and the definition of n. The assertion of (b) now follows by comparing dimensions.

Finally, $\mathrm{Lie}(U \cap nUn^{-1}) \subseteq \mathrm{Lie}(U) \cap \mathrm{Lie}(nUn^{-1}) = \mathfrak{g}_\alpha \cap \mathfrak{g}_{-\alpha} = 0$, so the group $U \cap nUn^{-1}$ is finite by Theorem 7.4(b), unipotent and normalized by T. Hence this group lies in $C_G(T) = T$ by Exercise 10.4. So $U \cap nUn^{-1} = 1$. $\quad\square$

The above considerations finally enable one to establish the following classification of rank 1 semisimple groups ([66, Thm. 7.2.4]):

Theorem 8.8 *Any semisimple group of rank 1 is isomorphic to* SL_2 *or* PGL_2.

Example 8.9 For later use, let's have a closer look at the case of SL_2. By Example 8.2(2) we get the same root space decomposition

$$\mathfrak{sl}_2 = (\mathfrak{sl}_2)_0 + (\mathfrak{sl}_2)_\alpha + (\mathfrak{sl}_2)_{-\alpha}$$

of $\mathrm{Lie}(\mathrm{SL}_2) = \mathfrak{sl}_2$ with respect to $T = \mathrm{D}_2 \cap \mathrm{SL}_2$ as for the action of PGL_2, where now $\alpha \in X(T)$ is defined by $\alpha \begin{pmatrix} c & \\ & c^{-1} \end{pmatrix} = c^2$. Note that $\langle \alpha \rangle = 2X(T)$ in this case.

Remark 8.10 In our calculations for PGL_2 above we also found two roots $\pm\beta$ and a Weyl group of order two interchanging these two. That is, the groups SL_2 and PGL_2 have the same Weyl group and "isomorphic root systems".

We finish this section with the following extension of our results on groups of rank 1:

Proposition 8.11 *Let G be connected reductive of semisimple rank 1, $T \leq G$ a maximal torus, $Z = Z(G)$ and $G' = [G, G]$. Then:*

(a) $C_G(T) = T$ *and thus* $Z \leq T$.

(b) G' *is semisimple of rank 1 with maximal torus* $T \cap G'$, *and* $G = G'Z$.

Proof By Proposition 6.20, the radical $R(G) = Z(G)^\circ$ is a torus, so contained in T, G' is semisimple and $G' \cap R(G)$ is finite. As $G/R(G)$ is assumed to be semisimple of rank 1 we have $\dim(G/R(G)) = 3$ by Proposition 8.6, and $G/R(G)$ is isomorphic to SL_2 or PGL_2 by Theorem 8.8. Now observe that a maximal torus of SL_2 or PGL_2 is self-centralizing. So T and $C_G(T)$ have the same image in $G/R(G)$ under the canonical epimorphism, and hence $T \leq C_G(T) \leq TR(G) = T$.

Since the product map $R(G) \times G' \to G$ has finite kernel we have

$$\dim G' \leq \dim G - \dim R(G) = \dim G/R(G) = 3.$$

On the other hand, G' is non-solvable (since $G/R(G)$ and thus G is), so $\dim G' = 3$ by Corollary 6.9. Now let T_1 be a maximal torus of G'. Comparing ranks we get that $T_1 R(G)$ is a maximal torus of G, thus, after conjugating we may assume $T = T_1 R(G)$. Using again that maximal tori of SL_2 or PGL_2 are self-centralizing we conclude that $T \cap G' = T_1$. □

8.3 Structure of connected reductive groups

We start with the following characterization of the unipotent radical of a general connected group, whose proof relies crucially on the structure results for groups of semisimple rank 1 ([32, Thm. 26.1]); although it is easy to check in the case of classical groups, its proof is rather involved:

Theorem 8.12 *Let G be a connected linear algebraic group with maximal torus T. Then $R_u(G)$ is the identity component of the intersection of the unipotent parts of the Borel subgroups which contain T.*

Note that one inclusion is given by Proposition 6.16. For a reductive group, in particular, we obtain:

Corollary 8.13 *Let G be connected reductive.*

(a) *If S is a subtorus of G, then $C_G(S)$ is connected and reductive.*
(b) *If T is a maximal torus of G, then $C_G(T) = T$.*

Proof Part (a) follows directly from the preceding theorem, applied to a maximal torus T of G containing S, and Theorem 6.19. The second assertion then follows from (a) with $S = T$, Proposition 6.18 and Example 6.17(2). □

Example 8.14 Let $G = \mathrm{GL}_n$, $S \le G$ a torus of dimension $m = \dim(S)$ (so without loss of generality $S \le \mathrm{D}_n$). Then $C_G(S) \cong S \times \mathrm{GL}_{n-m}$ is connected reductive, and $C_G(S) = S$ if $m = n$ (see Example 3.11).

We now study general reductive algebraic groups via their adjoint action on the Lie algebra. Let G be a non-solvable connected reductive algebraic group. We keep the notations T, $\mathfrak{g} = \mathrm{Lie}(G)$, Φ and W as in Section 8.2.

Let $\alpha \in \Phi$; then $T_\alpha := (\ker \alpha)^\circ \le T$ is a subtorus of T by Proposition 3.9, of codimension 1, and we let $C_\alpha := C_G(T_\alpha)$ be its centralizer.

Proposition 8.15 *Let G be non-solvable connected reductive. Then all C_α, $\alpha \in \Phi$, are connected reductive and non-solvable, and $G = \langle C_\alpha \mid \alpha \in \Phi \rangle$.*

Proof The C_α, $\alpha \in \Phi$, are connected reductive by Corollary 8.13(a). By Proposition 1.16, they generate a closed connected subgroup H of G. Moreover, by Proposition 7.14, $\text{Lie}(C_\alpha) = \text{Lie}(C_G(T_\alpha)) = C_{\mathfrak{g}}(T_\alpha)$, which contains both the root space $\mathfrak{g}_\alpha$ and $\text{Lie}(T)$ (as $\text{Lie}(T) = \text{Lie}(C_G(T)) = C_{\mathfrak{g}}(T)$). Since these subspaces span $\mathfrak{g}$, we have that $\text{Lie}(H) = \text{Lie}(G)$, hence $\dim(H) = \dim(G)$ and so $H = G$ by Proposition 1.22.

The reductive group C_α is solvable if and only if it is a torus. But we just saw that the dimension of the centralizer of T_α is bigger than $\dim(T)$. $\qquad\square$

Since C_α is connected reductive, non-solvable and $T_\alpha < R(C_\alpha)$, we see that C_α is of semisimple rank 1 and so Proposition 8.11 shows that its derived subgroup $G_\alpha := [C_\alpha, C_\alpha]$ is semisimple of rank 1 with maximal torus $T_1 := T \cap G_\alpha$. Moreover:

Proposition 8.16 *For $\alpha \in \Phi$ we have $\text{Lie}(C_\alpha) = \text{Lie}(T) \oplus \mathfrak{g}_\alpha \oplus \mathfrak{g}_{-\alpha}$ and* $\dim \mathfrak{g}_\alpha = 1$.

Proof We may decompose $\text{Lie}(C_\alpha) = \text{Lie}(T) \oplus \mathfrak{h}$ with $\mathfrak{h}$ a two-dimensional T-invariant subspace which contains the root space $\mathfrak{g}_\alpha$ of $\mathfrak{g}$. By Proposition 8.7(b) we have $\text{Lie}(G_\alpha) = \text{Lie}(T_1) \oplus \text{Lie}(G_\alpha)_\beta \oplus \text{Lie}(G_\alpha)_{-\beta}$ for some $\beta \in X(T_1) \setminus \{0\}$. Comparing with the T_1-eigenspaces on $\text{Lie}(C_\alpha)$, we see that $\mathfrak{h} = \mathfrak{g}_\beta \oplus \mathfrak{g}_{-\beta}$, whence $\mathfrak{h} = \mathfrak{g}_{\alpha_1} \oplus \mathfrak{g}_{\alpha_2}$ as T-module, with $\alpha_i \in X(T)$, and $\alpha_1|_{T_1} = \beta$, $\alpha_2|_{T_1} = -\beta$. Since both α_i have T_α in their kernel, they are proportional. As one of them must be α the claim follows. $\qquad\square$

We are now ready to prove the first major structure theorem for reductive groups (see also [66, Thm. 8.1.1, Cor. 8.1.2, 8.1.3] or [32, Thm. 26.3]):

Theorem 8.17 *Let G be a connected reductive group, $T \leq G$ a maximal torus of G, $\mathfrak{g} = \text{Lie}(G)$ and $\Phi = \Phi(G)$. Then:*

(a) $\mathfrak{g} = \text{Lie}(T) \oplus \bigoplus_{\alpha \in \Phi} \mathfrak{g}_\alpha$ *with* $\dim \mathfrak{g}_\alpha = 1$ *for all* $\alpha \in \Phi$, *and* $\text{Lie}(T) = \mathfrak{g}_0$.

(b) $\dim G = \dim \text{Lie}(G) = |\Phi| + \text{rk}(G)$.

(c) *For each* $\alpha \in \Phi$ *there exists a morphism of algebraic groups* $u_\alpha : \mathbf{G}_a \to G$, *which induces an isomorphism onto* $u_\alpha(\mathbf{G}_a)$ *such that* $t u_\alpha(c) t^{-1} = u_\alpha(\alpha(t)c)$, *for all* $t \in T, c \in k$. *If* u' *is a morphism with the same properties, there is a unique* $a \in k^\times$ *with* $u'(c) = u_\alpha(ac)$ *for* $c \in k$. *Moreover,* $\text{im}(du_\alpha) = \mathfrak{g}_\alpha$.

(d) $U_\alpha := \text{im}(u_\alpha)$ *is the unique one-dimensional connected unipotent subgroup of* G *normalized by* T *with* $\text{Lie}(U_\alpha) = \mathfrak{g}_\alpha$.

(e) *For* $w \in W$ *with preimage* $n \in N_G(T)$ *we have* $n U_\alpha n^{-1} = U_{w.\alpha}$.

(f) $[C_\alpha, C_\alpha] = \langle U_\alpha, U_{-\alpha} \rangle$.

(g) $G = \langle T, U_\alpha \mid \alpha \in \Phi \rangle$.

(h) $Z(G) = \bigcap_{\alpha \in \Phi} \ker \alpha$.

Proof Part (a) and thus (b) follows from Propositions 7.14 and 8.16. Now let $u_\alpha : \mathbf{G_a} \to G_\alpha$ be the morphism u defined in the proof of Proposition 8.7, relative to the torus T_1. Then for all $t \in T$, $c \in k$, we have $t u_\alpha(c) t^{-1} = u_\alpha(\alpha(t)c)$ as required, since $T = T_1 \cdot \ker(\alpha)$. The argument that $\mathrm{im}(du_\alpha) = \mathfrak{g}_\alpha$ is as in the proof of loc. cit. Let $U_\alpha := \mathrm{im}(u_\alpha)$. Suppose that $u' : \mathbf{G_a} \to G$ is a homomorphism as in (c), then clearly $V := \mathrm{im}(u') \leq C_\alpha$. Since

$$t u'(c) t^{-1} u'(c)^{-1} = u'(\alpha(t)c) u'(-c) = u'((\alpha(t) - 1)c) \quad \text{for } t \in T, \ c \in k,$$

we even have that $V \leq G_\alpha$, a one-dimensional connected unipotent subgroup normalized by T. Then $T_1 V$ is a Borel subgroup of G_α, containing T_1. But by Proposition 8.7 there are precisely two such Borel subgroups of G_α, namely $T_1 U_\alpha$ and $T_1 U_{-\alpha}$. (For, any Borel subgroup containing T gives rise to a corresponding weight space in $\mathrm{Lie}(G_\alpha)$.) Since T acts via α, we must have $V = U_\alpha$. Thus $u_\alpha^{-1} \circ u'$ is an automorphism of $\mathbf{G_a}$, and hence given by multiplication by a non-zero scalar. We now have the statement of (c).

If V is a one-dimensional closed connected unipotent subgroup normalized by T, with $\mathrm{Lie}(V) = \mathfrak{g}_\alpha$, then the action of T on V must be given by the character α and so $V \leq C_\alpha = C_G(T_\alpha)$, in which case the conclusion of (d) follows as above.

For $n \in N_G(T)$, $n U_\alpha n^{-1}$ is a one-dimensional closed connected unipotent subgroup normalized by T and T acts by $w\alpha$. So the uniqueness statement of (d) gives the result of (e). The statement of (f) follows from Theorem 8.8 applied to the rank 1 group G_α; (g) follows from (f) and Proposition 8.15.

Finally, by (c) and (g), the right-hand side of (h) lies in $Z(G)$. As $Z(G) \leq C_G(T) \leq T$ by Corollary 8.13, the other inclusion also holds by part (c). $\square$

Definition 8.18 The one-dimensional subgroup U_α, $\alpha \in \Phi$, in Theorem 8.17(d) is called the *root subgroup* of G (with respect to T) associated to α. Similarly, the one-dimensional subspace $\mathfrak{g}_\alpha$ of $\mathfrak{g}$ is called a *root subspace*.

8.4 Structure of semisimple groups

Let's now specialize to the situation of semisimple groups. For this, we study the action of W in some more detail. For $\alpha \in \Phi$, let $C_\alpha = C_G(T_\alpha)$ as in the previous section. Since T_α lies in the center of C_α, the canonical epimorphism $C_\alpha \to \bar{C}_\alpha := C_\alpha/T_\alpha$ induces a bijection $W_{C_\alpha}(T) \cong W_{\bar{C}_\alpha}(T/T_\alpha)$. By Proposition 8.5, this latter Weyl group is of order 2. Recall that $C_{C_\alpha}(T) =$

$T = C_G(T)$, by Corollary 8.13. Choose $n_\alpha \in N_{C_\alpha}(T) \setminus C_{C_\alpha}(T)$, and let s_α be the image of n_α in

$$N_{C_\alpha}(T)/C_{C_\alpha}(T) \le N_G(T)/C_G(T) = W.$$

We saw in Section 8.1 that s_α acts naturally on $X = X(T)$ and on $Y = Y(T)$. More precisely we have:

Lemma 8.19 *Let $\alpha \in \Phi$. Then:*

(a) *There exists a unique $\alpha^\vee \in Y$ such that $s_\alpha.\chi = \chi - \langle \chi, \alpha^\vee \rangle \alpha$ for all $\chi \in X$. In particular $\langle \alpha, \alpha^\vee \rangle = 2$.*
(b) *We have $s_\alpha.\gamma = \gamma - \langle \alpha, \gamma \rangle \alpha^\vee$ for all $\gamma \in Y$.*
(c) *$\mathrm{im}(\alpha^\vee) \subseteq [C_\alpha, C_\alpha]$.*

Proof For $\chi \in X$ and $t \in T$ we have

$$(\mathrm{id}_X - s_\alpha).\chi(t) = \chi(t)/(s_\alpha.\chi)(t) = \chi(t)/\chi(t^{s_\alpha}) = \chi(tt^{-s_\alpha}).$$

But tt^{-s_α} lies in the rank 1 group $G_\alpha = [C_\alpha, C_\alpha]$, isomorphic to SL_2 or PGL_2 by Theorem 8.8. Let $T_1 := T \cap G_\alpha$, a maximal torus of G_α by Proposition 8.11(b). Since $\alpha|_{T_1}$ is a root of T_1, we have $2X(T_1) \subseteq \langle \alpha|_{T_1} \rangle$ according to Example 8.9 (in case $G_\alpha \cong \mathrm{SL}_2$), respectively our calculations for PGL_2 in Section 8.2, so $2\chi|_{T_1} = c_\chi \, \alpha|_{T_1}$ for some $c_\chi \in \mathbb{Z}$. Thus

$$2\chi(tt^{-s_\alpha}) = c_\chi \alpha(tt^{-s_\alpha}) = c_\chi(\mathrm{id}_X - s_\alpha).\alpha(t) = 2c_\chi \alpha(t),$$

showing that $(\mathrm{id}_X - s_\alpha).\chi = c_\chi \alpha$. This defines a homomorphism $X \to \mathbb{Z}$, $\chi \mapsto c_\chi$. Evaluation on T_1, where s_α acts by inversion, shows that $c_\alpha = 2$. By Proposition 3.6 there exists a unique $\alpha^\vee \in Y$ such that $c_\chi = \langle \chi, \alpha^\vee \rangle$, whence $s_\alpha.\chi = \chi - \langle \chi, \alpha^\vee \rangle \alpha$ and $\langle \alpha, \alpha^\vee \rangle = c_\alpha = 2$, thus proving (a).

For $\chi \in X$ we have by Lemma 8.3 and part (a)

$$\langle \chi, s_\alpha.\gamma \rangle = \langle s_\alpha.\chi, \gamma \rangle = \langle \chi - \langle \chi, \alpha^\vee \rangle \alpha, \gamma \rangle$$
$$= \langle \chi, \gamma \rangle - \langle \alpha, \gamma \rangle \langle \chi, \alpha^\vee \rangle = \langle \chi, \gamma - \langle \alpha, \gamma \rangle \alpha^\vee \rangle.$$

Since this is true for all χ, Proposition 3.6 yields that $s_\alpha.\gamma = \gamma - \langle \alpha, \gamma \rangle \alpha^\vee$.

For (c) note that $\alpha^\vee(c) n_\alpha \alpha^\vee(c)^{-1} n_\alpha^{-1} \in [C_\alpha, C_\alpha]$ for $c \in k^\times$. But (a) and (b) give that $s_\alpha.\alpha^\vee = -\alpha^\vee$, so $n_\alpha \alpha^\vee(c)^{-1} n_\alpha^{-1} = \alpha^\vee(c)$. So we have $\mathrm{im}(\alpha^\vee) = \mathrm{im}(2\alpha^\vee) \subseteq [C_\alpha, C_\alpha]$. $\square$

We say that $\alpha^\vee$ is the *coroot corresponding to* α and set

$$\Phi^\vee := \{\alpha^\vee \mid \alpha \in \Phi\}.$$

Recall that an element $s \in \mathrm{GL}(V)$, where V is a finite-dimensional real vector space, is called a *reflection along $v \in V$*, if v is an eigenvector of s

with eigenvalue -1, and s fixes a hyperplane of V pointwise. Lemma 8.19 shows that the elements s_α act as reflections on $X(T) \otimes_{\mathbb{Z}} \mathbb{R}$. Let's note the following result on the s_α for future use (this is Exercise 10.31, or see [66, Thm. 8.2.8]):

Proposition 8.20 *The Weyl group W of a connected reductive group G is generated by the s_α, $\alpha \in \Phi$.*

The structure of semisimple groups is now given as follows:

Theorem 8.21 *Let G be a semisimple group, T, Φ, U_α as in Theorem 8.17. Then:*

(a) $G = \langle U_\alpha \mid \alpha \in \Phi \rangle$.
(b) $G = [G, G]$.
(c) G *has only finitely many minimal non-trivial closed connected normal subgroups $G_1, \ldots, G_r$. Moreover, $[G_i, G_j] = 1$ for all $i \neq j$ and $G_i \cap \prod_{i \neq j} G_j$ is finite.*
(d) $G = G_1 \cdot \ldots \cdot G_r$ *and each G_i is a simple algebraic group.*

Here a non-trivial semisimple algebraic group is called *simple* if it has no non-trivial proper closed connected normal subgroups.

Proof Let $Z = \bigcap_{\alpha \in \Phi} \ker \alpha$. Then by Theorem 8.17(c) and (g), Z lies in the center of G and hence is finite. Thus the roots Φ span a subgroup of finite index in $X(T)$ and in particular span the vector space $X(T) \otimes_{\mathbb{Z}} \mathbb{R}$. We claim that $S := \langle \mathrm{im}(\alpha^\vee) \mid \alpha \in \Phi \rangle$ equals T. Assume not. Then there exists $0 \neq \chi \in X(T)$ such that $\chi(S) = 1$. Thus $\chi \circ \alpha^\vee = \mathrm{id}$, whence $\langle \chi, \alpha^\vee \rangle = 0$ for all $\alpha \in \Phi$. Hence χ is fixed by all reflections s_α by Lemma 8.19(a). This contradicts the fact that the $\alpha \in \Phi$ span $X(T) \otimes_{\mathbb{Z}} \mathbb{R}$. The assertion of (a) now follows from Lemma 8.19(c) and Theorem 8.17(f) and (g). As shown in the proof of Theorem 8.17, $U_\alpha \leq [C_\alpha, C_\alpha]$, so (b) follows.

If $G \neq 1$ has no proper non-trivial closed connected normal subgroups, then G itself is a simple group and the statements of (c) and (d) are clear. So assume the contrary and let H be a proper non-trivial closed connected normal subgroup of G. As the radical of H is normal in G, it must be trivial and so H is semisimple. We now decompose Φ into $\Phi_1 := \{\alpha \in \Phi \mid U_\alpha \leq H\}$ and $\Phi_2 := \Phi \setminus \Phi_1$; as H is proper non-trivial and semisimple, $\Phi_1, \Phi_2 \neq \emptyset$ by (a).

We claim that the subgroup $H' := \langle U_\beta \mid \beta \in \Phi_2 \rangle$ commutes with H. For this, let T_H be a maximal torus of H; as all such are conjugate in H and H is normal in G, we may assume that $T_H \leq T$. For $\beta \in \Phi_2$ we have from Theorem 8.17(c) that $t u_\beta(c) t^{-1} u_\beta(-c) = u_\beta((\beta(t) - 1)c) \in H$ for all $t \in T_H$

and $c \in k$. Hence $\beta \in \Phi_2$ implies that $T_H \le \ker \beta$, so T_H centralizes U_β. Next, for $\alpha \in \Phi_1$ set

$$u_y(c) := [u_\beta(y), u_\alpha(c)]u_\alpha(c) = u_\beta(y)u_\alpha(c)u_\beta(-y) \quad \text{for } c, y \in k.$$

Then $u_y(c) \in H$ for all y and $tu_y(c)t^{-1} = u_y(\alpha(t)c)$ for all $t \in T_H$. Thus the map $\mathbf{G_a} \to G$, $c \mapsto u_y(c)$, satisfies the conditions of Theorem 8.17(c), and by the uniqueness result there exists $a_y \in k^\times$ such that $u_y(c) = u_\alpha(a_y c)$ for all $c \in k$. This defines a morphism of affine 1-space $a : k \to k$, $y \mapsto a_y$, with $a_0 = 1$, whose image does not contain 0. This can only be satisfied if a is the constant function 1 and hence U_β and U_α commute. Thus $[H, H'] = 1$ and $G = H \cdot H'$ by (a).

Finally we note that $H \cap H'$ must be finite as it is a closed normal subgroup of H which contains no U_α. One now proceeds by induction on the dimension of G, using the fact that H and H' are both semisimple. $\square$

Finally we can improve the statement of Exercise 10.21.

Corollary 8.22 *Let G be connected reductive. Then*

$$G = [G, G]R(G) = [G, G]Z(G)^\circ;$$

in particular, $\mathrm{rk_{ss}}(G) = \mathrm{rk}([G, G])$ *and* $\mathrm{rk}(G) = \mathrm{rk_{ss}}(G) + \dim Z(G)$.

Proof We already argued that $G/R(G)$ is semisimple (see Exercise 10.21). Thus by Theorem 8.21(b)

$$G/R(G) = [G/R(G), G/R(G)] = [G, G]R(G)/R(G). \qquad \square$$

This result can be used to show that $\ker(\mathrm{Ad}) = Z(G)$, as claimed in Theorem 7.12(c), see Exercise 10.32.

9

The classification of semisimple algebraic groups

The aim here is to achieve a classification of semisimple algebraic groups in terms of combinatorial data. It is clear from the previous section that the set of roots plays an essential role in the structure of reductive groups. We now formalize this concept.

9.1 Root systems

Let G be a connected reductive group and $T \leq G$ a maximal torus. Then associated to this we have a finite set of roots $\Phi \subset X := X(T)$ with the finite Weyl group W acting faithfully on X, preserving Φ (see Proposition 8.4). Recall the group $Y = Y(T)$ of cocharacters of T and the pairing $\langle\ ,\ \rangle : X \times Y \to \mathbb{Z}$ defined in Section 3.2. We identify X and Y with subgroups of $E := X \otimes_{\mathbb{Z}} \mathbb{R}$ and $E^\vee := Y \otimes_{\mathbb{Z}} \mathbb{R}$, respectively, and denote the induced pairing on $E \times E^\vee$ also by $\langle\ ,\ \rangle$. The actions of W on X and on Y may be extended to actions on E and $E^\vee$. Recall the reflections $s_\alpha \in W$ introduced in Section 8.4.

We first axiomatize the combinatorial properties satisfied by these data.

Definition 9.1 A subset Φ of a finite-dimensional real vector space E is called an *(abstract) root system* in E if the following properties are satisfied:

(R1) Φ is finite, $0 \notin \Phi$, $\langle \Phi \rangle = E$;

(R2) if $c \in \mathbb{R}$ is such that $\alpha, c\alpha \in \Phi$, then $c = \pm 1$;

(R3) for each $\alpha \in \Phi$ there exists a reflection $s_\alpha \in \mathrm{GL}(E)$ along α stabilizing Φ;

(R4) *(crystallographic condition)* for $\alpha, \beta \in \Phi$, $s_\alpha.\beta - \beta$ is an integral multiple of α.

The group $W = \langle s_\alpha \mid \alpha \in \Phi \rangle$ is called the *Weyl group* of Φ. The dimension of E is called the *rank* of Φ.

Since Φ is finite, generates E and is stabilized by W, the Weyl group of an abstract root system is always finite. Thus, there exists a positive definite W-invariant bilinear form $(\ ,\ )$ on E, which is unique up to non-zero scalars on each irreducible W-submodule of E. We'll always assume such a form to have been chosen so that we can speak of lengths of vectors and angles between them.

Let's return to our reductive group G with root system Φ. We have seen in Lemma 8.19 that s_α acts as a reflection on $E = X \otimes_{\mathbb{Z}} \mathbb{R}$. In fact, we have:

Proposition 9.2 *Let G and $\Phi \subset X$ be as above. View Φ as a subset of $E := X \otimes_{\mathbb{Z}} \mathbb{R}$. Then Φ, together with $\{s_\alpha \mid \alpha \in \Phi\}$ is an abstract root system in $\langle \Phi \rangle_{\mathbb{R}}$ with Weyl group W.*

Moreover, if G is semisimple then $\langle \Phi \rangle_{\mathbb{R}} = E$.

Proof We first verify the axioms (R1)–(R4) of a root system. The first axiom is clear (by our restriction of Euclidean space). (R2) follows from the fact that $C_{c\alpha} = C_\alpha$ for all $c \in \mathbb{R}^\times$. (R3) is just Proposition 8.4 and Lemma 8.19. The final axiom (R4) will be proved in Part II when we discuss the representation theory of semisimple groups. See Lemma 15.4 and Example 15.5. The fact that the s_α generate the finite group W is Proposition 8.20.

The last statement was established in the proof of Theorem 8.21(a). $\square$

Abstract root systems can be classified. A crucial ingredient in this classification is the notion of a base:

Definition 9.3 Let Φ be an abstract root system in E. A subset $\Delta \subseteq \Phi$ is called a *base of* Φ if it is a vector space basis of E and any $\beta \in \Phi$ is an integral linear combination $\beta = \sum_{\alpha \in \Delta} c_\alpha \alpha$ with either all $c_\alpha \geq 0$ or all $c_\alpha \leq 0$.

If Δ is a base of Φ, then the subset

$$\Phi^+ := \{\alpha \in \Phi \mid \alpha \text{ is a non-negative linear combination over } \Delta\}$$

is called the *system of positive roots* of Φ with respect to the base Δ.

We develop the basic theory of abstract root systems in the appendix. As one of the first results, one obtains that bases exist and are essentially unique:

Proposition 9.4 *Let Φ be an abstract root system.*

(a) *There exists a base Δ of Φ.*

(b) Let $\Delta_1, \Delta_2 \subseteq \Phi$ be two bases of Φ. Then there exists a unique $w \in W$ such that $w(\Delta_1) = \Delta_2$.

(c) If Δ is a base, then $W = \langle s_\alpha \mid \alpha \in \Delta \rangle$. Furthermore, for every $\alpha \in \Phi$ there is $w \in W$ such that $w\alpha \in \Delta$.

See Proposition A.7 for (a), Theorem A.22 for (b) and Proposition A.11 for (c).

Example 9.5 We classify the two-dimensional abstract root systems. Let $E = \mathbb{R}^2$ and $\Phi \subset E$ an abstract root system with $\alpha, \beta \in \Phi$ linearly independent. With respect to this basis, the reflections in α, respectively β, have the form

$$s_\alpha = \begin{pmatrix} -1 & a \\ 0 & 1 \end{pmatrix}, \quad s_\beta = \begin{pmatrix} 1 & 0 \\ b & -1 \end{pmatrix},$$

with $a, b \in \mathbb{Z}$ by (R4). Then

$$w := s_\alpha s_\beta = \begin{pmatrix} ab - 1 & -a \\ b & -1 \end{pmatrix}$$

lies in the finite Weyl group W of Φ, so its trace $ab - 2 \in \mathbb{Z}$ is the sum of two roots of unity. This implies that $ab \in \{0, 1, 2, 3, 4\}$. Moreover, w is not diagonalizable when $ab = 4$, so not of finite order. There remain four possibilities for ab, and in fact all lead to root systems. Writing ψ for the angle between α and β, we essentially have the following four cases (note that, with Exercise 10.35(b) or Proposition A.1, ab above is just $4\cos(\psi)^2$):

(1) $\alpha \perp \beta$, then $\Phi = \{\pm\alpha, \pm\beta\}$ with $W = Z_2 \times Z_2$ the Klein four group.
(2) $\psi = 2\pi/3$, then $\Phi^+ = \{\alpha, \beta, \alpha + \beta\}$. This gives the root system of SL_3 already encountered in Example 8.2(2), of type A_2 (see Table 9.1 below). Here $W = \mathfrak{S}_3$.
(3) $\psi = 3\pi/4$, then $\Phi^+ = \{\alpha, \beta, \alpha + \beta, 2\alpha + \beta\}$. This gives the root system of Sp_4, see Exercise 10.29(a), of type $B_2 = C_2$, with W the dihedral group of order 8.
(4) $\psi = 5\pi/6$, then $\Phi^+ = \{\alpha, \beta, \alpha + \beta, 2\alpha + \beta, 3\alpha + \beta, 3\alpha + 2\beta\}$, with W the dihedral group of order 12. This root system is said to be of type G_2. It does not occur as a root system of a *classical* algebraic group.

In all cases, $\Delta = \{\alpha, \beta\}$ is a base of Φ.

A root system can be recovered from a base by Proposition 9.4(c), so in order to describe a root system, it is sufficient to give its base. This is most conveniently done with the associated Dynkin diagram which we now describe. Its underlying graph has one node for each element of the base Δ, and two nodes corresponding to $\alpha, \beta \in \Delta$ are joined by an edge of multiplicity $m_{\alpha,\beta}$ as follows:

$$m_{\alpha,\beta} = \begin{cases} 0 & \text{if } |(\mathbb{Z}\alpha + \mathbb{Z}\beta) \cap \Phi^+| = 2, \\ 1 & \text{if } |(\mathbb{Z}\alpha + \mathbb{Z}\beta) \cap \Phi^+| = 3, \\ 2 & \text{if } |(\mathbb{Z}\alpha + \mathbb{Z}\beta) \cap \Phi^+| = 4, \\ 3 & \text{if } |(\mathbb{Z}\alpha + \mathbb{Z}\beta) \cap \Phi^+| = 6. \end{cases}$$

(By Example 9.5 these are the only possibilities. Note that $|(\mathbb{Z}\alpha+\mathbb{Z}\beta)\cap\Phi^+| = o(s_\alpha s_\beta)$, the order of the product of the corresponding simple reflections.) The resulting graph is called the *Coxeter diagram* associated to Φ or to W. It does not determine Φ uniquely. So in addition, whenever α, β are of different lengths and joined by at least one edge, then we put an arrow on this edge, pointing towards the shorter of the two (see Table 9.1 for examples). This is called the *Dynkin diagram* of Φ. It can be shown that the base can essentially be recovered from its Dynkin diagram (that is, up to changing lengths in different connected components of the diagram), see [33, 11.1].

There is an obvious notion of isomorphism of root systems, and then one has that two root systems are isomorphic if and only if their Dynkin diagrams agree.

A root system Φ with base Δ for which there exists a partition $\Delta = \Delta_1 \sqcup \Delta_2$ into non-empty mutually orthogonal subsets is called *decomposable*; by Proposition A.14 this is the case if and only if

$$\Phi = (\mathbb{Z}\Delta_1 \cap \Phi) \sqcup (\mathbb{Z}\Delta_2 \cap \Phi).$$

If $\Phi \neq \emptyset$ and no such decomposition exists, Φ is said to be *indecomposable*. It can easily be seen that a root system is indecomposable if and only if its Weyl group acts irreducibly on the ambient real vector space (see Proposition A.16). A root system is indecomposable if and only if its associated Dynkin diagram is connected. For example, the root system (1) in Example 9.5 is decomposable, while the other three are indecomposable.

It can be shown (see Exercise 10.33 or [32, Cor. 27.5]) that simple algebraic groups have indecomposable root systems (and conversely). Thus, as a first step in the determination of simple groups one needs a classification of indecomposable root systems (see [9, VI, §4] or [33, Thm. 11.4]):

Theorem 9.6 *Let Φ be an indecomposable root system in some real vector space $E \cong \mathbb{R}^n$. Then up to isomorphism Φ is of one of the following types:*

$A_n \ (n \geq 1), \quad B_n \ (n \geq 2), \quad C_n \ (n \geq 3), \quad D_n \ (n \geq 4), \quad E_6, \quad E_7, \quad E_8, \quad F_4, \quad G_2,$

with corresponding Dynkin diagrams as shown in Table 9.1.

Table 9.1 *Dynkin diagrams of indecomposable root systems*

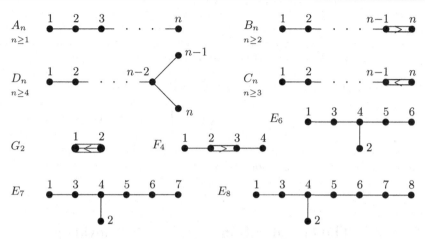

The numbering of nodes is as in [9, Pl. I–IX] and [33, p.58].

A root system or Dynkin diagram is said to be *simply laced* if all roots have the same length, or equivalently, if all $m_{\alpha,\beta} \in \{0,1\}$. Thus the indecomposable, simply laced Dynkin diagrams are those of types A_n, D_n, E_6, E_7, E_8.

Finite reflection groups arising from root systems only satisfying (R1)–(R3), but not necessarily the crystallographic condition (R4), are the so-called *finite Coxeter groups*. Interestingly, in addition to the Weyl groups arising from Theorem 9.6 above, only the two-dimensional dihedral groups and two further indecomposable cases, denoted H_3 and H_4, in dimension 3, respectively 4, occur. For more information on this and on the indecomposable root systems and their Weyl groups, see for example [34, §2.7–2.11].

Definition 9.7 We will often want to consider decomposable root systems. We will use the following notation: if Φ is the orthogonal union of indecomposable root systems $\Phi_1, \ldots, \Phi_t$, then we will refer to the root system Φ as of type $\Phi_1 \Phi_2 \ldots \Phi_t$. If several of the Φ_i are isomorphic, we will write $(\Phi_i)^j$ for the union $\Phi_i \sqcup \cdots \sqcup \Phi_i$ (j copies).

Example 9.8 We identify the root system of $G = \mathrm{SL}_n$ and determine a base and the corresponding positive roots. Here $T = \mathrm{D}_n \cap \mathrm{SL}_n$ is a maximal torus and $B = \mathrm{T}_n \cap \mathrm{SL}_n$ is a Borel subgroup (see Example 6.7(2)). Then $\Phi = \{\chi_{ij} \mid i \neq j\}$ where $\chi_{ij}(\mathrm{diag}(t_1, \ldots, t_n)) := t_i t_j^{-1}$ by Example 8.2(2). A base of Φ is given by

$$\Delta := \{\chi_{i,i+1} \mid 1 \le i \le n-1\}.$$

Indeed,

$$\chi_{ij} = \chi_{i,i+1} + \chi_{i+1,i+2} + \ldots + \chi_{j-1,j} \qquad \text{for } i < j,$$

so $\Phi^+ = \{\chi_{ij} \mid i < j\}$ and $\Phi = \Phi^+ \sqcup -(\Phi^+)$. In particular we have

$$|(\mathbb{Z}\chi_{i,i+1} + \mathbb{Z}\chi_{j,j+1}) \cap \Phi^+| = \begin{cases} 3 & j = i+1, \\ 2 & j > i+1, \end{cases}$$

from which we see that the root system of SL_n is of type A_{n-1}.

9.2 The classification theorem of Chevalley

The aim here is to achieve a classification of semisimple algebraic groups in terms of combinatorial data.

One might now hope that semisimple groups are already determined up to isomorphism by their root systems. But unlike the case of simple complex Lie algebras, there exist non-isomorphic simple algebraic groups having the same root system, e.g., as we have seen $\mathrm{SL}_2, \mathrm{PGL}_2$ both have root system of type A_1 (see Section 8.2). The additional piece of combinatorial data will be provided by the coroots and the coroot lattice.

Let G be semisimple, T, W, Φ, X be as in the previous section. In Lemma 8.19 we constructed for each $\alpha \in \Phi$ a unique cocharacter $\alpha^\vee \in Y := Y(T)$, the coroot corresponding to α, such that $\langle \alpha, \alpha^\vee \rangle = 2$, and moreover $\mathrm{im}(\alpha^\vee)$ is a maximal torus of the rank 1 semisimple group $[C_\alpha, C_\alpha]$.

Example 9.9 Let's compute the coroots for SL_2 and PGL_2.

(1) $G = \mathrm{SL}_2$. The character group is generated by χ, where $\chi\begin{pmatrix} t & \\ & t^{-1} \end{pmatrix} = t$.

We have seen in Example 8.9 that $\Phi = \{\pm\alpha\}$, where $\alpha\begin{pmatrix} t & \\ & t^{-1} \end{pmatrix} = t^2$. So $\mathbb{Z}\Phi = \langle 2\chi \rangle$ and $X = \mathbb{Z}\chi$. The coroot $\alpha^\vee$ is given by $\alpha^\vee : t \mapsto \begin{pmatrix} t & \\ & t^{-1} \end{pmatrix}$, thus $\mathbb{Z}\Phi^\vee = Y$.

(2) $G = \mathrm{PGL}_2$. By Section 8.2 we have $\Phi = \{\pm\beta\}$, where $\beta\overline{\begin{pmatrix} t & \\ & 1 \end{pmatrix}} = t$; in

particular, $\mathbb{Z}\Phi = X$. The coroot $\beta^\vee$ is then given by

$$\beta^\vee : t \mapsto \overline{\begin{pmatrix} t & \\ & t^{-1} \end{pmatrix}} = \overline{\begin{pmatrix} t^2 & \\ & 1 \end{pmatrix}}.$$

In this case $\mathbb{Z}\Phi^\vee = 2Y$.

We are led to introduce the following combinatorial structure:

Definition 9.10 A quadruple $(X, \Phi, Y, \Phi^\vee)$ is called a *root datum* if

(RD1) $X \cong \mathbb{Z}^n \cong Y$, with a perfect pairing $\langle\ ,\ \rangle : X \times Y \to \mathbb{Z}$ as in Proposition 3.6;

(RD2) $\Phi \subseteq X, \Phi^\vee \subseteq Y$ are abstract root systems in $\mathbb{Z}\Phi \otimes_{\mathbb{Z}} \mathbb{R}$, respectively $\mathbb{Z}\Phi^\vee \otimes_{\mathbb{Z}} \mathbb{R}$;

(RD3) there exists a bijection $\Phi \to \Phi^\vee$ such that $\langle \alpha, \alpha^\vee \rangle = 2$; and

(RD4) the reflections s_α of the root system Φ, respectively $s_\alpha^\vee$ of $\Phi^\vee$ are given by

$$s_\alpha . \chi = \chi - \langle \chi, \alpha^\vee \rangle \alpha \qquad \text{for all } \chi \in X,$$

$$s_{\alpha^\vee} . \gamma = \gamma - \langle \alpha, \gamma \rangle \alpha^\vee \qquad \text{for all } \gamma \in Y.$$

It follows easily that whenever $(X, \Phi, Y, \Phi^\vee)$ is a root datum then the Weyl groups of Φ and of $\Phi^\vee$ are isomorphic via $s_\alpha \mapsto s_{\alpha^\vee}$ (see Exercise 10.34).

The preceding definition is justified by the following result:

Proposition 9.11 *Let Φ be the root system of a connected reductive group G with respect to the maximal torus T, with Weyl group W, and set $\Phi^\vee = \{\alpha^\vee \mid \alpha \in \Phi\}$. Then $(X(T), \Phi, Y(T), \Phi^\vee)$ is a root datum.*

Proof We already saw in Proposition 9.2 that Φ is an abstract root system in $\langle \Phi \rangle_{\mathbb{R}} \leq X_{\mathbb{R}} := X(T) \otimes_{\mathbb{Z}} \mathbb{R}$. Furthermore in Lemma 8.19 we constructed a surjective map $\Phi \to \Phi^\vee$, $\alpha \mapsto \alpha^\vee$, satisfying properties (RD3) and (RD4) with respect to the pairing coming from Proposition 3.6. Suppose that for $\alpha, \beta \in \Phi$ we have $\alpha^\vee = \beta^\vee$. Then

$$s_\alpha s_\beta(\chi) = \chi + \langle \chi, \alpha^\vee \rangle (\alpha - \beta) \qquad \text{for all } \chi \in X(T).$$

Since $\langle \alpha - \beta, \alpha^\vee \rangle = 0$ we have $(s_\alpha s_\beta - \mathrm{id}_X)^2 = 0$, so all eigenvalues of the map $s_\alpha s_\beta : X_{\mathbb{R}} \to X_{\mathbb{R}}$ are 1. But this is a transformation of finite order on the real vector space $X_{\mathbb{R}}$ and so $s_\alpha s_\beta = 1$ and $\alpha = \beta$. Then $\Phi^\vee$ is an abstract root system by Exercise 10.35. $\qquad\square$

Example 9.12 Let G be a connected reductive algebraic group with root datum $(X, \Phi, Y, \Phi^\vee)$ with respect to the maximal torus T, and let $T' \leq T$ be a subtorus. Then T' has root datum $(X', \emptyset, Y', \emptyset)$, with

$$Y' := \{\gamma \in Y \mid \gamma(\mathbf{G}_m) \leq T'\} = Y(T')$$

and

$$X' := \{\chi|_{T'} \mid \chi \in X\} \cong X/\operatorname{Ann}(Y') = X(T'),$$

where, for any submodule $Z \leq Y$ we set

$$\operatorname{Ann}(Z) := \{\chi \in X \mid \langle \chi, \gamma \rangle = 0 \text{ for all } \gamma \in Z\}.$$

There is a natural notion of isomorphism of root data (see [41, II.1.13]). Then the following fundamental result classifies semisimple groups (see [66, 9.6.2,10.1.1]):

Theorem 9.13 (Chevalley Classification Theorem) *Two semisimple linear algebraic groups are isomorphic if and only if they have isomorphic root data. For each root datum there exists a semisimple algebraic group which realizes it. This group is simple if and only if its root system is indecomposable.*

The groups with root system of type A_n, B_n, C_n or D_n are called *groups of classical type*; the remaining simple groups are called *groups of exceptional type*.

One can determine precisely the possible root data. This involves introducing one further finite group.

By (R1) and Proposition 9.2, $\mathbb{Z}\Phi$ is of finite index in X, and $\mathbb{Z}\Phi^\vee$ is of finite index in Y, when G is a semisimple group. For simple groups G with root system Φ not of type D_{2n}, the root datum, and hence the isomorphism type of G, is determined up to isomorphism by Φ and the index $|X : \mathbb{Z}\Phi|$.

In general, let $\Omega = \operatorname{Hom}(\mathbb{Z}\Phi^\vee, \mathbb{Z})$. Restriction gives a natural homomorphism

$$X \cong \operatorname{Hom}(Y, \mathbb{Z}) \to \operatorname{Hom}(\mathbb{Z}\Phi^\vee, \mathbb{Z}) = \Omega$$

which is injective. Thus we may view $\mathbb{Z}\Phi \subseteq X \subseteq \Omega$.

The finite group $\Lambda := \Lambda(\Phi) := \Omega/\mathbb{Z}\Phi$ does not depend on X and is called the *fundamental group* of the root system Φ. The root data with fixed root system Φ are classified by subgroups X of Ω satisfying $\mathbb{Z}\Phi \subseteq X \subseteq \Omega$ (up to automorphisms of Ω stabilizing Φ) hence by subgroups $X/\mathbb{Z}\Phi \leq \Omega/\mathbb{Z}\Phi$ of the fundamental group.

Definition 9.14 Let G be a semisimple algebraic group, with X, Φ, Ω as above. Then $\Lambda(G) := \Omega/X$ is called the *fundamental group* of G. If $X = \Omega$, so $\Lambda(G) = 1$, then G is said to be *simply connected*; if $X = \mathbb{Z}\Phi$ then G is said to be of *adjoint type*. We write $G_{\mathrm{ad}}, G_{\mathrm{sc}}$ for an algebraic group with given root system Φ of adjoint, respectively simply connected type. (See Remark 9.17 for the origin of this terminology.)

A surjective homomorphism $\varphi : G \to H$ of algebraic groups with finite kernel is called an *isogeny*. If such a morphism φ exists, one says that G and H are *isogenous*. If G is connected then $\ker \varphi$ is central by Exercise 10.4, and if moreover G is reductive, then $\ker \varphi$ lies in all maximal tori. The various semisimple algebraic groups G with fixed root system Φ are called the *isogeny types* corresponding to Φ due to the following result:

Proposition 9.15 *Let G be semisimple with root system Φ. Then there exist natural isogenies*

$$G_{\mathrm{sc}} \xrightarrow{\pi_1} G \xrightarrow{\pi_2} G_{\mathrm{ad}}$$

from a simply connected group G_{sc} and to an adjoint group G_{ad}, each with root system Φ, with $\ker(\pi_1) \cong \Lambda(G)_{p'}$, $\ker(\pi_2) \cong (\Lambda(G_{\mathrm{ad}})/\Lambda(G))_{p'}$, where $p = \mathrm{char}(k)$, and such that $d\pi_i$ is an isomorphism for $i = 1, 2$.

Proof for the case that $\mathrm{char}(k)$ is prime to $|\Lambda(\Phi)|$ Let T be a maximal torus of G with corresponding root system Φ. Then $Z := Z(G) \le C_G(T) = T$, so $Z = \bigcap_\alpha \ker(\alpha)$. Since Z lies in the kernel of the adjoint representation, G acts via G/Z on $\mathrm{Lie}(G)$, so the roots of G and G/Z are the same under the natural inclusion $X(T/Z) \le X(T)$ induced by the surjection $T \to T/Z$ (see Exercise 10.13). In particular, for any subgroup $S \le Z$, we have inclusions $\mathbb{Z}\Phi \le X(T/Z) \le X(T/S) \le X(T)$. In fact, as $\gcd(\mathrm{char}(k), |\Lambda|) = 1$, Proposition 3.8 implies that $X(T)/\mathbb{Z}\Phi \cong Z$. Hence, $X(T/Z)/\mathbb{Z}\Phi = 0$ and G/Z is of adjoint type; so we take $\pi_2 : G \to G/Z$ to be the natural surjection.

On the other hand, starting with G_{sc} and $\Lambda_1 \le X(T_{\mathrm{sc}})/\mathbb{Z}\Phi$ the fundamental group of G, then with $S = \Lambda_1^\perp \le Z(G_{\mathrm{sc}})$ the quotient G_{sc}/S has root datum with fundamental group Λ_1. With $\pi_1 : G_{\mathrm{sc}} \to G_{\mathrm{sc}}/S$ we then have $G \cong G_{\mathrm{sc}}/S$. The assertion that $d\pi_i$ is an isomorphism follows from Theorem 7.9(b). $\qquad\qquad \square$

For the general case, the claim follows from the isogeny theorem [66, Thm. 9.6.5] which asserts that any morphism of root data induces an isogeny of corresponding semisimple groups.

Table 9.2 gives a list of the possible isogeny types for simple groups and an identification with various classical groups (see e.g. [73, p.45]).

Table 9.2 *Isogeny types of simple algebraic groups*

Φ	$\Lambda(\Phi)$	G_{sc}	G_{ad}	in between
$A_{n-1},\ n \geq 2$	Z_n	SL_n	PGL_n	$SL_n/Z_d\ (d\|n)$
$B_n,\ n \geq 2$	Z_2	$Spin_{2n+1}$	SO_{2n+1}	$-$
$C_n,\ n \geq 2$	Z_2	Sp_{2n}	$PCSp_{2n}$	$-$
$D_n,\ n \geq 3$ odd	Z_4	$Spin_{2n}$	PCO_{2n}°	SO_{2n}
$D_n,\ n \geq 4$ even	$Z_2 \times Z_2$	$Spin_{2n}$	PCO_{2n}°	$SO_{2n}, HSpin_{2n}$
G_2	1	G_2		$-$
F_4	1	F_4		$-$
E_6	Z_3	$(E_6)_{\text{sc}}$	$(E_6)_{\text{ad}}$	$-$
E_7	Z_2	$(E_7)_{\text{sc}}$	$(E_7)_{\text{ad}}$	$-$
E_8	1	E_8		$-$

Example 9.16 (1) In Example 9.9 we saw that $X = \mathbb{Z}\Phi$ for PGL_2, so PGL_2 is of adjoint type. On the other hand for SL_2, $X = \Omega$, so SL_2 is simply connected.

(2) For Φ of type E_6 we have $|\Omega/\mathbb{Z}\Phi| = 3$, so there exist two isogeny types of groups of type E_6.

(3) The groups SO_{2n} with root system Φ of type D_n are neither adjoint nor simply connected. Here, the fundamental group Λ of Φ is of order 4, and SO_{2n} corresponds to a subgroup of Λ of order 2. If n is even, then Λ contains two other subgroups of order 2 which correspond to isomorphic simple groups, the so-called *half-spin groups* $HSpin_{2n}$. If $n = 4$ both are also isomorphic to SO_8. (These isomorphisms are induced by graph automorphisms of the Dynkin diagram of type D_n, see Theorem 11.12 below.) The groups of adjoint type D_n are obtained as the quotient $PCO_{2n}^\circ := CO_{2n}^\circ/Z(CO_{2n}^\circ)$ of the conformal orthogonal group (see Section 1.2) modulo its one-dimensional central torus, or also as the quotient $SO_{2n}/Z(SO_{2n})$.

(4) The orthogonal groups of simply connected type are the so-called *spin-groups* $Spin_n$. Their smallest faithful matrix representation is of dimension 2^s, with $s = \lfloor \frac{n-1}{2} \rfloor$. They are best constructed and studied as subgroups of the units in Clifford algebras.

For more information on the various types of classical groups, see for example the books by Dieudonné [21], Grove [30], or Goodman and Wallach [27] (for k of characteristic 0).

Remark 9.17 Steinberg has shown that G_{sc} is the universal perfect central extension, in the category of abstract groups, of any semisimple group with root system Φ [73, Thm. 10]. There is no such universal central extension in the category of algebraic groups since there exist bijective morphisms of simple groups which are not isomorphisms. These isogenies will play a crucial role in Part III. Over the field $k = \mathbb{C}$ of complex numbers the groups G_{sc} *are* in fact simply connected in the complex topology, and for G of arbitrary type, $\Lambda(G)$ *is* the topological fundamental group [73, Thm. 13]. Finally we note that the group G_{ad} is isomorphic to its image under the adjoint representation, which explains the denomination. (It is not however the case that the image of the adjoint representation of any group is necessarily of adjoint type. See Exercise 10.37.)

There is another universal property of groups of simply connected type which will be needed later on:

Proposition 9.18 *Let G be semisimple of simply connected type. Then whenever $\pi : H_1 \to H_2$ is an isogeny of semisimple groups with $d\pi$ an isomorphism, any isogeny $\varphi : G \to H_2$ lifts to an isogeny $\psi : G \to H_1$ such that $\varphi = \pi \circ \psi$.*

Proof Let $\tilde{G} := \{(g, h) \in G \times H_1 \mid \varphi(g) = \pi(h)\}$, a closed subgroup of $G \times H_1$. The projection $\mathrm{pr}_1 : \tilde{G} \to G$ onto the first factor is a surjective morphism with finite central kernel $1 \times C$, where $C = \ker(\pi)$, hence its restriction to $G_1 := \tilde{G}^\circ$ is an isogeny.

$$
\begin{array}{ccc}
G_1 \subseteq G \times H_1 & \xrightarrow{\ \mathrm{pr}_2\ } & H_1 \\
{\scriptstyle \mathrm{pr}_1}\Big\downarrow \quad {\scriptstyle \psi} \nearrow & & \Big\downarrow{\scriptstyle \pi} \\
G & \xrightarrow[\ \varphi\ ]{} & H_2
\end{array}
$$

In particular G_1 is semisimple and as pr_1 has central kernel, it induces an isomorphism of root systems and so G_1 has the same type root system as G. By Chevalley's Classification Theorem 9.13 this shows that G_1 is also of simply connected type and thus $\mathrm{pr}_1|_{G_1}$ is injective. Moreover, the differential $d\mathrm{pr}_1$ (which is just the projection onto $\mathrm{Lie}(G)$) has kernel $\ker(d\pi)$ which is 0 by assumption, hence $d\mathrm{pr}_1$ is an isomorphism. By the criterion in Proposition 7.7(c), $\mathrm{pr}_1|_{G_1}$ is an isomorphism, so we may take $\psi = \mathrm{pr}_2 \circ (\mathrm{pr}_1|_{G_1})^{-1}$ which is surjective since H_1 is connected. $\qquad\square$

10
Exercises for Part I

Throughout we take k to be an algebraically closed field.

Exercise 10.1 Let X be a topological space. Show the following assertions of Proposition 1.10:

(a) $Z \subset X$ is irreducible if and only if its closure $\bar{Z}$ is irreducible.
(b) Let $f : X \to Y$ be a continuous map to a topological space Y. If X is irreducible then so is the image $f(X)$.

Exercise 10.2 Show that the set $\{(x, y) \in k^2 \mid xy = 0\}$ is not irreducible but is connected in the Zariski topology.

Exercise 10.3 Let G be a linear algebraic group.

(a) Show that $C_G(x)$ is a closed subgroup, for $x \in G$.
(b) Show that $Z(G)$ is a closed subgroup of G.
(c) If H is a subgroup of G, then so is its closure $\bar{H}$.
(d) If H is a subgroup of G containing a non-empty open subset of its closure, then H is closed.

[**Hint:** For (c) first show that $H\bar{H} \subseteq \bar{H}$. For (d), first show that if $U, V \subseteq G$ are dense open subsets, then $UV = G$.]

Exercise 10.4 Let G be a linear algebraic group and let $N \lhd G$ be a finite normal subgroup. Let $H \leq G$ be a closed connected subgroup. Show that N lies in $C_G(H)$, the centralizer of H in G. (See also Exercise 10.18.)
[**Hint:** For x in N consider the orbit map $H \to N$, $h \mapsto hxh^{-1}$.]

Exercise 10.5 Show that $\mathrm{CSp}_{2n} = \mathrm{Sp}_{2n} \cdot Z(\mathrm{CSp}_{2n})$.

Exercise 10.6 Show that a non-trivial connected nilpotent group has a center of dimension at least 1.

[**Hint**: Look at the last non-trivial term in the descending central series and use Proposition 1.18.]

Exercise 10.7 Show that each of the groups T_n, U_n, D_n and SL_n is connected.

Exercise 10.8 Show that GO_{2n} is not connected when $\mathrm{char}(k) \neq 2$.
[**Hint**: Determine GO_2 explicitly to find an element of determinant -1.]

Exercise 10.9 Show that $\dim(T_n) = \binom{n+1}{2}$, $\dim(U_n) = \binom{n}{2}$, $\dim(D_n) = n$.

Exercise 10.10 Show that the (algebraic group) automorphisms of $\mathbf{G_a}$ are the multiplications by non-zero elements of k.

Exercise 10.11 Show that

$$\mathrm{End}(\mathbf{G_m}) := \{\varphi : \mathbf{G_m} \to \mathbf{G_m} \mid \varphi \text{ a morphism of algebraic groups}\} \cong \mathbb{Z};$$

conclude that the group of algebraic group automorphisms of $\mathbf{G_a}$ is $\mathbb{Z}_2$.

Exercise 10.12 Let G be a linear algebraic group.

(a) Show that the set G_u of unipotent elements in G is closed.
(b) Show, by example, that the set G_s of semisimple elements in G is not necessarily closed nor open.
(c) Show that the conjugacy class of a semisimple element in GL_n is closed.

[**Hint**: For (a) consider the characteristic polynomials of unipotent elements.]

Exercise 10.13 Let $\varphi : S \to T$ be a surjective homomorphism of tori. Show that φ induces a natural injective group homomorphism $\varphi^* : X(T) \to X(S)$.

Exercise 10.14 (Dedekind's Lemma) Let G be a group, K a field. Then any finite subset of $\mathrm{Hom}(G, K^\times)$ is linearly independent.
[**Hint**: See [32, 16.1].]

Exercise 10.15 Let T be a torus, $H \leq T$ a closed subgroup, and $X_1 \leq X(T)$ a subgroup. Show the following:

(a) $X(H)$ forms a basis of $k[H]$.
(b) Restriction defines an isomorphism

$$X(H) \cong X(T)/H^\perp.$$

(c) If $H_1 \leq H$ is a subgroup of finite index, then $H_1^\perp/H^\perp$ is finite.
(d) $X_1^{\perp\perp}/X_1$ is a finite p-group, where $p = \mathrm{char}(k)$; in particular, $X_1^{\perp\perp} = X_1$ if $X(T)/X_1$ has no p-torsion.

[**Hint**: By Example 3.5, $X(T)$ is a basis of $k[T]$. The embedding $\varphi : H \to T$ defines an epimorphism $k[T] \to k[H]$, which is just restriction. Now restrictions of characters are again characters. Use Exercise 10.14. For (c) note that $X(T)$ is a finitely generated $\mathbb{Z}$-module. For (d) let $x \in X_1^{\perp\perp} \setminus X_1$. Apply the elementary divisor theorem to $\langle X_1, x \rangle$ and again use that characters are linearly independent.]

Exercise 10.16 The purpose of this exercise is to point out that the structure of connected unipotent groups is difficult to classify.

(a) Show that

$$H_1 = \left\{ \begin{pmatrix} 1 & a & b \\ 0 & 1 & a \\ 0 & 0 & 1 \end{pmatrix} \;\middle|\; a, b \in k \right\}$$

is a closed connected two-dimensional commutative unipotent algebraic group.

(b) Show that $H_1 \cong \mathbf{G}_a \times \mathbf{G}_a$ as algebraic groups if and only if k is a field of characteristic different from 2.

(c) Show that

$$H_2 = \left\{ \begin{pmatrix} 1 & t & u & s \\ 0 & 1 & 0 & u \\ 0 & 0 & 1 & -t \\ 0 & 0 & 0 & 1 \end{pmatrix} \;\middle|\; s, t, u \in k \right\} \leq \mathrm{Sp}_4$$

is a three-dimensional connected unipotent algebraic group.

(d) Determine $[H_2, H_2]$; conclude that H_2 is abelian if and only if $\mathrm{char}(k) = 2$.

[**Hint**: For (b) note that if $\mathrm{char}(k) = 2$, H_1 has elements of order 4.]

Exercise 10.17 The goal of this exercise is to see that PGL_n, the projective linear group in dimension n, is a linear algebraic group.

Set $\mathrm{PGL}_n = \mathrm{GL}_n/Z$, where $Z = \{cI_n \mid c \in k^\times\}$, as abstract group. Let V be the n-dimensional vector space over k on which GL_n naturally acts and $V^* = \mathrm{Hom}(V, k)$ the dual space. Consider the action of GL_n on $V \otimes V^*$. This defines a group homomorphism $\rho : \mathrm{GL}_n \to \mathrm{GL}_{n^2}$.

(a) Show that ρ is a morphism of algebraic groups and hence its image is a closed subgroup H of GL_{n^2}.

(b) Show that $\ker \rho = Z$.

Hence we conclude that PGL_n is isomorphic, as an abstract group, to the closed subgroup H of GL_{n^2} and thus can be given the structure of a linear algebraic group.

Exercise 10.18 Let G be a linear algebraic group, X a G-space.

(a) Show that the fixed point set $X^g := \{x \in X \mid g.x = x\}$ is closed for any $g \in G$. Conclude that $X^G := \{x \in X \mid g.x = x \text{ for all } g \in G\}$ is closed.

(b) Let Y and Z be closed subsets of X and set $\mathrm{Tran}_G(Y, Z) := \{x \in G \mid x.Y \subseteq Z\}$ (the *transporter* of Y into Z). Show that $\mathrm{Tran}_G(Y, Z)$ is a closed subset of G.

(c) Show that for H a closed subgroup of G, $N_G(H)$ is a closed subgroup.

[**Hint**: For (a), consider the morphism $\psi : X \to X \times X$, $x \mapsto (x, g.x)$, and use that the diagonal in $X \times X$ is closed. For (b), consider the maps $\varphi_y : G \to X$ given by $x \mapsto x.y$. Then $\mathrm{Tran}_G(Y, Z) = \bigcap_{y \in Y} (\varphi_y^{-1}(Z))$.]

Exercise 10.19 (Maximal tori and Borel subgroups of classical groups)

(a) Show that $\mathrm{rk}(\mathrm{Sp}_{2n}) = \mathrm{rk}(\mathrm{SO}_{2n+1}) = n$.

(b) Show that $\mathrm{Sp}_{2n} \cap \mathrm{T}_{2n}$ (respectively $\mathrm{SO}_{2n+1} \cap \mathrm{T}_{2n+1}$) is a Borel subgroup of Sp_{2n} (respectively of SO_{2n+1}).

Exercise 10.20 Show that a connected linear algebraic group with nilpotent Borel subgroup is solvable. Conclude that any two-dimensional connected group is solvable.

[**Hint**: Consider a counterexample of minimal dimension and use Exercise 10.6 and Proposition 6.8.]

Exercise 10.21 Show that for a connected group G, $G/R(G)$ is semisimple, and $G/R_u(G)$ is reductive.

Exercise 10.22 (A semisimplicity criterion)

(a) Let G be a linear algebraic group. Show that the radical $R(G)$ is equal to $(\bigcap_B B)^\circ$, where B runs over all Borel subgroups of G.

(b) Show that Sp_{2n} is semisimple.

Exercise 10.23 (Algebraic groups consisting of semisimple elements)

(a) Let G be a connected linear algebraic group whose elements are semisimple. Show that G is a torus.

(b) Give an example of an algebraic group (closed subgroup of GL_n) which consists of semisimple elements but which is not conjugate to a subgroup of D_n, the group of diagonal matrices.

[**Hint**: For (a) use Exercise 10.20.]

Exercise 10.24 Let V be the three-dimensional vector space on which SL_3 naturally acts, with a fixed basis e_1, e_2, e_3. Let $P = \mathrm{Stab}_{\mathrm{SL}_3}(\langle e_1, e_2 \rangle)$. Let

$$S = \{g \in P \mid g.e_i = \lambda_{g,i}\, e_i, \text{ for some } \lambda_{g,i} \in k^\times, 1 \leq i \leq 3\}$$

(so S is a torus of P). Find $N_P(S)$, $N_P(S)^\circ$, $C_P(S)$, and $C_P(S)^\circ$ and verify that $N_P(S)/C_P(S)$ is finite.

Exercise 10.25 Let G be a linear algebraic group. Show that the inversion morphism $i \colon G \to G$, $g \mapsto g^{-1}$, on G has differential $di(X) = -X$, for $X \in \mathrm{Lie}(G)$.
[**Hint**: Apply Example 7.8(1) and Proposition 7.7(a) to the morphism $\mu \circ (i, \mathrm{id}) \colon G \times G \to G$.]

Exercise 10.26 In this exercise, we identify the Lie algebra of various algebraic groups.

(a) The Lie algebra of T_n, as well as that of U_n and D_n, may be identified with a subalgebra of $\mathfrak{gl}_n$. Show that it equals the set of all upper triangular, strictly upper triangular, diagonal matrices, respectively.
(b) Show that the Lie algebra of SL_n may be identified with the Lie-subalgebra $\mathfrak{sl}_n$ of $\mathfrak{gl}_n$ of $n \times n$ trace zero matrices.
(c) Show that $\mathrm{Lie}(\mathrm{PGL}_n) \cong \mathrm{Lie}(\mathrm{SL}_n)$ if $\mathrm{char}(k)$ does not divide n.

Exercise 10.27 Assume k to be a field of characteristic $p > 0$, $G = \mathrm{SL}_3$, and let

$$\varphi \colon \mathbf{G}_a \to G, \qquad \varphi(t) = \begin{pmatrix} 1 & t & t^p \\ 0 & 1 & 0 \\ 0 & 0 & 1 \end{pmatrix}.$$

(a) Show that φ defines an isomorphism of algebraic groups $\mathbf{G}_a \cong \mathrm{im}(\varphi)$. Set $H = \mathrm{im}(\varphi)$.
(b) Determine $\mathrm{Lie}(H)$, as a subalgebra of $\mathfrak{gl}_3$.
(c) Show that $C_G(H)$ is not equal to

$$C_G(\mathrm{Lie}(H)) := \{g \in G \mid \mathrm{Ad}\,(g)X = X \text{ for all } X \in \mathrm{Lie}(H)\}.$$

(d) Show that if $\mathrm{char}(k) = 3$, then $\mathrm{Lie}(C_G(H))$ is not equal to

$$C_{\mathrm{Lie}(G)}(H) := \{X \in \mathrm{Lie}(G) \mid \mathrm{Ad}\,(h)X = X \text{ for all } h \in H\}.$$

Exercise 10.28 (a) Let G be a linear algebraic group. Show that $Z(G) \leq \ker(\mathrm{Ad}\,)$.
(b) Determine $\ker(\mathrm{Ad}\,)$ for each of the following groups: GL_n, U_n, T_n.

(c) This example shows that $\ker(\mathrm{Ad})$ may be larger than $Z(G)$. Let k be a field of characteristic $p > 0$. Let $G = \left\{ \begin{pmatrix} a & 0 & 0 \\ 0 & a^p & b \\ 0 & 0 & 1 \end{pmatrix} \;\middle|\; a \in k^\times, b \in k \right\}$.

Show that $Z(G) \subsetneq \ker(\mathrm{Ad}) \subsetneq G$.

Exercise 10.29 Let $G = \mathrm{Sp}_{2n}$. Then $\mathrm{Lie}(G) = \{X \in \mathfrak{gl}_{2n} \mid X^{tr} J_{2n} = -J_{2n} X\}$, where J_{2n} is as in the definition of symplectic groups in Section 1.2. Let $T = \mathrm{D}_{2n} \cap \mathrm{Sp}_{2n}$ be a maximal torus of Sp_{2n} (see Exercise 10.19).

(a) In the case $n = 2$, find the roots and root subspaces of $\mathrm{Lie}(G)$. For each root α, exhibit a one-dimensional closed subgroup $U_\alpha \le G$ whose Lie algebra is the corresponding root subspace.

(b) Generalize to the case of arbitrary n. Show that the root system of Sp_{2n} is of type C_n. Conclude that $\dim \mathrm{Sp}_{2n} = 2n^2 + n$.

Exercise 10.30 Let T be a maximal torus of a connected reductive group G, with character group X, cocharacter group Y and Weyl group W.

(a) Show that W acts faithfully on both X and Y.

(b) Prove that $\langle w.\chi, \gamma \rangle = \langle \chi, w^{-1}.\gamma \rangle$ for all $w \in W$, $\chi \in X$, $\gamma \in Y$.

Exercise 10.31 Show that the Weyl group W of a connected reductive group G is generated by the s_α, $\alpha \in \Phi$.

[**Hint**: Argue by induction on $\dim(G)$. For $w \in W$, with preimage $n \in N_G(T)$, consider the homomorphism $\psi : T \to T$, $t \mapsto ntn^{-1}t^{-1}$. If ψ is not surjective, then $S := \ker(\psi)^\circ$ is a non-trivial subtorus of T with $n \in C := C_G(S)$. If $C < G$, use induction for the maximal torus $C \cap T$ of C. If $C = G$, then $S \le Z(G)$ is normal, and induction applies to G/S. If ψ is surjective, use Exercise 10.13 and argue that there exists $x \in X(T) \otimes_{\mathbb{Z}} \mathbb{R}$ such that $(w - 1)x = \alpha$. Then show that $s_\alpha x = wx$; hence $s_\alpha x$ has eigenvalue 1 and we can apply the argument of the first case to this element.]

Exercise 10.32 Show that for a connected reductive group G, $\ker(\mathrm{Ad}) = Z(G)$.

[**Hint**: By Exercise 10.28 we have $Z(G) \le \ker(\mathrm{Ad})$. If $\ker(\mathrm{Ad})$ is finite, use Exercise 10.4. For G semisimple, apply Theorem 8.17 to see that a maximal torus of any simple factor G_i of G acts non-trivially on some root space in $\mathfrak{g}$. For the general case, $R(G) = Z(G)^\circ \le \ker(\mathrm{Ad})$, and Ad induces the adjoint representation of $G/R(G)$ with kernel $Z(G/R(G))$. Since $G/R(G)$ is semisimple, we have $Z(G/R(G))^\circ = 1$.]

Exercise 10.33 Let G be semisimple, $G = G_1 \cdots G_r$ its decomposition into simple components as in Theorem 8.21(d). Then the root system Φ of G decomposes into an orthogonal disjoint union $\Phi = \Phi_1 \sqcup \ldots \sqcup \Phi_r$ of indecomposable root systems Φ_i, such that G_i has root system Φ_i.

In particular, simple algebraic groups have indecomposable root systems.

Exercise 10.34 Let $(X, \Phi, Y, \Phi^\vee)$ be a root datum. Then the Weyl groups of Φ and of $\Phi^\vee$ are isomorphic via $s_\alpha \mapsto s_{\alpha^\vee}$.

[**Hint**: Define an action of the Weyl group W of Φ on Y by: $w.\gamma$ is the unique element such that $\langle \chi, w.\gamma \rangle = \langle w^{-1}.\chi, \gamma \rangle$ for all $\chi \in X$. Now mimic the proof of Lemma 8.19 to see that this sends s_α to $s_{\alpha^\vee}$.]

Exercise 10.35 (A realization of the dual root system inside $X_\mathbb{R}$) . Let X, respectively Y, be the character, respectively cocharacter, group of a connected reductive group, $\Phi \subset X$ the root system with Weyl group W and $\{\alpha^\vee \mid \alpha \in \Phi\} \subset Y$ the coroots as constructed in Lemma 8.19. We write $V := X_\mathbb{R} = X \otimes_\mathbb{Z} \mathbb{R}$.

(a) The pairing between X, Y induces a natural isomorphism $Y_\mathbb{R} \cong V^* := \mathrm{Hom}(V, \mathbb{R})$.

(b) Let $(,) : V \times V \to \mathbb{R}$ be a W-invariant scalar product on V. Show that the reflection s_α along α is given by $s_\alpha(x) = x - 2(x, \alpha)\alpha$.

(c) The scalar product $(,)$ induces an isomorphism $i : V^* \to V$, via $\langle x, y \rangle = (x, i(y))$ for $x \in X$, $y \in Y$, which satisfies $i(\alpha^\vee) = 2\alpha/(\alpha, \alpha)$ for $\alpha \in \Phi$.

(d) Show that $\tilde{\Phi} := \{i(\alpha^\vee) \mid \alpha \in \Phi\}$ is a root system in V, and hence $\Phi^\vee$ is a root system in $Y_\mathbb{R}$.

[**Hint**: For (c) compare $\alpha^\vee$ and $2\alpha/(\alpha, \alpha)$ using Lemma 8.19.]

Exercise 10.36 Show that

$$[\mathrm{GL}_n, \mathrm{GL}_n] = \mathrm{SL}_n, \quad [\mathrm{CSp}_{2n}, \mathrm{CSp}_{2n}] = \mathrm{Sp}_{2n}, \quad [\mathrm{CO}_n^\circ, \mathrm{CO}_n^\circ] = \mathrm{SO}_n.$$

[**Hint**: Compare the root systems of the two groups in each case and apply Proposition 6.20(c) and Table 9.2. For GL_n, this is also easy to show by direct matrix calculation.]

Exercise 10.37 Let $G = \mathrm{Sp}_2$, $\mathfrak{g} = \mathrm{Lie}(G)$, $\mathrm{char}(k) = 2$, and $\mathrm{Ad} : G \to \mathrm{GL}(\mathfrak{g})$ be the adjoint representation. Show that $\mathrm{Ad}\,(G)$ is a simply connected simple algebraic group of type A_1.

PART II

SUBGROUP STRUCTURE AND REPRESENTATION THEORY OF SEMISIMPLE ALGEBRAIC GROUPS

In this part we develop the basic structure and representation theory of semisimple and more generally of reductive algebraic groups. The aim of this part is twofold. First, we want to provide sufficient information in order to be able to formulate the classification results on maximal subgroups of simple algebraic groups presented in Chapters 18 and 19 and to sketch the important ingredients in their proof. Secondly, we prepare the notions and results necessary for the investigation of the finite groups of Lie type in Part III.

It has already been indicated in Chapter 8 that the structure and classification of reductive groups is somehow controlled by their maximal connected solvable subgroups, the Borel subgroups. This will become even more apparent here. The crucial starting point, on which most of the development in this part will be based, is the structure result for Borel subgroups of reductive groups in Theorem 11.1. From this, we first derive the so-called Bruhat decomposition, which is of a more topological nature, and then also the more combinatorial BN-pair structure of connected reductive groups. The latter leads to the notion of parabolic subgroups and their Levi decomposition, which are studied in Chapter 12. In the framework of classical groups, parabolic subgroups arise very naturally as the stabilizers of isotropic subspaces in the natural matrix representation.

Chapter 13 is devoted to the study of subsystem subgroups, that is, subgroups normalized by a maximal torus. These comprise the Levi subgroups already encountered before, but also the centralizers of semisimple elements.

We derive the classification of subsystem subgroups by the algorithm of Borel and de Siebenthal and also treat the case of small characteristic. In Chapter 14 we then go on to consider properties of centralizers and conjugacy classes. We give a parametrization of semisimple conjugacy classes, and prove some basic properties of regular semisimple elements. We introduce the important concepts of bad primes and torsion primes and indicate their relation to connectedness of centralizers.

In Chapters 15 and 16 we give an introduction to the representation theory of reductive algebraic groups over their underlying field, the so-called weight theory up to Steinberg's tensor product theorem. We also study two important ingredients for the determination of maximal subgroups: the characterization of the self-dual irreducible modules in terms of their highest weight and certain properties of restrictions to Levi subgroups.

In Chapter 17 we return to the structure theory of parabolic subgroups to study their action on the unipotent radical, which gives rise to an important class of naturally occurring representations, the so-called internal modules. Moreover, we sketch in Theorem 17.10 the proof of the important result of Borel and Tits that the normalizer of any non-trivial unipotent subgroup in a connected reductive group is contained in a proper parabolic subgroup, and give some applications to the study of maximal subgroups.

In Chapter 18 we state the reduction result for maximal subgroups of simple algebraic groups of classical type, sketch its proof and give some indications on the complete classification of maximal subgroups of positive dimension. The maximal subgroups of simple groups of exceptional type have to be investigated by different methods, as we explain in Chapter 19.

Throughout this part, we consider algebraic groups defined over a fixed algebraically closed field k of characteristic $p \geq 0$.

11

BN-pairs and Bruhat decomposition

We fix the following notation. Let G be a connected reductive algebraic group, $T \leq G$ a maximal torus with character group X and Weyl group W. Let Φ be the root system of G with respect to T. For each $\alpha \in \Phi$ there is an associated root subgroup $U_\alpha \leq G$ and a reflection $s_\alpha \in W$ (see Sections 8.3 and 8.4).

We start by refining the assertion that $G = \langle T, U_\alpha \mid \alpha \in \Phi \rangle$ from Theorem 8.17(g). There will be two such results, the first giving a product decomposition of the Borel subgroup B, the second a partition of G into double cosets with respect to B. This will also lead to the determination of the algebraic automorphism group of a semisimple group.

11.1 On the structure of B

Our first main result here not only gives structural information on Borel subgroups but also provides a strong connection between Borel subgroups containing the given maximal torus T and bases Δ of the root system with respect to T.

Theorem 11.1 *Let G be connected reductive, $B \geq T$ a Borel subgroup of G.*

(a) *There exists a base Δ of Φ with positive system $\Phi^+ \subseteq \Phi$ such that*

$$B = T \cdot \prod_{\alpha \in \Phi^+} U_\alpha$$

for any fixed order of the factors U_α; moreover, we have uniqueness of expression with respect to the product in the fixed order.

(b) *If G is semisimple then $|\Delta| = \dim(X \otimes_{\mathbb{Z}} \mathbb{R}) = \dim(T) = \operatorname{rk} G$.*

(c) $W = \langle s_\alpha \mid \alpha \in \Delta \rangle$.

(d) $G = \langle T, U_\alpha \mid \alpha \in \pm\Delta \rangle$.

Proof of (b)–(d)   Parts (b) and (c) are just Proposition 9.4(c) together with Proposition 9.2. For (d) note that $\langle U_\alpha, U_{-\alpha} \rangle$ contains a preimage of the simple reflection s_α, for $\alpha \in \Delta$, so $H := \langle U_\alpha \mid \alpha \in \pm\Delta \rangle$ contains preimages of all $w \in W$ by (c). Hence $U_\beta \leq H$ for all $\beta \in \Phi$ by Proposition 9.4(c) and Theorem 8.17(e). Then (d) follows from Theorem 8.17(g). See [66, Prop. 8.2.1] for a proof of (a).   □

Definition 11.2   In the situation of Theorem 11.1, Δ is called the set of *simple roots* with respect to $T \leq B$, and $\{s_\alpha \mid \alpha \in \Delta\}$ are called the *simple reflections*.

Remark 11.3   Recall from Theorem 8.17(f) and Proposition 8.15 that the subgroup $\langle U_\alpha, U_{-\alpha} \rangle$ is non-solvable, so it follows from Theorem 11.1(a) that $U_\alpha \nleq B$ for $\alpha \in \Phi^-$.

Example 11.4   We verify Theorem 11.1 in the case of $G = \mathrm{SL}_n$. Here, $T = D_n \cap \mathrm{SL}_n$ is a maximal torus contained in the Borel subgroup $B = T_n \cap \mathrm{SL}_n$. By Example 9.8 the root system of G with respect to T is given by $\Phi = \{\chi_{ij} \mid i \neq j\}$ with $\chi_{ij}(\mathrm{diag}(t_1, \ldots, t_n)) = t_i t_j^{-1}$.

Thus, for $i \neq j$ we have $U_{\chi_{ij}} = U_{ij}$ where

$$U_{ij} := I_n + \langle E_{ij} \rangle.$$

Now, clearly B can be decomposed as

$$B = \left\{ \begin{pmatrix} * & \cdots & * \\ & \ddots & \vdots \\ 0 & & * \end{pmatrix} \in \mathrm{SL}_n \right\} = T \cdot \prod_{i<j} U_{ij}.$$

Moreover, with respect to the base $\Delta = \{\chi_{i,i+1} \mid 1 \leq i \leq n-1\}$ the set of positive roots equals $\Phi^+ = \{\chi_{ij} \mid i < j\}$, by Example 9.8. So indeed

$$B = T \cdot \prod_{\alpha \in \Phi^+} U_\alpha$$

as claimed in Theorem 11.1, and $|\Delta| = n - 1 = \dim(T) = \mathrm{rk}(G)$. Finally, the algorithm of Gaussian elimination shows that SL_n is generated by the U_α together with diagonal matrices.

We investigate the structure of B and $R_u(B)$ a bit further; for this, let's choose and fix isomorphisms $u_\alpha : \mathbf{G}_a \to U_\alpha$, for all $\alpha \in \Phi$.

Proposition 11.5 *Let G be reductive, and let $U \leq R_u(B)$ be a T-stable subgroup. Then U is the product of the root subgroups U_α it contains, hence closed and connected. In particular, the U_α, $\alpha \in \Phi^+$, are the minimal non-trivial T-invariant subgroups of $R_u(B)$.*

Proof Let $U \leq R_u(B)$ be T-stable. Since U is unipotent, $U' := [U, U]$ is a T-stable subgroup of smaller dimension, hence a product of root subgroups by induction. Thus we may argue in the abelian group U/U'. Clearly, if $u_\alpha(c) \in U$ with $c \neq 0$, then $tu_\alpha(c)t^{-1} = u_\alpha(\alpha(t)c) \in U$ for all $t \in T$, so $U_\alpha \leq U$ since $\alpha \neq 0$.

Assume the result fails. Then by Theorem 11.1(a) there is $1 \neq u \in U$ which is a product $u = \prod_{\alpha \in M} u_\alpha(c_\alpha)$, for some $M \subseteq \Phi^+$, such that $U_\alpha \not\subseteq U$ for all $\alpha \in M$. Take u such that $|M|$ is minimal, and let $\beta, \gamma \in M$ be different. Since β, γ are linearly independent, there is $t \in T$ with $\beta(t) = 1$, $\gamma(t) \neq 1$. Then modulo U' we have that

$$tut^{-1}u^{-1} \equiv \prod_{\alpha \in M} u_\alpha(\alpha(t)c_\alpha) \prod_{\alpha \in M} u_\alpha(-c_\alpha) \equiv \prod_{\alpha \in M} u_\alpha(c_\alpha(\alpha(t) - 1))$$

does not involve u_β any more, but a non-trivial element from U_γ occurs in the expression. So it is a non-trivial product of smaller length, contradicting our minimal choice. $\square$

The following consequence will be used in the investigation of maximal rank subgroups:

Corollary 11.6 *Let G be connected reductive with maximal torus T and $H \leq G$ a connected reductive subgroup normalized by T. Then*

$$H = \langle T \cap H, U_\alpha \mid U_\alpha \leq H \rangle.$$

Proof Note that HT is a reductive subgroup of G. Let B_H be a Borel subgroup of HT containing T. It lies in a Borel subgroup B of G, hence so does $U := R_u(B_H)$. Application of Proposition 11.5 shows that $U = \prod_{\alpha \in M} U_\alpha$ for some subset $M \subseteq \Phi$. But then $HT = \langle T, U_\alpha \mid \alpha \in \pm M \rangle$ by Theorems 8.17(g) and 11.1. As HT is reductive, we have $U \leq [HT, HT] \leq H$ and the claim follows. $\square$

Example 11.7 (Root system and root subgroups of SO_{2n}) Let $G = SO_{2n}$ with Borel subgroup $B = G \cap T_{2n}$ and maximal torus $T = G \cap D_{2n}$ (see Example 6.7). Here,

$$U_{ij} := I_{2n} + \langle E_{ij} - E_{2n-j+1, 2n-i+1} \rangle, \qquad 1 \leq i < j \leq 2n - i,$$

are one-dimensional connected subgroups of the unipotent radical $R_u(B) =$

$G \cap U_{2n}$ normalized by T. By the description of B in Example 6.7(4) we see that $R_u(B)$ is the product of the U_{ij}. Thus, the U_{ij} are the positive root subgroups with respect to B and T, by Proposition 11.5.

A short calculation shows that T acts via the character $\epsilon_i - \epsilon_j$ on U_{ij}, where

$$\epsilon_i : T \to \mathbf{G}_m, \qquad \mathrm{diag}(t_1, \ldots, t_n, t_n^{-1}, \ldots, t_1^{-1}) \mapsto t_i.$$

Now $\epsilon_i = -\epsilon_{2n-i+1}$ on T, so we see that $\Phi^+ = \{\epsilon_i \pm \epsilon_j \mid 1 \le i < j \le n\}$, with base $\Delta := \{\epsilon_1 - \epsilon_2, \epsilon_2 - \epsilon_3, \ldots, \epsilon_{n-1} - \epsilon_n, \epsilon_{n-1} + \epsilon_n\}$. Computation of $|(\mathbb{Z}\alpha + \mathbb{Z}\beta) \cap \Phi^+|$ for $\alpha, \beta \in \Delta$ now shows that the root system Φ of SO_{2n} has type D_n.

The following result essentially describes multiplication in the unipotent radical of B (see [32, Lemma 32.5], and [73, Thm. 8]).

Theorem 11.8 (Commutator formula) *Given a root system* Φ, *with a fixed total ordering compatible with addition, there exist integers* $c_{\alpha\beta}^{mn}$ *such that for any connected reductive group* G *with root system* Φ *over an algebraically closed field* k, *the morphisms* u_α *can be chosen so that for all roots* $\alpha \ne \pm\beta$ *we have*

$$[u_\alpha(t), u_\beta(u)] = \prod_{m,n>0} u_{m\alpha+n\beta}(c_{\alpha\beta}^{mn} t^m u^n) \qquad \text{for all } t, u \in k,$$

where the product is over all integers $m, n > 0$ *such that* $m\alpha + n\beta \in \Phi$, *taken according to the chosen ordering.*

The $c_{\alpha\beta}^{mn}$ above are called *structure constants*. Note that all unipotent elements of a connected reductive group G are contained in its derived subgroup, which is semisimple, and also that any semisimple group is a quotient of a group of simply connected type by a subgroup consisting of semisimple elements. So it suffices to prove the above statement for semisimple groups of simply connected type. The above, together with more precise versions of Theorem 8.17(c) and (e), yields the very important *Steinberg presentation* of a semisimple group G, see [73, §6] or [13, Thm. 12.1.1].

From now on, we fix a total ordering on Φ and, for each $\alpha \in \Phi$, an isomorphism $u_\alpha : \mathbf{G}_a \to U_\alpha$, $c \mapsto u_\alpha(c)$, for which Theorem 11.8 holds.

Example 11.9 We determine the commutator relations for SL_3 and Sp_4.

(1) For the group $G = \mathrm{SL}_3$, with B and T as in Example 9.8, by Example 11.4 the root groups of G with respect to T are

$$U_\alpha = I_3 + \langle E_{12} \rangle, \ U_\beta = I_3 + \langle E_{23} \rangle, \ U_{\alpha+\beta} = I_3 + \langle E_{13} \rangle,$$

for the roots $\alpha = \chi_{12}$ and $\beta = \chi_{23}$. A direct computation gives

$$[u_\alpha(t), u_\beta(u)] = u_{\alpha+\beta}(tu)$$

and all other commutators between root subgroups corresponding to positive roots are trivial.

(2) For $G = \mathrm{Sp}_4$, we have by Exercise 10.19 that $B = G \cap T_4$ is a Borel subgroup with maximal torus $T = G \cap D_4$. We write (t_1, t_2) to denote the element $\mathrm{diag}(t_1, t_2, t_2^{-1}, t_1^{-1}) \in T$. A similar calculation to the one in Example 11.7 shows that the root subgroups of G in B are given by

$$U_\alpha = I_4 + \langle E_{12} - E_{34} \rangle, \qquad U_\beta = I_4 + \langle E_{23} \rangle,$$
$$U_{\alpha+\beta} = I_4 + \langle E_{13} + E_{24} \rangle, \qquad U_{2\alpha+\beta} = I_4 + \langle E_{14} \rangle,$$

for the roots $\alpha, \beta \in X(T)$ with $\alpha(t_1, t_2) = t_1 t_2^{-1}$, $\beta(t_1, t_2) = t_2^2$. In particular the root system of Sp_4 is of type C_2 (compare with Exercise 10.29). A direct computation gives

$$[u_\alpha(t), u_\beta(u)] \quad = u_{\alpha+\beta}(-tu) u_{2\alpha+\beta}(-t^2 u),$$
$$[u_\alpha(t), u_{\alpha+\beta}(u)] \quad = u_{2\alpha+\beta}(-2tu),$$

and all other commutators between root subgroups corresponding to positive roots are trivial. It becomes apparent from these relations that the structure of the unipotent radical of B in characteristic 2 is considerably different from the case of odd (or zero) characteristic. See also Example 13.16 where the effect of this on maximal rank subgroups will emerge.

We can now describe the automorphism group of a semisimple group G. Let's fix a pair $T \le B$ consisting of a maximal torus of G contained in a Borel subgroup, with corresponding root system Φ, set of positive roots Φ^+ and base Δ. We first prove a general result about endomorphisms which will also be used in Part III in the investigation of Steinberg endomorphisms.

Lemma 11.10 *Let G be connected reductive with T, B, Φ^+ as above. Let $\sigma : G \to G$ be an endomorphism stabilizing T and B. Then there exists a permutation ρ of Φ^+ stabilizing Δ such that for all $\alpha \in \Phi^+$ the following conditions hold:*

(a) *There exists a positive integer q_α, equal to 1 or to a power of $\mathrm{char}(k) > 0$ such that $\sigma(\rho(\alpha)) := \rho(\alpha) \circ \sigma|_T = q_\alpha \alpha$.*

(b) *There exists $a_\alpha \in k^\times$ such that $\sigma(u_\alpha(c)) = u_{\rho(\alpha)}(a_\alpha c^{q_\alpha})$ for all $c \in k$.*

If moreover σ is an automorphism of algebraic groups, then $q_\alpha = 1$ for all $\alpha \in \Phi^+$.

Proof As σ stabilizes both B and T, it permutes the minimal T-stable subgroups of the unipotent radical $R_u(B)$ of B, viz., the root subgroups U_α, $\alpha \in \Phi^+$, see Proposition 11.5. This induces a permutation ρ of Φ^+, via $\sigma(U_\alpha) = U_{\rho(\alpha)}$, which must preserve Δ since any positive system contains a unique base (see Proposition A.7). In particular for any $\alpha \in \Phi^+$ there exists an endomorphism $\nu : \mathbf{G_a} \to \mathbf{G_a}$ such that $\sigma(u_\alpha(c)) = u_{\rho(\alpha)}(\nu(c))$ for $c \in k$. For $t \in T$ we have $tu_\alpha(c)t^{-1} = u_\alpha(\alpha(t)c)$ by Theorem 8.17(c). Application of σ yields

$$\sigma(tu_\alpha(c)t^{-1}) = \sigma(t)\sigma(u_\alpha(c))\sigma(t)^{-1} = u_{\rho(\alpha)}\big(\rho(\alpha)(\sigma(t))\,\nu(c)\big)$$

on the left-hand side, and

$$\sigma(u_\alpha(\alpha(t)c)) = u_{\rho(\alpha)}\big(\nu(\alpha(t)c)\big)$$

on the right, whence

$$\nu(\alpha(t)c) = \rho(\alpha)(\sigma(t))\,\nu(c) \qquad \text{for all } c \in k, \, t \in T.$$

Thus, ν is a monomial, of degree q_α say. Clearly, this defines a group morphism only if q_α is a power of $\mathrm{char}(k)$, showing (b). If σ is an isomorphism, then so is ν, but then $q_\alpha = 1$ by Exercise 10.10.

Note that as T is stabilized by σ, by composition σ also defines a linear map $X(T) \to X(T)$, $\chi \mapsto \chi \circ \sigma$. Putting $c = 1$ above we get

$$\sigma(\rho(\alpha))(t) = \rho(\alpha)(\sigma(t)) = \nu(\alpha(t))/\nu(1) = \alpha(t)^{q_\alpha}$$

for all $t \in T$, which shows (a). $\qquad\qquad\square$

Let's now write $\mathrm{Aut_a}(G)$ for the group of (algebraic group) automorphisms of G, $\mathrm{Inn}(G)$ for the subgroup of inner automorphisms, and let Γ_G denote the subgroup of $\mathrm{Aut_a}(G)$ of automorphisms of G which fix a pair $T \le B$ consisting of a maximal torus of G contained in a Borel subgroup.

Theorem 11.11 *Let G be semisimple. Then:*

(a) $\mathrm{Aut_a}(G) = \mathrm{Inn}(G).\Gamma_G$.
(b) *The elements of Γ_G induce diagram automorphisms of the Dynkin diagram of G, which determine them uniquely modulo $\mathrm{Inn}(G) \cap \Gamma_G$.*

In particular, $\mathrm{Out_a}(G) := \mathrm{Aut_a}(G)/\mathrm{Inn}(G)$ is finite.

Proof Let $\sigma \in \mathrm{Aut_a}(G)$. Since G acts transitively on its Borel subgroups by conjugation, and B acts transitively on its maximal tori, there is $g \in G$ such that σ composed with conjugation by g fixes both T and B. This proves the first part.

For part (b), let $\sigma \in \Gamma_G$. Then by Lemma 11.10 it induces a permutation ρ of the positive system Φ^+ such that $\sigma(\rho(\alpha)) = \alpha$, that is, the homomorphism of $X(T)$ induced by σ permutes the set of positive roots. Exercise C.1 then shows that σ must induce a diagram automorphism of the Dynkin diagram with respect to the base Δ in Φ^+.

For the uniqueness assertion it suffices to show that any $\sigma \in \Gamma_G$ which induces the identity on Δ is contained in $\mathrm{Inn}(G)$. Any such σ fixes all root subgroups $U_\alpha = u_\alpha(\mathbf{G}_a)$ for $\alpha \in \Delta$, thus by Lemma 11.10 there exist $a_\alpha \in k^\times$ such that $\sigma(u_\alpha(c)) = u_\alpha(a_\alpha c)$ for $c \in k$. As Δ is linearly independent, there exists $s \in T$ such that $\alpha(s) = a_\alpha$, for all $\alpha \in \Delta$. Replacing σ by $s^{-1}\sigma$ we may assume by Theorem 8.17(c) that σ acts as the identity on all U_α, $\alpha \in \Delta$.

Furthermore, as σ acts trivially on Δ, for $t \in T$ we have $\alpha(\sigma(t)) = \alpha(t)$ for $\alpha \in \Delta$, so $\alpha(\sigma(t)t^{-1}) = 1$. But G is semisimple, so $\bigcap_{\alpha \in \Delta} \ker(\alpha)$ is finite. Hence the morphism $T \to T$, $t \mapsto \sigma(t)t^{-1}$, has finite, connected image, forcing $\sigma(t) = t$ for all $t \in T$. Thus σ centralizes T_α, hence stabilizes $C_\alpha = C_G(T_\alpha)$ for all $\alpha \in \Delta$. As it centralizes the Borel subgroup TU_α of the latter, it must be trivial on C_α by Proposition 6.8. Since G is generated by the C_α (see Proposition 8.15), the claim follows. $\qquad\square$

The non-trivial groups of diagram automorphisms of connected Dynkin diagrams are collected in Table 11.1. It then follows from the previous result that the outer (algebraic) automorphism group of a simple algebraic group is solvable, of order at most 6.

Table 11.1 *Graph automorphisms of Dynkin diagrams*

Φ	$A_n\ (n \geq 2)$	$D_n\ (n \geq 5)$	D_4	E_6
Γ_G	Z_2	Z_2	$\mathfrak{S}_3$	Z_2

Theorem 11.11 is complemented by the following existence result for automorphisms (see [15, 23.7, Cor. 3] or [73, Cor. on p.156]):

Theorem 11.12 (Chevalley) *Let G be semisimple of simply connected or adjoint type, ρ a permutation of the set of simple roots Δ of G inducing a symmetry of the Dynkin diagram. Then there exists an automorphism $\sigma \in \mathrm{Aut}_a(G)$ with*

$$\sigma(u_\alpha(c)) = u_{\rho(\alpha)}(c) \qquad \text{for all } \alpha \in \Delta,\ c \in k.$$

In particular, $\mathrm{Out}_a(G)$ is isomorphic to the group of graph automorphisms of the Dynkin diagram of G.

Proof for type A_{n-1} For A_{n-1}, $n \geq 3$, there is a unique non-trivial graph automorphism of the Dynkin diagram, which sends α_i to α_{n-i}, for $1 \leq i \leq n-1$ (with the simple roots labeled as in Table 9.1). An easy calculation with the root subgroups in Example 11.4 shows that the product of the transpose-inverse automorphism with conjugation by $\begin{pmatrix} 0 & & .1 \\ & \cdot & \\ 1 & & 0 \end{pmatrix} \cdot \mathrm{diag}(1, -1, 1, -1, \ldots)$ is an automorphism of SL_n with the required property. Factoring out the center we obtain a corresponding automorphism on the adjoint group PGL_n. $\square$

The automorphisms in Theorem 11.12 are called *graph automorphisms* of G. For SO_{2n}, of type D_n, elements of $\mathrm{GO}_{2n} \setminus \mathrm{SO}_{2n}$ induce a non-trivial graph automorphism of order 2 (see Exercise 20.1). The exceptional graph automorphisms for D_4, of order 3, and for E_6, of order 2, are realized inside suitable larger groups, see Examples 12.12 and 13.9.

Remark 11.13 There exist further abstract group automorphisms of a simple algebraic group G, but which are not invertible as morphisms. For example, any field automorphism of the underlying field k extends to an automorphism of G. Also, if G is of type B_2 or F_4 in characteristic 2, or of type G_2 in characteristic 3, and k is perfect, there exists a bijective endomorphism which comes from the non-trivial symmetry of the corresponding Coxeter diagram (see [73, Thm. 29 and Cor.]). These will be important in Part III.

11.2 Bruhat decomposition

We now discuss the partition of a connected reductive group G into double cosets of a Borel subgroup. Throughout, we fix a maximal torus T lying in a Borel subgroup B, with root system Φ, positive roots Φ^+ and base Δ determined by B. We write $R_u(B) = \prod_{\alpha \in \Phi^+} U_\alpha$ according to Theorem 11.1. For any $w \in W = N_G(T)/T$ let $\dot{w}$ denote an arbitrary fixed preimage in $N_G(T)$. We first need to study the multiplication of double cosets with respect to B.

Lemma 11.14 *Let $\alpha \in \Delta$ be a simple root with corresponding simple reflection $s \in W$. Then for all $w \in W$ we have*

$$B\dot{w}B \cdot B\dot{s}B \subseteq B\dot{w}\dot{s}B \cup B\dot{w}B.$$

Proof It can easily be checked inside the homomorphic image $\langle U_\alpha, U_{-\alpha} \rangle$ of SL_2 that $\dot{s}(U_\alpha \setminus \{1\})\dot{s} \subseteq TU_\alpha \dot{s} U_\alpha$. So clearly

$$\dot{s}U_\alpha \dot{s} \subseteq TU_\alpha \dot{s} U_\alpha \cup U_\alpha \subseteq B\dot{s}B \cup B.$$

Moreover, by Theorem 11.1 we have $B = (\prod_{\beta \in \Phi^+ \setminus \{\alpha\}} U_\beta) \cdot U_\alpha \cdot T$. Recall that $\dot{v} U_\beta \dot{v}^{-1} = U_{v\beta}$ for all $v \in W$ and $\beta \in \Phi$ by Theorem 8.17(e), and $s_\alpha . \beta \in \Phi^+$ for $\alpha \neq \beta \in \Phi^+$ by Lemma A.8. Now first assume that $w\alpha \in \Phi^+$. Then

$$B \dot{w} B \cdot B \dot{s} B = B \dot{w} U_\alpha \dot{s} B = B U_{w\alpha} \dot{w} \dot{s} B = B \dot{w} \dot{s} B.$$

On the other hand, if $w\alpha \in -\Phi^+$ then $ws\alpha = w(-\alpha) = -w\alpha \in \Phi^+$. So, with $v := ws$ we have $B\dot{v}B \cdot B\dot{s}B = B\dot{w}B$ by the above. Then

$$B\dot{w}B \cdot B\dot{s}B = B\dot{v}\dot{s}B\dot{s}B \subseteq B\dot{v}(B\dot{s}B \cup B) = B\dot{v}B\dot{s}B \cup B\dot{v}B = B\dot{w}B \cup B\dot{w}\dot{s}B$$

as claimed, where the inclusion follows from the first part of the proof. $\square$

We now introduce a crucial combinatorial group theoretical structure which is particularly well adapted to reductive algebraic groups.

Definition 11.15 A pair B, N of subgroups of a group G is called a *BN-pair* (for G) if the following axioms are satisfied:

(BN1) G is generated by B and N.
(BN2) $B \cap N$ is a normal subgroup of N.
(BN3) The group $W := N/(B \cap N)$ is generated by a set S of involutions.
(BN4) If $\dot{s} \in N$ maps to $s \in S$ under the natural homomorphism $N \to W$, and $n \in N$, then $BnB \cdot B\dot{s}B \subseteq Bn\dot{s}B \cup BnB$.
(BN5) If $\dot{s}$ is as before then $\dot{s}B\dot{s} \neq B$.

The group W is called the *Weyl group* of the BN-pair.

Theorem 11.16 (Tits) *Let G be a connected reductive algebraic group with Borel subgroup B and $N := N_G(T)$ for some maximal torus $T \leq B$. Then B, N is a BN-pair in G whose Weyl group is equal to that of G.*

Proof By Theorem 4.4 and Corollary 8.13(b), $B \cap N = N_B(T) = C_B(T) = T$ and so $B \cap N$ is normal in N, giving (BN2). Moreover $W = N/(B \cap N)$ is the Weyl group of G, hence generated by the set S of simple reflections by Theorem 11.1(c), whence we have (BN3). By Theorem 8.17(g), G is generated by T and the U_α for $\alpha \in \Phi$. But for $\alpha \in \Phi^+$, $\dot{s}_\alpha U_\alpha \dot{s}_\alpha^{-1} = U_{-\alpha} \leq \langle B, N \rangle$ by Theorem 8.17(e), showing (BN1). We just proved (BN4) in Lemma 11.14. Finally, writing α for the root of a simple reflection $s \in S$, $\dot{s}B\dot{s}$ contains $\dot{s}U_\alpha\dot{s} = U_{-\alpha}$, which does not lie in B by Remark 11.3. This shows (BN5). $\square$

Let's put $\Phi^- := -\Phi^+$ whenever Φ^+ is a positive system inside Φ, and

$$U_w^- := \prod_{\alpha \in \Phi^+, w.\alpha \in \Phi^-} U_\alpha$$

for any $w \in W$ (note that by Theorem 11.8 this is a subgroup of G). With this we have the following extension of Theorem 8.17(g) (see [32, Thm. 28.3 and 28.4]):

Theorem 11.17 (Bruhat decomposition)   *Let G be a group with a BN-pair. Then*

$$G = \bigsqcup_{w \in W} B\dot{w}B$$

for any choice of preimages $\dot{w} \in N$ mapping to $w \in W = B/N$.

If G is connected reductive, $T \leq B$ a maximal torus in a Borel subgroup of G, W the Weyl group of G with respect to T, then more precisely, every $g \in G$ can be uniquely written as $g = u\dot{w}b$, where $b \in B$, $w \in W$ and $u \in U_w^-$ with respect to the positive system $\Phi^+ \subseteq \Phi$ determined by $T \leq B$.

Proof of the double coset decomposition   We first show that $B\dot{v}B \cap B\dot{w}B = \emptyset$ if $v \neq w$. Write v as a product of minimal length $\ell(v)$ in the generators from S. We argue by induction on $\ell(v)$, which we may assume to be not larger than $\ell(w)$. If $\ell(v) = 0$, then $v = 1$. Now $B = B\dot{w}B$ implies $\dot{w} \in B \cap N$, so $w = 1$. If $\ell(v) > 0$, we may write $v = v's$ with $s \in S$ and $\ell(v') < \ell(v)$. Then $B\dot{v}'\dot{s} = B\dot{v} \subseteq B\dot{v}B = B\dot{w}B$. By (BN4) this gives

$$B\dot{v}' \subseteq B\dot{w}B\dot{s} \subseteq B\dot{w}\dot{s}B \cup B\dot{w}B.$$

So $B\dot{v}'B = B\dot{w}\dot{s}B$ or $B\dot{v}'B = B\dot{w}B$. As $\ell(v') < \ell(v)$ induction yields that either $v' = ws$ or $v' = w$. The second is not possible since $\ell(v') < \ell(v) \leq \ell(w)$ by assumption. So $v = v's = w$.

On the other hand, an easy induction on $\ell(w)$ shows from (BN4) that the union of the $B\dot{w}B$, where w runs over W, is closed under multiplication. Since it contains B and N, it is all of G by (BN1).   □

The proof of the uniqueness assertion uses in an essential way that G is reductive.

Corollary 11.18   *In the notation of Theorem 11.17, there exists a unique element $w_0 \in W$ such that $w_0(\Delta) = -\Delta$. Moreover $w_0^2 = 1$ and $B^{\dot{w}_0} \cap B = T$.*

Proof   The first part follows directly from the simple transitivity of W on the set of bases, see Proposition 9.4(b). Now assume that $g \in B^{\dot{w}_0} \cap B$, so $g = \dot{w}_0 u t \dot{w}_0 = t'u'$ for some $u, u' \in U = R_u(B)$, $t, t' \in T$. Thus

$$\dot{w}_0^{-1} t' u' = u \dot{w}_0 t^{\dot{w}_0}.$$

As $U_{w_0}^- = U$ by the defining property of w_0, the uniqueness statement in Theorem 11.17 shows that $u' = 1 = u$, whence $g = t' \in T$. The other inclusion is clear since $\dot{w}_0$ normalizes T.   □

The element w_0 above is called the *longest element* of W with respect to S since its length $\ell(w_0)$ as introduced in the proof of the Bruhat decomposition can be shown to be maximal possible (see also Proposition A.21 for another characterization of the length function on W). The subgroup $B^{\dot{w}_0}$ is called the Borel subgroup *opposite to B.*

Corollary 11.19 *Let G be reductive. Then the intersection of any two Borel subgroups of G contains a maximal torus.*

Proof Since all Borel subgroups and maximal tori of G lie in G°, we may assume that G is connected. Let B, B_1 be Borel subgroups, so $B_1 = gBg^{-1}$ for some $g \in G$ (by Theorem 6.4). Choose a maximal torus $T \le B$, with Weyl group W. Then $g = b\dot{w}b'$ with $b, b' \in B$, $w \in W$, as in Theorem 11.17; hence

$$B_1 = gBg^{-1} = b\dot{w}b'B(b')^{-1}\dot{w}^{-1}b^{-1} = b\dot{w}B\dot{w}^{-1}b^{-1}.$$

Now $\dot{w}T\dot{w}^{-1} = T$, so $b\dot{w}T\dot{w}^{-1}b^{-1} = bTb^{-1} \le B$ is a maximal torus in $B \cap B_1$. $\qquad\square$

We end this chapter with the following topological aspect of the Bruhat decomposition which will be needed for the investigation and construction of representations in Section 15.2.

Theorem 11.20 *Let G be connected reductive, $T \le B$ a maximal torus inside a Borel subgroup, with corresponding positive system Φ^+, $U^- := \langle U_\alpha \mid \alpha \in \Phi^- \rangle$. Then the product map $\pi\colon U^- \times B \to G$ is a bijective morphism of $U^- \times B$ onto a dense open subset of G (called the big cell).*

Proof By definition of U^- and of w_0 we have $U^- = \dot{w}_0 U \dot{w}_0$, so the product map $U^- \times B = \dot{w}_0 U \dot{w}_0 \times B \to G$ is a bijective morphism onto the subset $\dot{w}_0 U \dot{w}_0 B$ by the uniqueness of expression in the Bruhat decomposition (Theorem 11.17). Now

$$\dim(U^- \times B) = \dim(U^-) + \dim(B) = |\Phi^-| + \mathrm{rk}(G) + |\Phi^+| = \dim(G)$$

by Theorem 8.17(b). Moreover, $U^- \times B$ is irreducible, so the closure of its image is an irreducible subset of G of the same dimension, hence equal to G by Proposition 1.22.

It remains to show that the image is open; we use the fact that it is a translate of the coset $B\dot{w}_0 B = U\dot{w}_0 B$ by $\dot{w}_0$. As $B\dot{w}_0 B$ is the B-orbit of $\dot{w}_0 B$ (in the action of G on G/B by left multiplication), Proposition 5.4(a) implies that $B\dot{w}_0 B$ is open in its closure. But we have just observed that the closure is G/B. So $B\dot{w}_0 B$ is open in G/B. This then implies that $B\dot{w}_0 B$ is open in G because the quotient map is continuous. $\qquad\square$

It can be shown that the map π above is even an isomorphism onto its image, see [32, 28.5].

12

Structure of parabolic subgroups, I

We now come to a natural collection of subgroups of a reductive group, the parabolic subgroups; in fact, one can define parabolic subgroups in any group having a BN-pair. It will turn out that these are precisely the closed subgroups containing a Borel subgroup of G. Moreover, their structure as a semidirect product of a reductive group, the so-called Levi complement, with a unipotent normal subgroup enables one to argue inductively for many questions concerning the subgroup structure and representation theory of reductive groups. We'll see that Levi complements can alternatively be characterized as the centralizers of subtori.

12.1 Parabolic subgroups

Let G be a connected reductive algebraic group, $T \leq G$ a maximal torus contained in a Borel subgroup B of G. Let Φ be the root system of G, Δ the set of simple roots with respect to $T \leq B$, $S = \{s_\alpha \mid \alpha \in \Delta\}$ the corresponding set of generating reflections of the Weyl group $W = N_G(T)$ (see Theorem 11.1(c)).

We define a collection of natural subgroups of W. For a subset $I \subseteq S$, $W_I := \langle s \in I \rangle$ is called a *standard parabolic subgroup* of W. A *parabolic subgroup* of W is any conjugate of a standard parabolic subgroup. We let $\Delta_I := \{\alpha \in \Delta \mid s_\alpha \in I\}$ and

$$\Phi_I := \Phi \cap \sum_{\alpha \in \Delta_I} \mathbb{Z}\alpha$$

be the corresponding *parabolic subsystem* of roots.

Proposition 12.1 *Let* $I \subseteq S$. *Then* Φ_I *is a root system in* $\mathbb{R}\Phi_I$ *with base* Δ_I *and Weyl group* W_I.

This is proved in Proposition A.25.

The W_I are, up to conjugation, precisely the subspace centralizers of W in its natural reflection representation on $X_{\mathbb{R}} := X \otimes_{\mathbb{Z}} \mathbb{R}$, see Corollary A.29. The group G now contains similar natural subgroups:

Proposition 12.2 *Let G be connected reductive, $T \leq B$ a maximal torus in a Borel subgroup of G, with root system Φ and set of simple reflections S.*

(a) *Let $I \subseteq S$. Then $P_I := BW_I B = \bigsqcup_{w \in W_I} B\dot{w}B$ is a closed, connected, self-normalizing subgroup of G which contains B.*

(b) *The P_I are mutually non-conjugate; in particular, $P_I = P_J$ implies $I = J$.*

(c) *$P_I = \langle T, U_\alpha \mid \alpha \in \Phi^+ \cup \Phi_I \rangle$.*

Moreover, all overgroups of B in G arise in this way.

Proof (a) By Lemma 11.14 we have $B\dot{w}B \cdot B\dot{s}B \subseteq P_I$ for all $w \in W_I$ and all $s \in I$. Hence, by an easy induction, $P_I \leq G$. Now assume that $g \in N_G(P_I)$. Then B, B^g are two Borel subgroups of P_I, hence conjugate by some $p \in P_I$ by Theorem 6.4(a). Then $gp \in N_G(B) = B$, by Theorem 6.12, so $g \in P_I$.

In (b), if $P_J = P_I^g$ with $g \in G$ then B, B^g are Borel subgroups of P_J, whence $g \in P_J$ as before, so $P_I = P_J$. Next, if $P_I = P_J$ for $I, J \subseteq S$ then by (a) we must have $W_I = W_J$. This implies that $W_I = W_{I \cup J}$, so we may assume that $I \subseteq J$. But the elements of S are reflections on $X_{\mathbb{R}}$, so the fixed space of W_I on $X_{\mathbb{R}}$ has codimension at most $|I|$. On the other hand, by Corollary 8.22 and the proof of Theorem 8.21, the fixed space of $W = W_S$ on $X_{\mathbb{R}}$ has codimension $\mathrm{rk}_{\mathrm{ss}}(G)$, which equals $|\Delta|$ by Theorem 11.1(b), so we must in fact have equality. Thus $W_I = W_J$ implies that $I = J$.

For (c), write $\Phi_I^{\pm} := \Phi_I \cap \Phi^{\pm}$. Since $B \subseteq P_I$ we have $U_\alpha \subseteq P_I$ for all $\alpha \in \Phi^+$. Since Φ_I is a root system by Proposition 12.1, the longest element $w \in W_I$ satisfies $w(\Phi_I^+) = \Phi_I^-$ (see Corollary 11.18). Thus, for $\beta \in \Phi_I^-$ there exists $\alpha \in \Phi_I^+$ such that $w\alpha = \beta$. Then $U_\beta = U_{w\alpha} = \dot{w}U_\alpha \dot{w}^{-1} \subseteq P_I$, which proves the inclusion "$\supseteq$". For the converse note that if $\alpha \in \Delta$ is the simple root corresponding to $s \in S$, then we may choose $\dot{s} \in \langle U_{\pm\alpha} \rangle$. So $\langle T, U_\alpha \mid \alpha \in \Phi_I \rangle$ contains preimages of all $s \in I$, hence of all $w \in W_I$, which in view of the decomposition in (a) gives the other inclusion. The connectedness of P_I now follows by Proposition 1.16.

Finally, all overgroups of B are of the form P_I for some $I \subseteq S$ by Exercise 20.3. $\qquad\square$

Note that all statements except for (c) and the connectedness of P_I remain valid for arbitrary BN-pairs, see Exercise 20.3.

Definition 12.3 The P_I ($I \subseteq S$) are called *standard parabolic subgroups* of G. A *parabolic subgroup* of G is any subgroup containing a Borel subgroup. Note that by Proposition 12.2 these are just the conjugates of standard parabolic subgroups.

Example 12.4 Let $G = \mathrm{SL}_n$. Then $T = \mathrm{D}_n \cap \mathrm{SL}_n$ is a maximal torus contained in the Borel subgroup $B = \mathrm{T}_n \cap \mathrm{SL}_n$, with root system $\Phi = \{\chi_{ij} \mid 1 \le i, j \le n,\ i \ne j\}$, set of positive roots $\Phi^+ = \{\chi_{ij} \mid i < j\}$, and base $\Delta = \{\chi_{i,i+1} \mid i < n\}$ (see Example 9.8). Furthermore,

$$
s_i := \begin{pmatrix} I_{i-1} & & & \\ & 0 & 1 & \\ & -1 & 0 & \\ & & & I_{n-i-1} \end{pmatrix} T \in N_G(T)/T
$$

is the simple reflection corresponding to the simple root $\chi_{i,i+1}$. We identify $W = N_G(T)/T$ with $\mathfrak{S}_n$ via $s_i \mapsto (i, i+1)$, so $S = \{s_i \mid i < n\} = \{(12), (23), \ldots, (n-1, n)\}$. Let $I = S \setminus \{s_a\}$ for some $a < n$. The associated maximal parabolic subgroup of W is $W_I = \langle s_1, \ldots, s_{a-1}, s_{a+1}, \ldots, s_{n-1} \rangle \cong \mathfrak{S}_a \times \mathfrak{S}_{n-a}$. And so the corresponding parabolic subgroup of G consists of block diagonal matrices

$$
P_a := P_I = \left\{ \begin{pmatrix} A & * \\ 0 & B \end{pmatrix} \ \middle| \ A \in \mathrm{GL}_a,\ B \in \mathrm{GL}_{n-a} \right\} \cap \mathrm{SL}_n,
$$

the stabilizer in the natural representation of G of the subspace generated by the first a standard basis vectors.

More generally, parabolic subgroups in GL_n and in SL_n are just the stabilizers of flags of subspaces in their natural n-dimensional representation space. In fact, the parabolic subgroups of all classical groups (groups with root system of type A_n, B_n, C_n or D_n) have an interpretation in terms of the natural module for the group. We give a precise statement and proof in the next section (see Proposition 12.13).

The BN-pair setting also allows one to give an easy proof that simple algebraic groups (as defined in Theorem 8.21) are simple modulo center as abstract groups; for this recall from the discussion before Theorem 9.6 that the Weyl group of a simple algebraic group is irreducible.

Proposition 12.5 *Let G be a simple linear algebraic group. Then $G/Z(G)$ is simple as an abstract group.*

Proof Let $H \trianglelefteq G$ be a normal subgroup and B some Borel subgroup of G.

Since H is normal in G, the product BH is a subgroup of G containing B. Then

$$BH = BW_I B = \bigsqcup_{w \in W_I} B\dot{w}B \qquad \text{for some } I \subseteq S$$

by Proposition 12.2. We claim that also W_I is normal in W. For this, let $s \in S \setminus I$. For any $w \in W_I$, $B\dot{w}B \subseteq BH$, so there exists $b \in B$ with $\dot{w}b \in H$ by the normality of H. Thus

$$s\dot{w}b\dot{s} \in (Bs\dot{w}sB \cup Bs\dot{w}B) \cap BH$$

by Lemma 11.14. By the choice of s, clearly $sw \notin W_I$, so $sws \in W_I$, for all $s \in S \setminus I$, $w \in W_I$. Since S generates W by Theorem 11.1(c) this shows that $W_I \trianglelefteq W$ as claimed.

Now first assume that $I \neq S$. Then W_I centralizes the non-zero subspace $\bigcap_{s \in I} \ker(s - \mathrm{id})$ of $X_{\mathbb{R}}$, while W acts irreducibly on $X_{\mathbb{R}}$. But a normal subgroup of an irreducible group which fixes a non-zero vector is trivial, whence $I = \emptyset$ and $W_I = 1$. Then $BH = B$, so $H \leq B$. As H is normal in G, it also lies in all conjugates of B. Since

$$\left(\bigcap_{g \in G} B^g \right)^{\circ} = R(G) = 1$$

by Proposition 6.16, H is finite, and hence $H \leq Z(G)$ by Exercise 10.4.

Otherwise, if $I = S$ then $BH = G$, so $B/(B \cap H) \cong BH/H = G/H$. Now, since B is solvable, $B/(B \cap H)$ is solvable while G/H is perfect since G is. Thus $G/H = 1$ and hence $H = G$. □

12.2 Levi decomposition

We continue the investigation of the structure of parabolic subgroups of a connected reductive group G. For $I \subseteq S$ define

$$U_I := \prod_{\alpha \in \Phi^+ \setminus \Phi_I} U_\alpha = \langle U_\alpha \mid \alpha \in \Phi^+ \setminus \Phi_I \rangle$$

with Φ_I as above. In this situation,

$$L_I := \langle T, U_\alpha \mid \alpha \in \Phi_I \rangle \leq P_I$$

is a complement to U_I in the parabolic subgroup P_I:

Proposition 12.6 *Let $I \subseteq S$. Then $R_u(P_I) = U_I$, and L_I is a complement to U_I, so $P_I = U_I \rtimes L_I$. In particular, L_I is reductive with root system Φ_I.*

Furthermore, all closed complements to U_I are conjugate to L_I in P_I and $L_I = C_G(Z(L_I)^\circ)$.

Observe the special case when $I = \emptyset$, so $P_I = B$, $U_I = R_u(B)$ and $L_I = T$ where this is just Theorem 4.4.

Proof We see that $U_I \trianglelefteq P_I$ and $P_I = \langle U_I, L_I \rangle$ by the commutator relations (Theorem 11.8) and the generation property of P_I in Proposition 12.2(c). To show that $L_I \cap U_I = 1$ let $Z := (\bigcap_{\alpha \in \Phi_I} \ker \alpha)^\circ \leq T$ and $L := C_G(Z)$. Note that $L_I \leq L$ by Theorem 8.17(c). We claim that $L = L_I$.

By Corollary 8.13(a), L is connected reductive, so by Corollary 11.6 it is generated by T and the U_β it contains. Now assume that $U_\beta \leq L$. Then the restriction of β to Z is trivial by Theorem 8.17(c), so $\beta \in Z^\perp$ (see Definition 3.7). The identity component Z is of finite index in $\bigcap_{\alpha \in \Phi_I} \ker \alpha = \langle \Phi_I \rangle^\perp$, so some multiple of β lies in $\langle \Phi_I \rangle^{\perp\perp}$ (see Exercise 10.15). Furthermore, by Proposition 3.8(b), $\langle \Phi_I \rangle^{\perp\perp}/\langle \Phi_I \rangle$ is a finite group, so some non-zero multiple of β even lies in $\langle \Phi_I \rangle$. By (R2) this shows that $\beta \in \Phi_I$, so $L = L_I$. But then $U_I \cap L_I$ is normal in the reductive group $L_I = L$ and unipotent, whence trivial, so $P_I = U_I \rtimes L$. As L_I is reductive, we moreover have $Z(L_I)^\circ = Z$, so $L_I = C_G(Z(L_I)^\circ)$, and Z is a maximal torus of $R(P_I)$.

In general, if $L' \leq P_I$ is any closed complement to U_I, then by the first part $L' \cong L_I$ as abstract groups, hence also $Z' := Z(L')^\circ \cong Z(L_I)^\circ = Z$ by Theorem 8.17(h). Since $R_u(P_I) = U_I$ we have $R_u(L') \cong R_u(P_I/U_I) = 1$ and so $R_u(Z') = 1$. Thus, Z' is a torus by Example 6.17(2). Now tori of different dimensions cannot be isomorphic as abstract groups (count elements of fixed finite order), so Z' is a torus of the same dimension as Z, hence also a maximal torus of $R(P_I)$. But all maximal tori of $R(P_I)$ are conjugate, by Theorem 4.4(b), so $L' = C_{P_I}(Z')$ is conjugate to $L_I = C_{P_I}(Z)$. $\square$

Definition 12.7 The decomposition $P_I = U_I \rtimes L_I$ in the preceding proposition is called the *Levi decomposition* of the parabolic subgroup P_I, and L_I is called the *(standard) Levi complement* of P_I. The conjugates of standard Levi complements are called *Levi subgroups* of G.

Corollary 12.8 *In the setting of Proposition 12.6, $P_I = N_G(U_I)$ for all $I \subseteq S$.*

Proof Clearly, $P_I \leq N := N_G(U_I) = N_G(R_u(P_I))$, so $N = P_J$ is parabolic for some subset $J \supseteq I$ of S by Proposition 12.2. Now for $s = s_\alpha \in J \setminus I$ we have $s \in N$ and $U_\alpha \leq U_I$, so $\dot{s} U_\alpha \dot{s}^{-1} = U_{-\alpha} \leq U_I$, which is not the case, so $J = I$, $N = P_I$. $\square$

In contrast to the situation for parabolic subgroups in Proposition 12.2(b), Levi subgroups L_I for different subsets $I \subseteq S$ may be conjugate in G:

Example 12.9 Let $G = \mathrm{SL}_n$ and P_a the parabolic subgroup of G defined in Example 12.4, where $1 \leq a \leq n - 1$. It has Levi complement

$$L_a = \left\{ \begin{pmatrix} A & 0 \\ 0 & B \end{pmatrix} \ \middle| \ A \in \mathrm{GL}_a, B \in \mathrm{GL}_{n-a} \right\} \cap \mathrm{SL}_n,$$

and clearly L_a is conjugate to L_{n-a} in G.

More generally, for a flag $0 = V_0 \subset V_1 \subset \cdots \subset V_r = k^n$ with $\dim V_i/V_{i-1} = n_i$, so $n = n_1 + \cdots + n_r$, the stabilizer in GL_n is a parabolic subgroup (see Proposition 12.13) with Levi complement $\mathrm{GL}_{n_1} \times \cdots \times \mathrm{GL}_{n_r}$ consisting of block diagonal matrices, where $\mathrm{GL}_{n_i} \cong \mathrm{GL}(V_i/V_{i-1})$.

The preceding proposition shows in particular that Levi subgroups L are centralizers of their central torus $Z(L)^\circ$. In fact, the converse is also true:

Proposition 12.10 *Let G be connected reductive, $Z \leq G$ a torus. Then $C_G(Z)$ is a Levi subgroup of G.*

Proof Let $C := C_G(Z)$, a connected reductive group by Proposition 8.13. A Borel subgroup B_C of C lies in some Borel subgroup B of G. As $Z \leq Z(C)^\circ \leq B_C \leq B$, Z is contained in a maximal torus T of B, hence of G. Clearly, $T \leq C$. So by Theorem 8.17(g), $C = \langle T, V_\alpha \mid \alpha \in \Phi_C \rangle$, where Φ_C denotes the root system of C with respect to T, and V_α are the T-root subgroups of C. In particular, for $\alpha \in \Phi_C$ we have $\mathrm{Lie}(V_\alpha) \subseteq \mathfrak{g}_\alpha \neq 0$, so $\Phi_C \subseteq \Phi$. Moreover, V_α is the unique one-dimensional subgroup of C with $\mathrm{Lie}(V_\alpha) = (\mathrm{Lie}(C))_\alpha = \mathfrak{g}_\alpha$. Hence by Theorem 8.17(d), $V_\alpha = U_\alpha$, the T-root subgroup of G corresponding to α.

Now clearly $U_\alpha \leq C$ if and only if $Z \leq \ker \alpha$, whence $C = \langle T, U_\alpha \mid Z \leq \ker(\alpha) \rangle$. So $\Phi_C = \Phi \cap \langle \alpha \mid Z \leq \ker(\alpha) \rangle$, the intersection of Φ with a subspace of $\mathbb{R}\Phi$. By an elementary property of root systems, see Corollary A.29, Φ_C is a parabolic subsystem of Φ, and thus C is the Levi complement of the corresponding parabolic subgroup of G. $\square$

For later use let's also point out the following (see Exercise 20.4):

Corollary 12.11 *Let $I \subseteq S$, $W_I = \langle I \rangle$, L_I the corresponding standard Levi subgroup. Then there is a natural isomorphism $N_G(L_I)/L_I \cong N_W(W_I)/W_I$.*

Example 12.12 From the Dynkin diagram one sees that a simple group G of type E_6 has a Levi subgroup L with derived subgroup $L' = [L, L]$ of type D_4. By Corollary 12.11 the normalizer of L in G can be computed inside the Weyl group; it is an extension of L by the symmetric group $\mathfrak{S}_3$, which necessarily leaves L' invariant. An element σ therein of order 3 induces the *triality* automorphism σ of the root system D_4 of order 3, hence a corresponding triality graph automorphism of L' whose existence was asserted in Theorem 11.12:

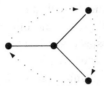

See also Example 13.9 for another realization of triality.

Similarly, the non-trivial graph automorphism of E_6 may be realized by embedding this group as a Levi subgroup inside E_7 with non-trivial normalizer.

We now consider the case of parabolic subgroups in classical groups and show that the situation of Examples 12.4 and 12.9 generalizes. Let V be a finite-dimensional vector space over k equipped with the 0 form β, a symplectic (i.e., non-degenerate skew-symmetric) bilinear form β, or a non-degenerate quadratic form $\mathcal{Q}$ (with associated bilinear form $\beta_{\mathcal{Q}}$). In the case where V of odd dimension is equipped with a quadratic form and $\mathrm{char}(k) = 2$, we say the quadratic form is non-degenerate if the radical of the form $\beta_{\mathcal{Q}}$ is a 1-space $\langle w \rangle$ with $\mathcal{Q}(w) \neq 0$. Recall that $v \in V$ is called *isotropic* if $\beta(v, v) = 0$, and in the orthogonal case it's said to be *singular* if $\mathcal{Q}(v) = 0$. A subspace W of V is said to be *totally isotropic* if $\beta(v, w) = 0$ (or $\beta_{\mathcal{Q}}(v, w) = 0$) for all $v, w \in W$, and in the orthogonal case it is said to be *totally singular* if $\mathcal{Q}(v) = 0$ for all $v \in W$. Note that (totally) singular implies (totally) isotropic and that the two notions are equivalent if $\mathrm{char}(k) \neq 2$. When the space is equipped only with a bilinear form, we may use the term "singular" in place of "isotropic".

Let $\mathrm{Isom}(V)$ denote the full group of isometries of V and $\mathrm{Cl}(V) \leq \mathrm{Isom}(V)$ be the group $\mathrm{SL}(V)$, $\mathrm{Sp}(V)$ or $\mathrm{SO}(V)$ respectively, as introduced in Section 1.2.

Proposition 12.13 *The parabolic subgroups of a simple classical group* $\mathrm{Cl}(V)$ *are precisely the subgroups of the form*

$$P = \mathrm{Stab}_{\mathrm{Cl}(V)}(0 \subset V_1 \subset \ldots \subset V_r \subset V),$$

where V_i *is a totally isotropic (totally singular if* V *is equipped with a quadratic form) subspace of* V *for* $1 \leq i \leq r$.

Proof To simplify the exposition, throughout this proof we will use the terms "singular" and "totally singular" for isotropic vectors and subspaces in the case where the space is equipped with a symplectic form or the zero form.

Given a flag $\mathcal{F}$ of totally singular subspaces, refine this to obtain a maximal flag inside a maximal totally singular subspace of V. By Witt's lemma (see [2, §20]), all maximal singular subspaces, and moreover all maximal flags inside a given maximal singular subspace are $\text{Isom}(V)$-conjugate. By the explicit constructions in Example 6.7, there exists a Borel subgroup of G stabilizing a maximal flag of a maximal singular subspace of V. So the stabilizer of $\mathcal{F}$ contains a Borel subgroup and hence is parabolic by Proposition 12.2.

In addition to the above remarks, we see that the explicit constructions of Borel subgroups in Example 6.7 also show that the Borel subgroups of a simple classical group $\text{Cl}(V)$ act indecomposably on V. Indeed, one first checks that except when V is an odd-dimensional orthogonal space and $\text{char}(k) = 2$, the unipotent radical of the given Borel subgroup B has a one-dimensional fixed point space on V; as $\text{Cl}(V)$ acts irreducibly this also follows easily from the theory of highest weights, see Corollary 15.10. In the exceptional case, one quotients out by the radical of the bilinear form and passes to the symplectic group, where B acts indecomposably. Then the result follows from the fact that the radical of the form has no B-invariant complement in V.

Now let $P \leq \text{Cl}(V)$ be a parabolic subgroup with Levi decomposition $P = QL$, containing a fixed Borel subgroup B. We first show that if P stabilizes a flag of subspaces on V with Q acting as the identity on the quotient spaces, then P is the full stabilizer in G of this flag. Let $\mathcal{F}$ be the flag $0 = V_0 \subset V_1 \subset \ldots \subset V_{r+1} = V$ with $P \leq \text{Stab}_G(\mathcal{F})$ and with Q acting as the identity on each of the quotients V_i/V_{i-1}. By Proposition 12.2, $\text{Stab}_G(\mathcal{F})$ is a parabolic subgroup, containing B, say $\hat{P}$. By Proposition 12.6, $R_u(\hat{P}) \leq Q$. Now let $U \leq \hat{P}$ be the subgroup of elements which act as the identity on all quotients in $\mathcal{F}$. Then U is a closed unipotent, normal subgroup of $\hat{P}$, and so $U^\circ \leq R_u(\hat{P})$. But $Q \leq U$ and so $Q \leq U^\circ \leq R_u(\hat{P}) \leq Q$, whence $Q = R_u(\hat{P})$. By Corollary 12.8, this implies $P = \hat{P}$.

To complete the proof, we must find a flag of totally singular subspaces of V stabilized by P such that Q acts as the identity on the quotients. By Proposition 2.9, Q has a non-zero fixed point on any representation space, so P stabilizes the flag $0 = V_0 \subset V_1 \subset \ldots \subset V_{r+1} = V$, where $V_i/V_{i-1} := (V/V_{i-1})^Q$, for $i \geq 1$. If V is equipped with the trivial form (so $G = \text{SL}(V)$), this flag satisfies the criteria.

Now suppose V is equipped with a non-degenerate form $\mathcal{Q}$, and assume for now that $\text{char}(k) \neq 2$ if V is equipped with a quadratic form. Set $W :=$

V^Q. Note that $W \cap W^\perp \neq 0$, else W is a non-degenerate subspace and $P \subseteq \mathrm{Isom}(W) \times \mathrm{Isom}(W^\perp)$, contradicting the indecomposability of B on V. Hence $X := W \cap W^\perp \neq 0$ and P stabilizes X, a totally singular subspace of V. If $X^\perp = X$, then X is a maximal totally singular subspace of V and $P \subseteq \mathrm{Stab}(0 \subset X \subset V)$. As the dual space X^* is isomorphic as Q-module to $V/X^\perp = V/X$, Q also acts trivially on V/X and we have the desired result. If $X^\perp \neq X$, P acts on the non-trivial quotient space $X^\perp/X$ on which Q has a non-trivial fixed point, and we proceed by induction on $\dim V$.

Finally, look at the case of $G = \mathrm{SO}(V)$ and $p = 2$. Set $W = V^Q$; if $W \cap W^\perp = 0$, we proceed as above to get a contradiction. So $X := W \cap W^\perp \neq 0$. If this space contains non-zero singular vectors, then set $U := \{v \in X \mid \mathcal{Q}(v) = 0\}$, a totally singular subspace, stabilized by P, fixed pointwise by Q. The remainder of the argument goes through as in the preceding case.

Now suppose $W \cap W^\perp$ contains no non-zero singular vectors, so $\dim X = 1$. If $\dim V = 2n$, so that the bilinear form is non-degenerate, P acts on $X^\perp/X$ a non-degenerate symplectic space. Argue using the quadratic form that the kernel of this action is trivial. So P embeds in Sp_{2n-2} and we have a contradiction by comparing the dimensions of Borel subgroups.

Now suppose $\dim V = 2n + 1$. Then $X = \mathrm{rad}(\beta_{\mathcal{Q}})$, and in V/X, we have the direct sum decomposition $W/X \oplus W^\perp/X = W/X \oplus (W/X)^\perp$, stabilized by P, and hence by a Borel subgroup of $\mathrm{Isom}(V/X)$, contradicting the indecomposability of the latter. $\qquad\square$

We end our discussion of Levi complements by an observation on isogeny types.

Proposition 12.14 *Let G be semisimple of simply connected type. Then for any Levi subgroup L of G, the derived subgroup $[L, L]$ is again of simply connected type.*

Proof Let $T \leq G$ be a maximal torus. Assume that L is the standard Levi subgroup corresponding to $I \subseteq S$ and let $T' := T \cap [L, L]$, a maximal torus of $[L, L]$. We have to show that $\mathrm{Hom}(\mathbb{Z}\Phi_I^\vee, \mathbb{Z}) \leq X(T')$. For this, let $\chi' \in \mathrm{Hom}(\mathbb{Z}\Phi_I^\vee, \mathbb{Z})$. Extend it to $\chi \in \mathrm{Hom}(\mathbb{Z}\Phi^\vee, \mathbb{Z})$ by setting $\chi(\alpha) = 0$ for $\alpha \in \Delta \setminus \Delta_I$. Since G is simply connected, $\chi \in X(T)$. But then $\chi' = \chi|_{T'} \in X(T')$ as claimed. $\qquad\square$

See Exercise 20.7 for one generalization of this statement, and Theorem 14.16(b) for a generalization in a different direction.

We'll have more to say about the structure of unipotent radicals of parabolic subgroups in Chapter 17, once we've prepared the necessary tools from representation theory.

13
Subgroups of maximal rank

Let G be a semisimple linear algebraic group. Many naturally defined closed subgroups of G are normalized by a maximal torus T of G, for example those containing a maximal torus. The following are some examples of such subgroups:

- parabolic subgroups containing T;
- Levi complements of parabolic subgroups;
- centralizers $C_G(s)$ of elements $s \in T$.

The aim of this chapter is a classification of all semisimple such subgroups in terms of data from the root system of G.

13.1 Subsystem subgroups

We are led to make the following definition:

Definition 13.1 A *subsystem subgroup* of a connected reductive group G is a semisimple subgroup normalized by a maximal torus of G.

One family of subsystem subgroups of a group G arises from certain subsets of the root system of G.

Definition 13.2 Let Φ be a root system. A subset $\Psi \subseteq \Phi$ is said to be *closed* if

(C1) for all $\alpha, \beta \in \Psi$ we have $s_\alpha.\beta \in \Psi$, and

(C2) for $\alpha, \beta \in \Psi$ with $\alpha + \beta \in \Phi$, we have $\alpha + \beta \in \Psi$.

The subset Ψ is called *p-closed* if it satisfies (C1) and moreover

(C2p) for $\alpha, \beta \in \Psi$, $m, n > 0$, we have $m\alpha + n\beta \in \Psi$ whenever in a semisimple group of type Φ over a field k of characteristic p the structure constant $c_{\alpha\beta}^{mn}$ (from the commutator formula in Theorem 11.8) is non-zero in k.

Note that by Theorem 11.8 the integers $c_{\alpha\beta}^{mn}$ only depend on the root system, not on the isogeny type of a corresponding semisimple group G.

It is clear from (C1) that closed and p-closed subsets are root systems in their own right, in the appropriate Euclidean spaces.

Example 13.3 If Δ_I is a subset of a base Δ of Φ, then the corresponding parabolic subsystem Φ_I is a closed subset. Also, for Φ indecomposable, any subsystem consisting of long roots can be seen to be closed (see Exercise B.2).

It can be seen from the classification of two-dimensional root systems that a closed subsystem is p-closed for all $p \geq 0$. In order to discuss the converse, one needs a knowledge of the values of the structure constants $c_{\alpha\beta}^{mn}$ occurring in the commutator relations, as given for example in [13, Thm. 5.2.2]. From this information, the following is easily verified:

Proposition 13.4 *Let Φ be an indecomposable root system, $\Psi \subseteq \Phi$ a p-closed subset. Then Ψ is closed unless possibly if Ψ contains some short roots and either Φ is of type B_n, C_n, F_4 and $p = 2$, or Φ is of type G_2 and $p = 3$.*

We now discuss the relationship between subsystem subgroups of a reductive group and p-closed subsystems of its root system.

Proposition 13.5 *Let G be connected reductive with root system Φ and $H \leq G$ a subsystem subgroup. Then the root system of H can be naturally regarded as a subsystem of Φ, and as such it is char(k)-closed.*

Proof By assumption H is normalized by a maximal torus T of G, so $\tilde{H} := HT$ is connected reductive. Let B_1 be a Borel subgroup of $\tilde{H}$ containing T, and B a Borel subgroup of G containing B_1. The root subgroups in B_1 are one-dimensional connected unipotent T-invariant subgroups of B, hence of the form U_α for some $\alpha \in \Phi$ by Proposition 11.5. This defines an embedding of the root system Ψ of $H = [\tilde{H}, \tilde{H}]$ with respect to $T \cap H$ into Φ.

Clearly Ψ satisfies (C1). Now let $\alpha, \beta \in \Psi^+$ (without loss: see Exercise A.5). Then $U_\alpha, U_\beta \leq B_1$, hence also $[U_\alpha, U_\beta] \subseteq B_1$. Again by Proposition 11.5 and Theorem 11.8, it follows that B_1 will contain all U_γ where $\gamma = m\alpha + n\beta$ with $c_{\alpha\beta}^{mn} \neq 0$ in k. Thus Ψ is char(k)-closed. $\square$

Conversely, we obtain the following generalization of Proposition 12.6 for Levi complements:

Theorem 13.6 *Let G be connected reductive with root system Φ, and $\Psi \subseteq$*
Φ a p-closed subset, where $p = \mathrm{char}(k)$. Then $G(\Psi) := \langle T, U_\alpha \mid \alpha \in \Psi \rangle$ is a
connected reductive subgroup with root system Ψ and Weyl group $W(\Psi) :=$
$\langle s_\alpha \mid \alpha \in \Psi \rangle$.

Proof As remarked above, the p-closed subset Ψ is a root system in $\mathbb{R}\Psi$. Let
Δ_1 be a base of Ψ with positive system Ψ^+. By the commutator relations
in Theorem 11.8, $[U_\alpha, U_\beta] \subseteq \prod_{m,n>0} U_{m\alpha+n\beta}$, so $B_1 := T \prod_{\alpha \in \Psi^+} U_\alpha$ is a
connected solvable subgroup of $G(\Psi)$ by p-closedness. Let $N_1 \leq N_{G(\Psi)}(T)$
so that $N_1/T = \langle s_\alpha \mid \alpha \in \Psi \rangle = W(\Psi)$ is the Weyl group of Ψ. We claim
that B_1, N_1 form a BN-pair in $G(\Psi)$ with Weyl group $W(\Psi)$. (Note that at
this point we don't yet know that $G(\Psi)$ is reductive.)

Axioms (BN1)–(BN3) are immediate. Now let $w \in W(\Psi)$ and $s = s_\alpha$ with
$\alpha \in \Delta_1$. Using that $\dot{v}U_\beta \dot{v}^{-1} = U_{v\beta}$ for $v \in W = N_G(T)/T$ we have

$$B_1 \dot{w} B_1 \dot{s} B_1 \subseteq B_1 \dot{w} U_\alpha \dot{s} B_1.$$

If $w\alpha \in \Psi^+$ then this lies in $B_1 \dot{w}\dot{s} B_1$. Otherwise $ws\alpha = -w\alpha \in \Psi^+$, so using
$U_{-\alpha} = \dot{s} U_\alpha \dot{s} \subseteq T U_\alpha \dot{s} U_\alpha \cup U_\alpha$ from the proof of Lemma 11.14 we have

$$B_1 \dot{w} U_\alpha \dot{s} B_1 = B_1 \dot{w}\dot{s} U_{-\alpha} B_1 \subseteq B_1 \dot{w}\dot{s}(U_\alpha \dot{s} U_\alpha \cup U_\alpha) B_1 \subseteq B_1 \dot{w} B_1 \cup B_1 \dot{w}\dot{s} B_1,$$

showing (BN4). Finally, $\dot{s} B_1 \dot{s}$ contains $U_{-\alpha}$, which together with U_α generates the non-solvable group $\langle U_\alpha, U_{-\alpha} \rangle$, so cannot be contained in B_1, as
required for (BN5).

We next claim that B_1 is a Borel subgroup of $G(\Psi)$. The weak form of the
Bruhat decomposition in Theorem 11.17, valid for any BN-pair, gives

$$G(\Psi) = \bigsqcup_{w \in W(\Psi)} B_1 \dot{w} B_1.$$

Now let $g \in G(\Psi) \setminus B_1$, so $g = b\dot{w}b'$ for some $b, b' \in B_1$ and $w \in W(\Psi) \setminus \{1\}$.
Let $\alpha \in \Psi^+$ with $\beta := w\alpha \in -\Psi^+$, which exists by Proposition A.21. Then
$\langle B_1, g \rangle$ contains $\dot{w} U_\alpha \dot{w}^{-1} = U_\beta$, hence the non-solvable group $\langle U_\beta, U_{-\beta} \rangle$.
Thus B_1 is a maximal connected solvable subgroup of $G(\Psi)$, that is, a Borel
subgroup.

Hence B_1 can be embedded into a Borel subgroup B of G, with maximal
torus T. The unipotent radical $R_u(G(\Psi))$ is then a T-invariant subgroup of
B_1, hence of B, hence generated by suitable root subgroups U_β, $\beta \in \Phi$, by
Proposition 11.5. Assume that $U_\beta \leq R_u(G(\Psi)) \leq B_1$. Then $\beta \in \Psi$ by the
definition of B_1 and Theorem 11.1(a), so $U_{-\beta} \leq G(\Psi)$. But $\langle U_\beta, U_{-\beta} \rangle$ has
trivial radical, so $R_u(G(\Psi)) = 1$. Finally, as $B_1 = T \prod_{\alpha \in \Psi^+} U_\alpha$ is a Borel
subgroup of $G(\Psi)$, it follows from Theorem 11.1 that Ψ is the root system
of $G(\Psi)$. $\square$

The preceding two results can be summarized as follows:

Corollary 13.7 *Let G be connected reductive with maximal torus T. The map $\Psi \mapsto [G(\Psi), G(\Psi)]$ establishes a bijection between char(k)-closed subsets $\Psi \subseteq \Phi$ of the root system with respect to T, and subsystem subgroups of G normalized by T.*

We have the following generalization of Corollary 12.11 (see Exercise 20.8):

Proposition 13.8 *Let G be semisimple with a maximal torus T with Weyl group W, $H \leq G$ a subsystem subgroup normalized by T with Weyl group W_H. Then there is a natural isomorphism $N_G(HT)/HT \cong N_W(W_H)/W_H$.*

Example 13.9 Let G be a simple algebraic group of type F_4, and H the subsystem subgroup of type D_4 generated by the long root subgroups (corresponding to the closed subsystem consisting of all long roots). It can be calculated with Proposition 13.8 that the normalizer $N_G(H)$ is an extension of H by the symmetric group $\mathfrak{S}_3$; as in Example 12.12 this realizes the full group of graph automorphisms of a group of type D_4 described in Theorem 11.12.

13.2 The algorithm of Borel and de Siebenthal

The preceding results reduce the question of determining all subsystem subgroups to the purely combinatorial task of finding all p-closed subsets of the root system. Clearly it is sufficient to do this in the indecomposable case. For the case of closed subsets the algorithm described by Borel and de Siebenthal does precisely this. In order to formulate it we need a further ingredient from the theory of root systems, see Proposition B.5:

Proposition 13.10 *Let Φ be an indecomposable root system with base Δ. Then there exists a unique root $\alpha_0 \in \Phi^+$ with the following property: Writing $\alpha_0 = \sum_{\alpha \in \Delta} n_\alpha \, \alpha$, then for every root $\beta = \sum_{\alpha \in \Delta} c_\alpha \, \alpha \in \Phi^+$ we have $c_\alpha \leq n_\alpha$ for all $\alpha \in \Delta$.*

Definition 13.11 The *height* of a root $\beta = \sum_{\alpha \in \Delta} c_\alpha \alpha \in \Phi$ is ht(β) := $\sum_{\alpha \in \Delta} c_\alpha$. The root α_0 above is called the *highest root* of Φ with respect to Δ (since, clearly it has the largest height of all roots in Φ). The *extended Dynkin diagram* of an indecomposable root system Φ is the diagram obtained from the set $\Delta \cup \{-\alpha_0\}$ by the same procedure by which the ordinary Dynkin diagram is obtained from Δ.

See Table B.1 for the list of highest roots in indecomposable root systems. The extended Dynkin diagrams for the indecomposable root systems are shown in Table 13.1 (see [34, §4.7]). It is clear from this list that the highest root is always a long root, if Φ has two root lengths. This can also be shown without using the classification of root systems, see Corollary B.7.

Table 13.1 *Extended Dynkin diagrams*

The following result (see Theorem B.18), which was first shown in [5, §7] in the context of compact Lie groups exhibits the role played by the extended Dynkin diagrams in the study of subsystem subgroups:

Theorem 13.12 (Borel–de Siebenthal)   *Let Φ be an indecomposable root system with base Δ and highest root $\alpha_0 = \sum_{\alpha \in \Delta} n_\alpha \alpha$ with respect to Δ. Then the maximal closed subsystems of Φ up to conjugation by W are those with bases:*

(1) $\Delta \setminus \{\alpha\} \cup \{-\alpha_0\}$, *for $\alpha \in \Delta$ with n_α a prime, and*
(2) $\Delta \setminus \{\alpha\}$ *for $\alpha \in \Delta$ with $n_\alpha = 1$.*

It is thus easy to write down explicitly the possibilities for maximal closed subsystems in any of the various types of indecomposable root systems, see Exercise 20.9.

The *algorithm of Borel–de Siebenthal* for the determination of all closed subsystems of a root system Φ now proceeds as follows. For any proper subset of $\Delta \cup \{-\alpha_0\}$, corresponding to a subdiagram of the extended Dynkin diagram, form the extended Dynkin diagram of each indecomposable part of that subdiagram and repeat the process. At any stage of the process, the set of nodes of the current diagram is a subset $J \subseteq \Phi$ and $\Psi := \mathbb{Z}J \cap \Phi$ is a closed subset of Φ.

Example 13.13 We start with a root system of type E_8. The corresponding extended Dynkin diagram is as shown in Figure 13.1.

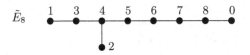

Figure 13.1 The extended Dynkin diagram for E_8.

Removal of the first node leads to the Dynkin diagram of type D_8, so the E_8-root system contains a closed subsystem of type D_8. Since the highest root for E_8 is given by $\alpha_0 = 2\alpha_1 + 3\alpha_2 + 4\alpha_3 + 6\alpha_4 + 5\alpha_5 + 4\alpha_6 + 3\alpha_7 + 2\alpha_8$ (see Table B.1), this is in fact maximal in E_8 by Theorem 13.12.

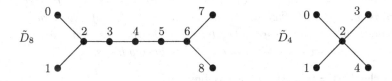

Figure 13.2 The extended Dynkin diagrams for D_8 and D_4.

The corresponding extended Dynkin diagram $\tilde{D}_8$ is shown in Figure 13.2. Removing the node with label 4, we reach a diagram of type $D_4 D_4$. Therefore the E_8 root system has a closed subsystem of type $D_4 D_4$. Clearly, this is not maximal inside E_8, but it is so inside the root system of type D_8, since the corresponding coefficient in the highest root for D_8 equals 2.

We now take the extended Dynkin diagram of the $D_4 D_4$ diagram. We see that there is a subsystem subgroup with root system of type $(A_1)^4$ of each of the D_4 subgroups and hence E_8 has a subsystem subgroup with root system of type $(A_1)^8$. In particular we have inclusions of subsystem subgroups of E_8 as follows:

$$(A_1)^8 \leq D_4 D_4 \leq D_8 \leq E_8.$$

Here, by abuse of notation, we identify the group with its root system.

We next discuss the extent to which closed subsystems account for all subsystem subgroups.

Theorem 13.14 *Let G be a simple algebraic group with root system Φ. The algorithm of Borel–de Siebenthal described above gives all subsystem subgroups of G, except when*

(1) *Φ is of type B_n, C_n, F_4 and $\mathrm{char}(k) = 2$, or*
(2) *Φ is of type G_2 and $\mathrm{char}(k) = 3$.*

Proof By Corollary 13.7 subsystem subgroups correspond bijectively to p-closed subsystems of Φ. By Proposition 13.4 these are even closed in Φ unless Φ and p are as in the exceptions. The result then follows from Theorem 13.12 by induction.   □

In the exceptional cases there exist subsystem subgroups corresponding to non-closed subsystems; the latter can also be enumerated, see Proposition B.21 and Example B.22, and we arrive at the following:

Proposition 13.15 *Let G be a simple algebraic group with root system Φ and assume further that (Φ, p) are as in (1) or (2) of Theorem 13.14. Then any non-closed p-closed subsystem Ψ of Φ contains short roots and if Ψ contains all short roots of Φ, one of the following holds:*

(1) *$(G, \mathrm{char}(k)) = (C_n, 2)$ and Ψ has type D_n,*
(2) *$(G, \mathrm{char}(k)) = (B_n, 2)$ and Ψ has type $B_{m_1} \cdots B_{m_r}$ with $\sum m_i = n$,*
(3) *$(G, \mathrm{char}(k)) = (F_4, 2)$ and Ψ has type C_4 or D_4, or*
(4) *$(G, \mathrm{char}(k)) = (G_2, 3)$ and Ψ has type A_2.*

We illustrate the exceptional behavior for small primes in one example:

Example 13.16 Let $G = \mathrm{Sp}_4$, with root system of type C_2, with base $\{\alpha, \beta\}$, where α is short. Here $\langle U_{\pm\beta}\rangle \times \langle U_{\pm(2\alpha+\beta)}\rangle$ is a subsystem subgroup of type $A_1 A_1$ corresponding to the subdiagram of the extended Dynkin diagram of G where the middle node is removed, see Table 13.1.

Now consider the short roots, $\alpha, \alpha+\beta$. If $p \neq 2$ we have $[U_\alpha, U_{\alpha+\beta}] \neq 1$ and indeed the group generated by the root subgroups corresponding to the short roots is all of G. However, if $p = 2$ we have $[U_{\pm\alpha}, U_{\pm(\alpha+\beta)}] = 1$ (see Example 11.9(2)) and the subgroup $\langle U_{\pm\alpha}\rangle \times \langle U_{\pm(\alpha+\beta)}\rangle$ is also a proper subsystem subgroup of type $D_2 = A_1A_1$, which arises from the fact that the group SO_4 is naturally a subgroup of the symplectic group Sp_4 in characteristic 2.

14

Centralizers and conjugacy classes

We now consider properties like generation, conjugacy, classification, connectedness and dimension of centralizers in connected reductive groups. It turns out that the situation is easiest for semisimple elements.

14.1 Semisimple elements

Recall from Corollary 6.11(a) that every semisimple element of a connected group lies in some maximal torus. More precisely we have:

Proposition 14.1 *Let G be connected, $s \in G$ semisimple, $T \leq G$ a maximal torus. Then $s \in T$ if and only if $T \leq C_G(s)^\circ$. In particular, $s \in C_G(s)^\circ$.*

Proof As T is abelian, $s \in T$ if and only if $T \leq C_G(s)$, which is equivalent to $T \leq C_G(s)^\circ$ as T is connected. □

We remark that in contrast, for $u \in G$ unipotent, u may not be in $C_G(u)^\circ$. See Exercise 20.10 for an example in Sp_4 over a field of characteristic 2.

We now determine the structure of centralizers of semisimple elements:

Theorem 14.2 *Let G be connected reductive, $s \in G$ semisimple, $T \leq G$ a maximal torus with corresponding root system Φ. Let $s \in T$ and $\Psi := \{\alpha \in \Phi \mid \alpha(s) = 1\}$. Then:*

(a) $C_G(s)^\circ = \langle T, U_\alpha \mid \alpha \in \Psi \rangle$.
(b) $C_G(s) = \langle T, U_\alpha, \dot{w} \mid \alpha \in \Psi, \ w \in W \text{ with } s^w = s \rangle$.

Moreover, $C_G(s)^\circ$ is reductive with root system Ψ and Weyl group $W_1 = \langle s_\alpha \mid \alpha \in \Psi \rangle$.

Proof The inclusion "⊇" is clear in both (a) and (b) (see Theorem 8.17(c) for the action of s on U_α). Now let $g \in C_G(s)$. By the Bruhat decomposition in Theorem 11.17, g can be uniquely expressed as a product $g = u_1 \dot{w} t u_2$ with $u_1 \in U_w^-$, $w \in W$, $t \in T$ and $u_2 \in U = R_u(B)$. Now

$$sgs^{-1} = su_1 s^{-1} \cdot \dot{w} \cdot s^w t s^{-1} \cdot su_2 s^{-1} = g = u_1 \dot{w} t u_2$$

implies $u_1, u_2 \in C_G(s)$, $s = s^w$. But, if $u = \prod_{\alpha \in \Phi^+} u_\alpha(c_\alpha) \in U$ with $c_\alpha \in k$ then

$$sus^{-1} = \prod_{\alpha \in \Phi^+} u_\alpha(\alpha(s) c_\alpha),$$

so $u \in C_G(s)$ implies that $\alpha(s) = 1$ whenever $c_\alpha \neq 0$, by Theorem 11.1. This shows the inclusion "⊆" in (b). Now $\langle T, U_\alpha \mid \alpha \in \Psi \rangle$ is a connected subgroup of $C_G(s)$ of index at most $|W|$, hence the identity component of $C_G(s)$, which completes the proof of (a).

As Ψ satisfies (C1) (by Lemma 8.19(a)) and (C2) from Definition 13.2, it is a closed subsystem of Φ. The last claim then follows from Theorem 13.6. □

Corollary 14.3 *Let G be connected reductive. Then up to conjugation, there exist only finitely many different centralizers of semisimple elements in G.*

Proof Let $s \in G$ be semisimple. After conjugation we may assume that $s \in T$. Then by Theorem 14.2 the centralizer $C_G(s)$ is determined by $\Psi \subseteq \Phi$ and the subgroup $\{w \in W \mid s^w = s\}$. But for both of these, there are just finitely many possibilities. □

Example 14.4 (1) Let $G = \mathrm{GL}_n$, $s = \mathrm{diag}(t, u, \dots, u) \in G$ with $t \neq u$. Then $C_G(s) = \mathrm{GL}_1 \times \mathrm{GL}_{n-1}$, a Levi subgroup of the parabolic subgroup

$$\left\{ \left(\begin{array}{c|ccc} * & * \dots * \\ \hline 0 & \\ \vdots & & A \\ 0 & \end{array} \right) \;\middle|\; A \in \mathrm{GL}_{n-1} \right\}.$$

(2) In characteristic char$(k) \neq 2$ the image s of diag$(1, -1)$ in $G = \mathrm{PGL}_2$ is a non-central semisimple element whose centralizer (see Exercise 20.14) is not contained in a proper Levi subgroup; such semisimple elements are called *quasi-isolated*. In this example at least $C_G(s)^\circ$ is a torus, hence a proper Levi subgroup.

(3) Let $G = \mathrm{Sp}_4$, char$(k) \neq 2$, $s = \mathrm{diag}(1, -1, -1, 1)$. Then $C_G(s) = \mathrm{Sp}_2 \times \mathrm{Sp}_2$, which is a subgroup of maximal semisimple rank, hence not even

$C_G(s)^\circ$ lies inside a proper Levi subgroup. Such semisimple elements are called *isolated*.

Remark 14.5 There is a criterion established by Carter which allows one to decide combinatorially via the root datum which maximal rank subgroups actually do occur as centralizers of semisimple elements, see for example [35, 2.12]. Deriziotis [20] has shown that up to conjugation the root systems of centralizers of semisimple elements always have a base contained in $\Delta \cup \{-\alpha_0\}$, see Exercise 20.13.

The latter result relies on the following parametrization of semisimple classes (see Exercise 20.12):

Proposition 14.6 *The set of semisimple conjugacy classes of a connected reductive group G is in bijection with the orbits of its Weyl group on a maximal torus.*

General conjugacy classes of a reductive group G can be described as follows. Let $g \in G$, with Jordan decomposition $g = us$. If $h \in C_G(g)$, then h centralizes both s and u, by the uniqueness of Jordan decomposition. Hence $C_G(g) = C_H(u)$, where $H := C_G(s)$ is reductive by Theorem 14.2. Thus, the centralizer of g is known once the centralizers of unipotent elements in certain reductive (but not necessarily connected) subgroups are known. Moreover we have:

Proposition 14.7 *Let G be connected reductive, $g \in G$ with Jordan decomposition $g = su$. Then $u \in C_G(s)^\circ$.*

Proof Let B be a Borel subgroup containing g, and $T \le B$ a maximal torus containing s (see Theorem 6.10 and Corollary 4.5). Then u lies in $C_B(s) \cap R_u(B)$, a unipotent subgroup of B normalized by T, hence connected by Proposition 11.5. $\qquad\square$

In particular, for s semisimple, $|C_G(s) : C_G(s)^\circ|$ is prime to char(k). See Proposition 14.20 below for a much stronger assertion.

Unfortunately, there is no systematic description of unipotent elements and their centralizers comparable to that given above for semisimple elements. At present, the conjugacy classes and centralizers of unipotent elements in simple algebraic groups have to be determined case-by-case, using the natural matrix representation (and refined Jordan canonical forms) for the classical groups and explicit computations with root elements for the exceptional types. In principle this then allows one to parametrize arbitrary conjugacy classes. See Springer and Steinberg [67], Steinberg [75], and Spaltenstein [64] for further information.

We conclude our investigation of conjugacy classes by introducing an important type of element, which in a sense comprises most elements in a connected group (see Corollary 14.10 below).

Definition 14.8 An element x of a linear algebraic group G is called *regular* if $\dim C_G(x)$ is smallest possible among all elements of G.

Proposition 14.9 *Let G be connected, $x \in G$. Then $\dim C_G(x) \geq \mathrm{rk}(G)$.*

Proof Let B be a Borel subgroup of G containing x (according to Theorem 6.10), with unipotent radical $U = R_u(B)$. As B/U is abelian by Theorem 4.4(a), the conjugacy class $[x]_B$ of x in B is contained in the coset xU, whence of dimension at most $\dim U = \dim B - \dim T$, where T denotes a maximal torus of B, hence of G. Consider the surjective morphism $B \to [x]_B$, $g \mapsto gxg^{-1}$, with fibers the cosets of $C_B(x)$, of dimension $\dim C_B(x)$. This shows that $\dim[x]_B = \dim B - \dim C_B(x)$ by Proposition 1.19, so

$$\dim B - \dim C_B(x) = \dim[x]_B \leq \dim B - \dim T = \dim B - \mathrm{rk}(G),$$

whence $\dim C_G(x) \geq \dim C_B(x) \geq \mathrm{rk}(G)$ as claimed. □

Regular elements play an important role in the study of reductive groups. For semisimple elements, we have the following characterization:

Corollary 14.10 *Let G be connected reductive with maximal torus T and root system Φ. For $s \in T$ the following are equivalent:*

(i) *s is regular;*
(ii) *$\alpha(s) \neq 1$ for all $\alpha \in \Phi$;*
(iii) *$C_G(s)^\circ = T$.*

Moreover, regular semisimple elements are dense in G.

Proof By Theorem 14.2(a) the conditions (ii) and (iii) are equivalent. As $\dim T = \mathrm{rk}(G)$, any element with $C_G(s)^\circ = T$ is necessarily regular by Proposition 14.9, so (ii), (iii) imply (i). Now $\ker(\alpha)$ is a (closed) subgroup of T of codimension 1 for any $\alpha \in \Phi$, and $|\Phi|$ is finite, so the complement of $\cup_\alpha \ker(\alpha)$ is non-empty open in T and thus dense by Proposition 1.9. In particular, again by Theorem 14.2(a), the bound in Proposition 14.9 is attained and so (i) implies (iii).

Now let B denote a Borel subgroup of G containing T, with unipotent radical U. Let T_{reg} denote the open set of regular semisimple elements of G in T. We claim that $T_{\mathrm{reg}} \cdot U$ consists of regular semisimple elements of

G, and thus regular semisimple elements (of G) are dense in B. Indeed, let $su \in T_{\mathrm{reg}}U$ with Jordan decomposition tv, where t is semisimple and $v \in U$. As U acts transitively by conjugation on the set of maximal tori of B, the semisimple part t of su is conjugate to s, hence also regular in G. By the first part, $C_B(t)^\circ$ is a torus, so $v = 1$ by Proposition 14.7, and hence $su = t$ is regular semisimple.

Since G is the union of conjugates of B (see Theorem 6.10), regular semisimple elements are also dense in G.   □

For example in GL_n, regular elements are those whose characteristic polynomial and minimal polynomial agree. This shows that GL_n and SL_n also do contain regular unipotent elements. The proof of this fact for other types of semisimple groups is much more delicate, see [71, Thm. 4.6].

14.2 Connectedness of centralizers

We have seen in Example 14.4 that, in general, centralizers of semisimple elements need not be connected. In order to investigate the extent to which this fails, we'll need the following important notions related to closed subsystems of root systems:

Definition 14.11   Let Φ be a root system. A prime r is called *bad for* Φ if $\mathbb{Z}\Phi/\mathbb{Z}\Psi$ has r-torsion for some closed subsystem $\Psi \subseteq \Phi$.

Let $\Phi^\vee$ denote the dual root system (see Exercise 10.35 or Definition A.3). The prime r is said to be a *torsion prime for* Φ if $\mathbb{Z}\Phi^\vee/\mathbb{Z}\Psi^\vee$ has r-torsion for some closed subsystem $\Psi \subseteq \Phi$.

Example 14.12   Let Φ be a root system of type G_2, with base $\{\alpha, \beta\}$ where α is short (see Example 9.5). Then $\Psi_1 := \pm\{\alpha, 3\alpha + 2\beta\}$ is a closed subsystem of type $A_1 A_1$, and clearly $|\mathbb{Z}\Phi/\mathbb{Z}\Psi_1| = 2$; furthermore, $\Psi_2 = \pm\{\beta, 3\alpha + \beta, 3\alpha + 2\beta\}$ is a closed subsystem of type A_2 with $|\mathbb{Z}\Phi/\mathbb{Z}\Psi_2| = 3$, so 2 and 3 are certainly bad primes for G_2.

In the dual root system, $\Psi_1^\vee = \pm\{\alpha^\vee, \alpha^\vee + 2\beta^\vee\}$ still has $|\mathbb{Z}\Phi^\vee/\mathbb{Z}\Psi_1^\vee| = 2$, but $\Psi_2^\vee = \pm\{\beta^\vee, \alpha^\vee + \beta^\vee, \alpha^\vee + 2\beta^\vee\}$ now spans $\mathbb{Z}\Phi^\vee$, so we only find the torsion prime 2. Note that $\Psi_2^\vee$ is not a closed subsystem of $\Phi^\vee$.

Proposition 14.13   *The bad primes, respectively torsion primes, for the indecomposable root systems are as given in Table 14.1. The bad primes and torsion primes for a decomposable root system are those of its indecomposable components.*

Table 14.1 *Bad primes and torsion primes of root systems*

Φ	bad primes	torsion primes
A_n	$-$	$-$
$B_n\ (n \geq 3), D_n(n \geq 4)$	2	2
$C_n\ (n \geq 2)$	2	$-$
G_2	2, 3	2
F_4, E_6, E_7	2, 3	2, 3
E_8	2, 3, 5	2, 3, 5

Proof This can be verified from the criteria involving coefficients of highest roots in Corollary B.28 and Proposition B.32; see also Exercise B.4. □

Definition 14.14 Let G be a connected reductive group. A prime r is called *bad for G* if it is bad for the root system of G. It is called a *torsion prime for G* if the fundamental group of some subsystem subgroup of G has r-torsion.

Here, the fundamental group of a reductive group H is by definition that of its semisimple part $[H, H]$, see Definition 9.14.

Proposition 14.15 *The torsion primes of a connected reductive group G are precisely the torsion primes of its root system Φ together with the prime divisors of the order of its fundamental group.*

Proof This is Exercise 20.15. □

The fundamental groups for the various simple groups can be read off from Table 9.2. Comparison with that list shows that for simple groups G the torsion primes for G are those of its root system Φ, except possibly for prime divisors of $n + 1$ for type A_n, and for 2 for type C_n.

We can now characterize an important situation in which centralizers of semisimple elements are always connected (see [14, Thm. 3.5.6] for (a), [74, Thm. 0.1] for (b)):

Theorem 14.16 (Steinberg) *Let G be connected reductive such that the derived group $[G, G]$ is simply connected, and $s \in G$ a semisimple element. Then:*

(a) *$C_G(s)$ is connected.*

(b) *If the order of s in $G/Z(G)$ is finite, but not divisible by any torsion prime of G then $[C_G(s), C_G(s)]$ is again simply connected.*

Note that part (b) generalizes Proposition 12.14: by Proposition 12.6 any Levi subgroup L of G is the centralizer of its connected center $Z(L)^\circ$; by Exercise 20.11 we can choose a semisimple element $s \in Z(L)^\circ$ of order prime to all torsion primes with $L = C_G(s)$.

Corollary 14.17 *Let G be connected reductive with simply connected derived subgroup $[G,G]$, and $s_1, \ldots, s_r \in G$ mutually commuting semisimple elements of finite order. Let n be the number of s_i whose order is not prime to all torsion primes of G. If $n \leq 2$ then there exists a maximal torus of G containing all s_i.*

Proof Assume that $s_1, \ldots, s_{r-2}$ have order not divisible by any torsion prime of G, and for $i = 1, \ldots, r$ put $G_i := C_{G_{i-1}}(s_i)$, with $G_0 := G$. By Theorem 14.16(a) and (b) applied inductively we see that G_i is connected and $[G_i, G_i]$ is simply connected for $i = 1, \ldots, r - 2$. Furthermore, G_{r-1} is still connected and contains s_r. Now any maximal torus of G_{r-1} containing s_r (see Corollary 6.11(a)) is as claimed. $\qquad\square$

Corollary 14.18 *Let G be a semisimple group, $r \neq \mathrm{char}(k)$ a prime which is not a torsion prime for G. Then the maximal rank of an elementary abelian r-subgroup of G equals $\mathrm{rk}(G)$.*

Proof Let R be an elementary abelian r-subgroup of G and consider the natural isogeny $\pi : G_{\mathrm{sc}} \to G$ from a simply connected group of the same type (see Proposition 9.15). Then $\ker(\pi)$ is a subgroup of the fundamental group of G, so of order prime to r. Hence, the preimage $\pi^{-1}(R) \leq G_{\mathrm{sc}}$ has a Sylow r-subgroup $\tilde{R}$ isomorphic to R. Since r is not a torsion prime, by Corollary 14.17 there exists a maximal torus $T \leq G_{\mathrm{sc}}$ with $\tilde{R} \leq T$. Since $T \cong (k^\times)^l$, with $l = \mathrm{rk}(G)$, we see that $R \cong \tilde{R}$ has r-rank at most l. Conversely, T clearly contains an elementary abelian r-subgroup of rank l. $\qquad\square$

Example 14.19 We illustrate Theorem 14.16 and its corollaries.

(1) For $G = \mathrm{GL}_n$ or SL_n the derived group $G' = \mathrm{SL}_n$ (by Exercise 10.36) is simply connected, and indeed, in both groups any set of commuting semisimple elements can be simultaneously diagonalized, hence embedded into a maximal torus.

(2) The assumption on torsion primes in Theorem 14.16(b) is necessary: Let $G = G_2$ with $\mathrm{char}(k) \neq 2$, a group of simply connected type, with root system Φ with respect to a maximal torus $T \leq G$. Then the closed subsystem Ψ_1 of type $A_1 A_1$ defined in Example 14.12 satisfies $|\mathbb{Z}\Phi/\mathbb{Z}\Psi_1| = 2$. Let H denote the subsystem subgroup of G of type $A_1 A_1$ corresponding to Ψ_1 and containing T (see Theorem 13.6). As G

is also of adjoint type, $X(T) = \mathbb{Z}\Phi$, hence H has $|X(T) : \mathbb{Z}\Psi_1| = 2$, while its fundamental group is of order 4 (see Table 9.2). So it is a quotient of the simply connected group $SL_2 \times SL_2$ of type $A_1 A_1$ by a central subgroup of order 2. Thus $|Z(H)| = 2$, and by maximality of Ψ_1 in Φ, H is the centralizer of $Z(H)$, and it is not of simply connected type.

On the other hand, when $\mathrm{char}(k) \neq 3$, the closed subsystem Ψ_2 from Example 14.12 belongs to the simply connected centralizer SL_3 of a semisimple element of order 3, consistent with the fact that 3 is not a torsion prime for G_2.

(3) The assumption that G is simply connected in Corollary 14.17 is necessary: the quaternion group Q_8 of order 8 has a faithful two-dimensional representation over k when $\mathrm{char}(k) \neq 2$, embedding it into SL_2. The image of Q_8 in the quotient $PGL_2 = SL_2/Z(SL_2)$ is elementary abelian of order 4, hence clearly does not embed into a (one-dimensional) maximal torus of PGL_2. Note that the group PGL_2 is not simply connected (see Table 9.2)

Cases where the conclusion of Corollary 14.18 fails for torsion primes lead to interesting maximal subgroups, see Theorem 29.3 below.

For semisimple groups, Theorem 14.16 can be generalized as follows: the extent to which centralizers of semisimple elements can fail to be connected is controlled by the fundamental group.

Proposition 14.20 *Let G be a semisimple algebraic group, $\pi : G_{\mathrm{sc}} \to G$ the natural isogeny from a simply connected group of the same type as G. Then for any semisimple $s \in G$, the group of components $C_G(s)/C_G(s)^\circ$ is isomorphic to a subgroup of $\ker(\pi) \leq Z(G_{\mathrm{sc}})$. Moreover, if s is of finite order then the exponent of $C_G(s)/C_G(s)^\circ$ divides the order of s.*

Proof This follows from Theorem 14.16(a), see Exercise 20.16. □

Example 14.21 (Some isolated elements in E_6)   Let G be simple of simply connected type E_6 and assume $\mathrm{char}(k) \neq 3$. The subsystem subgroup H of G corresponding to the subset $J = \{0, 1, 2, 3, 5, 6\}$ of the extended Dynkin diagram of E_6 (see Table 13.1) has type A_2^3.

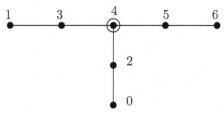

An application of Proposition 13.8 yields that $N_G(H)/H \cong \mathfrak{S}_3$. Direct calculation with the root datum shows that H is a quotient of SL_3^3 by a central subgroup of order 3, so has center $C_3 \times C_3$. By Theorem 13.12, H is a subsystem subgroup for a maximal subsystem, so it is the connected component of the centralizer $C := C_G(s)$ of any semisimple element $s \in Z(H) \setminus Z(G)$. In fact, $C = C^\circ = H$ is connected by Theorem 14.16. In particular, $N_G(H)/H$ acts fixed point freely, hence transitively, on the six elements of $Z(H) \setminus Z(G)$, so there is a unique conjugacy class of elements of order 3 in G with centralizer H.

The image $\bar{H} = H/Z(G)$ of H in the group $\bar{G} = G/Z(G)$ of adjoint type has center of order 3. Let $1 \neq s \in Z(\bar{H})$, with centralizer $C = C_{\bar{G}}(s)$. Then, as above we have $C^\circ = \bar{H}$, and C/C° is a subgroup of the fundamental group Z_3 by Proposition 14.20. Since clearly the elements of order 3 in $N_{\bar{G}}(\bar{H})/\bar{H} \cong \mathfrak{S}_3$ must act trivially on $Z(\bar{H})$, we see that $C/C^\circ \cong Z_3$. Again, the quotient $N_{\bar{G}}(C)/C \cong Z_2$ has just one orbit on the elements of $Z(\bar{H}) \setminus Z(\bar{G})$, of length 2.

15

Representations of algebraic groups

In Chapter 18 we will consider the subgroup structure of the classical algebraic groups SL_n, Sp_{2n} and SO_n. Hence we are interested in morphisms $\rho : G \to \mathrm{GL}_n$ of algebraic groups, that is rational representations of a linear algebraic group G. As with finite groups, the representation theory of linear algebraic groups is a tool in the study of a wide variety of questions, not only those directly concerned with the subgroup structure of classical groups.

We begin by developing the theory necessary for describing the irreducible representations of semisimple groups. It turns out that the irreducible representations are parametrized by "abstract weights" as in the representation theory of complex semisimple Lie algebras.

15.1 Weight theory

Let $\rho : G \to \mathrm{GL}(V)$ be a rational representation of the linear algebraic group G. The following result shows that if we are interested in irreducible representations, we may as well assume that G is semisimple:

Proposition 15.1 *Let $\rho : G \to \mathrm{GL}(V)$ be an irreducible rational representation of a connected linear algebraic group G. Then:*

(a) $R_u(G) \le \ker(\rho)$.
(b) $[G/R_u(G), G/R_u(G)]$ *acts irreducibly on V.*

Proof By Proposition 2.9, the fixed point space $V^{\rho(R_u(G))}$ of the unipotent radical $R_u(G)$ of G is a non-trivial G-invariant subspace of V, so all of V by irreducibility. That is, $R_u(G) \le \ker(\rho)$.

So ρ induces an irreducible representation $\bar{\rho} : \bar{G} \to \mathrm{GL}(V)$ of the connected reductive group $\bar{G} := G/R_u(G)$. Moreover, as G acts irreducibly on

V, $R(\bar{G}) \le Z(\bar{G})$ acts by scalars by Schur's Lemma. As $\bar{G} = R(\bar{G})[\bar{G}, \bar{G}]$ by Corollary 8.22, the semisimple group $[\bar{G}, \bar{G}]$ (see Proposition 6.20(c)) also acts irreducibly on V.   □

This seems to lead us to concentrate on semisimple groups; however for inductive arguments (where we pass to Levi factors of parabolic subgroups), it will be convenient to consider the more general setting of reductive groups. So let G be a connected reductive algebraic group. Let T be a maximal torus of G contained in a Borel subgroup B of G, Φ the root system with respect to T and Φ^+ a positive system corresponding to B with base $\Delta \subset \Phi^+$. Let $U = \langle U_\alpha \mid \alpha \in \Phi^+ \rangle$ and $U^- = \langle U_{-\alpha} \mid \alpha \in \Phi^+ \rangle$. Let $W = N_G(T)/T$ be the Weyl group of G.

Let $\rho : G \to \mathrm{GL}(V)$ be a rational representation. As $\rho(T)$ is a set of commuting semisimple elements in $\mathrm{GL}(V)$, $\rho(T)$ is a diagonalizable subgroup and V can be decomposed as a direct sum of common eigenspaces for T,

$$V = \bigoplus_{\chi \in X(T)} V_\chi, \quad \text{where} \quad V_\chi = \{v \in V \mid \rho(t)v = \chi(t)v \text{ for all } t \in T\}.$$

Definition 15.2 Let $\chi \in X(T)$ be such that $V_\chi \ne 0$. Then χ is said to be a *weight of V* (or a weight of the corresponding representation ρ) with respect to the torus T and V_χ a *T-weight space* of V. (If it is clear in the given context, we may omit the reference to the torus T.) The *multiplicity* of χ in V is $\dim V_\chi$.

Thus, for example, the weights in the adjoint representation of G are just 0 and the roots of G. The multiplicity of a non-zero weight (that is, a root) is 1 and the multiplicity of the 0 weight is the rank of G by Theorem 8.17(a). The weights of the natural representation of GL_n with respect to the maximal torus D_n of diagonal matrices are just the n coordinate functions, again with one-dimensional weight spaces.

We consider the action of certain elements and subgroups on weight spaces of V, starting with $N_G(T)$. As seen in Section 8.1, the Weyl group W acts on $X(T)$: for $w \in W$ and $\chi \in X(T)$ we have $w.\chi(t) = \chi(\dot{w}^{-1}t\dot{w})$, where $\dot{w} \in N_G(T)$ is an arbitrary preimage of w under the natural map $N_G(T) \to N_G(T)/T = W$. This leads to the following generalization of Proposition 8.4:

Lemma 15.3 Let $w \in W$. Then $\dot{w}(V_\chi) = V_{w.\chi}$; in particular weights in the same W-orbit have equal multiplicities.

Proof Let $v \in V_\chi$ and $t \in T$. Then

$$t(\dot{w}v) = \dot{w}(\dot{w}^{-1}t\dot{w})v = \dot{w}\chi(\dot{w}^{-1}t\dot{w})v = (w.\chi)(t)\dot{w}v$$

so $\dot{w}v \in V_{w.\chi}$. Hence $\dot{w}(V_\chi) \subseteq V_{w.\chi}$ and $\dot{w}^{-1}(V_{w.\chi}) \subseteq V_\chi$, and the result follows. $\qquad\square$

Recall from Theorem 8.17(g) that a connected reductive group G is generated by a maximal torus and its root subgroups U_α, for $\alpha \in \Phi$. We describe the action of the root groups U_α on the V_χ.

Lemma 15.4 *Let G, T, Φ be as above, $\rho : G \to \mathrm{GL}(V)$ a rational representation. Let $\alpha \in \Phi$ and $\lambda \in X(T)$ with T-weight space V_λ. Then for all $v \in V_\lambda$ we have*

$$U_\alpha.v \subseteq v + \sum_{m \in \mathbb{N}} V_{\lambda+m\alpha},$$

so $U_\alpha V_\lambda \subseteq \sum_{m \in \mathbb{N}_0} V_{\lambda+m\alpha}$.

Proof Fix an isomorphism $u_\alpha : \mathbf{G_a} \to U_\alpha$, $c \mapsto u_\alpha(c)$. Since G acts rationally on V, $u_\alpha(c).v$ is given by a polynomial function in c with respect to some basis for V and coordinates with respect to that basis. Collecting terms with equal c-power, we find $v_i \in V$ such that $u_\alpha(c).v = \sum_{i=0}^n c^i v_i$, for some $n \geq 0$. For $t \in T$ we have on the one hand

$$t u_\alpha(c) t^{-1}.v = \lambda(t^{-1}) t u_\alpha(c).v = \lambda(t^{-1}) \sum_{i=0}^n c^i t.v_i,$$

but (using Theorem 8.17(c)) also

$$t u_\alpha(c) t^{-1}.v = u_\alpha(\alpha(t)c).v = \sum_{i=0}^n (\alpha(t)c)^i v_i.$$

Since this is true for all $c \in k$, the coefficients at equal powers of c have to agree: $\lambda(t^{-1}) t.v_i = \alpha(t)^i v_i$, that is, $t.v_i = (\lambda + i\alpha)(t)v_i$ for all i. Hence $v_i \in V_{\lambda+i\alpha}$ for all i, and setting $c = 0$ we also find that $v_0 = v$. $\qquad\square$

As a consequence, $\langle U_\alpha, U_{-\alpha} \rangle$ preserves the submodule $\sum_{m \in \mathbb{Z}} V_{\lambda+m\alpha}$ of V.

Example 15.5 Let G be connected reductive with root system Φ. For $\alpha \in \Phi$ let $n_\alpha \in \langle U_\alpha, U_{-\alpha} \rangle$ as defined in Section 8.4. Applying the above result to the adjoint action of G on its Lie algebra $\mathfrak{g}$, we see that n_α preserves the sum of weight spaces $\sum_m \mathfrak{g}_{\beta+m\alpha}$ for all $\beta \in \Phi$, so $s_\alpha.\beta = \beta + m\alpha$ for some $m \in \mathbb{Z}$. This shows that the root system axiom (R4) is satisfied for Φ, as claimed in Proposition 9.2; that is, we have $s_\alpha.\beta - \beta$ an integral multiple of α.

We continue our study of rational representations $\rho : G \to \mathrm{GL}(V)$ of a connected reductive group G, where $V \neq 0$. By Lie–Kolchin (Theorem 4.1), there exists $v^+ \in V \setminus \{0\}$ such that $\langle v^+ \rangle$ is invariant under the image $\rho(B)$ of the Borel subgroup B of G. Any such vector v^+ is called a *maximal vector of V* (with respect to B). As $\langle v^+ \rangle$ is stabilized by any maximal torus T of B, $v^+ \in V_\lambda$ for some $\lambda \in X(T)$.

Proposition 15.6 *In the notation introduced above, set $V' = \langle Gv^+ \rangle$. Then the weights of V' are of the form*

$$\lambda - \sum_{\alpha \in \Delta} m_\alpha \alpha \quad \text{with all } m_\alpha \in \mathbb{N}_0.$$

Moreover, λ has multiplicity 1 in V', and V' has a unique maximal proper kG-submodule.

Proof For $\alpha \in \Phi$, $u \in U_\alpha$, $v \in V_\lambda$, we have

$$u.v \in v + \sum_{m>0} V_{\lambda+m\alpha}$$

by Lemma 15.4. So by successive applications of elements from $U^- = \langle U_\alpha \mid \alpha \in \Phi^- \rangle$ we obtain

$$U^-.v^+ \subseteq v^+ + \sum_{0 \neq \mu \in \mathbb{N}_0 \Phi^-} V_{\lambda+\mu}.$$

On the other hand, $U_\alpha v^+ = v^+$ for $\alpha \in \Phi^+$. So the weights occurring in $\langle U^- B v^+ \rangle$ are of the form $\lambda - \sum_{\alpha \in \Delta} m_\alpha \alpha$ with all $m_\alpha \geq 0$ and the multiplicity of the weight λ is 1. But by Theorem 11.20, $U^- B$ is dense in G, so $\langle U^- B v^+ \rangle = \langle \overline{U^- B} v^+ \rangle = \langle G v^+ \rangle = V'$. For the final statement of the proposition, let X be a proper kG-submodule of V'. Then $v^+ \notin X$, and so λ is not a weight of X. Hence, if we set M to be the sum of all proper kG-submodules of V', then M is proper and clearly the unique maximal proper kG-submodule of V'. $\qquad\square$

We will see that the weight of a maximal vector must satisfy certain conditions. To describe this we first need the following notions:

Definition 15.7 We say that $\lambda \in X(T)$ is *dominant* if $\langle \lambda, \alpha^\vee \rangle \geq 0$ for all $\alpha \in \Delta$. Given $\mu, \nu \in X(T)$, we write $\mu \leq \nu$ if $\nu - \mu = \sum_{\alpha \in \Phi^+} n_\alpha \alpha$, with $n_\alpha \in \mathbb{N}_0$ for all α.

For example, if $G = T$ is a torus, then $\Phi = \emptyset$ and all $\lambda \in X(T)$ are dominant. As the following result shows, it is enough to understand the dominant weights (in terms of the action of the Weyl group on $X(T)$):

Proposition 15.8 *For $\lambda \in X(T)$, there exists a unique dominant $\lambda^+ \in X(T)$ such that $\lambda \in W.\lambda^+$. If λ is dominant, then $w\lambda \leq \lambda$ for all $w \in W$.*

See Lemma B.4 for the proof.

Proposition 15.9 *Let $\rho : G \to \mathrm{GL}(V)$ be a rational representation of a connected reductive group G. Let v^+ be a maximal vector of V (with respect to B) of weight λ, for some $\lambda \in X(T)$. Then λ is dominant.*

Proof By Proposition 15.8 there exists $w \in W$ such that $\mu := w\lambda$ is dominant, and moreover $w'\mu \le \mu$ for all $w' \in W$. Thus

$$\lambda = w^{-1}(w\lambda) = w^{-1}\mu \le \mu = w\lambda.$$

So $w\lambda - \lambda$ is a sum of positive roots. As $\langle Gv^+ \rangle$ is a G-invariant subspace of V, the set of weights occurring in this subspace is invariant under the action of W. (See Lemma 15.3.) So $w\lambda$ is also a weight in $\langle Gv^+ \rangle$. But by Proposition 15.6 weights in $\langle Gv^+ \rangle$ are of the form $\lambda - \sum_{\alpha \in \Delta} c_\alpha \alpha$ with $c_\alpha \ge 0$. So we must have $w\lambda = \lambda$ and hence λ is itself dominant. □

15.2 Irreducible highest weight modules

The following result shows that the situation for irreducible representations of connected reductive linear algebraic groups is very similar to that for semisimple complex Lie algebras:

Corollary 15.10 *Let G be connected reductive, V an irreducible rational kG-module. Then all maximal vectors of V have the same weight λ, which is dominant, $\dim V_\lambda = 1$, and all weights in V are of the form $\lambda - \sum_{\alpha \in \Delta} c_\alpha \alpha$ with $c_\alpha \in \mathbb{N}_0$.*

Proof Let v^+ be a maximal vector for V, of weight λ. Then λ is dominant by Proposition 15.9, and $V' := \langle Gv^+ \rangle$ is a kG-submodule, hence all of V. If v is a maximal vector for some other weight μ, then $\mu \le \lambda$ by Proposition 15.6, but also $\lambda \le \mu$, and hence $\lambda = \mu$. The remaining statements are also immediate from Proposition 15.6. □

Definition 15.11 Let V be a rational kG-module generated by a maximal vector $v^+ \in V$ (with respect to the fixed Borel subgroup B), and assume that $v^+ \in V_\lambda$ for $\lambda \in X(T)$. Then V is called a *highest weight module* of G, with *highest weight* λ.

Let H be an algebraic group and V_1 and V_2 two rational kH-modules with corresponding rational representations $\rho_i : H \to \mathrm{GL}(V_i)$, $i = 1, 2$. We then define an action of H on $V_1 \otimes V_2$ as follows: for $v_i \in V_i$ we set $g(v_1 \otimes v_2) := gv_1 \otimes gv_2$. It is easy to see that this extends to an action on

$V_1 \otimes V_2$. We then obtain a homomorphism $\rho : H \to \mathrm{GL}(V_1 \otimes V_2)$ whose coordinate functions are algebraic expressions in terms of the coordinate functions of ρ_1 and ρ_2. Hence ρ is a rational representation of H.

Proposition 15.12 *Let V_1, V_2 be highest weight modules of highest weights μ_1, respectively μ_2. Then $V_1 \otimes V_2$ has an irreducible subquotient of highest weight $\mu_1 + \mu_2$.*

Proof Let v_i be maximal vectors for V_i, $i = 1, 2$. Then a straightforward calculation shows that $v_1 \otimes v_2$ is a maximal vector of $V_1 \otimes V_2$ of weight $\mu_1 + \mu_2$. Hence, by Proposition 15.6, the kG-submodule $\langle G(v_1 \otimes v_2) \rangle$ has an irreducible quotient of highest weight $\mu_1 + \mu_2$. $\square$

We now show existence and uniqueness of irreducible highest weight modules of a given dominant weight. While uniqueness is rather straightforward, the existence statement is more difficult to prove; it is solved by explicitly constructing, for semisimple groups G, a kG-module with a maximal vector of weight λ within the infinite-dimensional vector space $k[G]$.

Assume that G is semisimple, V an irreducible highest weight module for G of highest weight λ, and $v^+ \in V$ a maximal vector. Let V' denote the sum of the weight spaces for weights different from λ. Then $V = \langle v^+ \rangle \oplus V'$ by Corollary 15.10. Let $\varphi^+ \in V^* := \mathrm{Hom}(V, k)$ be defined by $\varphi^+(v^+) = 1$, $\varphi^+|_{V'} = 0$, and set

$$f_\lambda : G \to k, \qquad x \mapsto f_\lambda(x) := \varphi^+(x.v^+).$$

Note that $f_\lambda \in k[G]$ since G acts rationally on V and φ^+ is a morphism.

Lemma 15.13 *The function f_λ satisfies $f_\lambda(ub) = \lambda(b)$ for all $u \in U^-$, $b \in B$, where λ is considered as a character of B via the natural epimorphism $B \to T$.*

Proof By the definition of φ^+ we have

$$f_\lambda(ub) = \varphi^+(ub.v^+) = \lambda(b)\,\varphi^+(u.v^+) = \lambda(b)\,\varphi^+(v^+ + v') = \lambda(b)$$

for some $v' \in V'$, as claimed. $\square$

Let $\lambda \in X(T)$ be a weight of a rational representation $\rho : G \to \mathrm{GL}(V)$. Then as in Example 15.5, $s_\alpha.\lambda - \lambda$ is an integral multiple of α. With respect to a W-invariant non-degenerate bilinear form $(\ ,\ )$ on the Euclidean space $E = X(T) \otimes_{\mathbb{Z}} \mathbb{R}$, s_α acts as the reflection in α. So we have $s_\alpha.\lambda = \lambda - 2\frac{(\lambda, \alpha)}{(\alpha, \alpha)}\alpha$ which shows that the number $2(\lambda, \alpha)/(\alpha, \alpha)$ is an integer for all $\alpha \in \Phi$. Now any element $\lambda \in E$ with this property is called an *abstract (integral) weight* in the sense of root systems. Since Φ spans E by Proposition 9.2, the set of abstract weights is a lattice in E. From now on we will identify this lattice with $\Omega = \mathrm{Hom}(\mathbb{Z}\Phi^\vee, \mathbb{Z})$ via $\lambda \mapsto (\gamma^\vee \mapsto \langle \lambda, \gamma^\vee \rangle = 2\frac{(\lambda, \gamma)}{(\gamma, \gamma)})$, for all abstract weights $\lambda \in E$ and for all $\gamma^\vee \in \mathbb{Z}\Phi^\vee$. As explained in Section 9.2, $X(T)$ can be viewed as a sublattice of Ω of finite index.

The notion of dominant weight and the partial order $\leq$ from Definition 15.7 extend to Ω in an obvious way.

Definition 15.14   For G semisimple with base $\Delta = \{\alpha_1, \ldots, \alpha_l\}$ we let $\{\lambda_1, \ldots, \lambda_l\} \subseteq \Omega$ denote the dual basis of $\Delta^\vee = \{\alpha_1^\vee, \ldots, \alpha_l^\vee\}$ with respect to the perfect pairing $\langle \, , \, \rangle : X(T) \times Y(T) \to \mathbb{Z}$ from Proposition 3.6 (that is, $\langle \lambda_i, \alpha_j^\vee \rangle = \delta_{ij}$). The λ_i are called the *fundamental dominant weights* of T with respect to Δ. (The notation ϖ_i is also used in the literature.)

Thus $\lambda \in \Omega$ is dominant if and only if $\lambda = \sum_{i=1}^l a_i \lambda_i$ with $a_i \in \mathbb{N}_0$.

We start by constructing highest weight modules for multiples of the fundamental dominant weights:

Lemma 15.15   *Let G be semisimple, λ_m a fundamental dominant weight. Then there exists an irreducible highest weight module for G of highest weight $a_m \lambda_m$ for some $a_m \in \mathbb{N}$.*

Proof   Let $P \leq G$ denote the standard parabolic subgroup of G corresponding to the subset $\Delta \setminus \{\alpha_m\}$, see Proposition 12.2. By Theorem 5.5 there exists a rational representation $\rho : G \to \mathrm{GL}(V)$ with $v \in V$ such that P is the stabilizer of the line $\langle v \rangle$. Since P contains the Borel subgroup B, v is a maximal vector of V, say with dominant weight $\lambda = \sum_i a_i \lambda_i$, $a_i \geq 0$. But P also contains representatives $\dot{s}_i$ of the simple reflections $s_i \in S$, $i \neq m$, which thus also stabilize $\langle v \rangle$, hence fix λ by Lemma 15.3. But

$$s_i(\lambda_i) = \lambda_i - \alpha_i, \qquad s_i(\lambda_j) = \lambda_j - \langle \lambda_j, \alpha_i^\vee \rangle \alpha_i = \lambda_j \ \text{ for } j \neq i,$$

so we have $a_i = 0$ for $i \neq m$, whence $\lambda = a_m \lambda_m$. The kG-submodule of V generated by v now has highest weight λ, and by Proposition 15.6 it has an irreducible quotient of the same highest weight. $\qquad\Box$

To continue, we need a property of the field of fractions $k(G) := \mathrm{Frac}(k[G])$ of the coordinate ring $k[G]$. Note that $k[G]$ is an integral domain by Proposition 1.9, so it embeds into $k(G)$:

Proposition 15.16   *Let G be a connected linear algebraic group. Then $k[G]$ is integrally closed in $k(G)$.*

See for example [73, Lemma 70] for the proof.

We are now ready to prove the following existence and uniqueness result (see [15, Exposés 15,16]).

Theorem 15.17 (Chevalley) *Let G be connected reductive.*

(a) *Two irreducible kG-modules V_1, V_2 of highest weights μ_1, μ_2, are isomorphic if and only if $\mu_1 = \mu_2$.*

(b) *For each dominant weight $\lambda \in X(T)$, there exists an irreducible kG-module with highest weight λ.*

Proof (a) It is clear that an isomorphism of kG-modules $V_1 \to V_2$ will send maximal vectors to maximal vectors of the same weight. Hence, $V_1 \cong V_2$ implies that $\mu_1 = \mu_2$. Now assume that V_1, V_2 both have highest weight λ. Let v_i be a maximal vector of V_i, of weight λ, for $i = 1, 2$. Then $v := v_1 \oplus v_2$ is a maximal vector of the kG-module $V := V_1 \oplus V_2$, also of weight λ. By Proposition 15.6, λ has multiplicity 1 in $V_0 := \langle Gv \rangle$. As the v_i are not proportional to v, they cannot lie in V_0. So $V_i \cap V_0$ is a proper submodule of V_i, hence 0 by irreducibility, for $i = 1, 2$. Thus the projection $V \to V_i$ maps V_0 onto a non-zero submodule of V_i, so onto all of V_i. But then $V_1 \cong V_0 \cong V_2$ as claimed.

For (b), first assume that G is semisimple. Let $\lambda = \sum_{\alpha \in \Delta} c_\alpha \lambda_\alpha \in X(T)$ be dominant, with λ_α the fundamental dominant weight corresponding to the simple root α, so $c_\alpha \geq 0$. Let $a_\alpha > 0$ such that there exists a highest weight module V_α of highest weight $a_\alpha \lambda_\alpha$, as in Lemma 15.15, and set $a := \prod_\alpha a_\alpha$. Now $\mu := a\lambda$ is a non-negative integral linear combination of the $a_\alpha \lambda_\alpha$, so Proposition 15.12 and an easy induction show that $\bigotimes_{\alpha \in \Delta} V_\alpha^{\otimes c_\alpha d_\alpha}$ has an irreducible subquotient of highest weight μ, where $d_\alpha = a/a_\alpha$. Let $f_\mu \in k[G]$ be the corresponding function constructed in the discussion preceding Lemma 15.13.

Recall the dense open subset $O := U^- B$ of G from Theorem 11.20. We define $f : O \to k$ by $f(xz) := \lambda(z)$ for $x \in U^-$, $z \in B$. Then $f \in \mathcal{O}_G(O)$, so $f \in \mathrm{Frac}(\mathcal{O}_G(O)) = k(G)$, as pointed out at the beginning of Section 5.1. Since $\mu = a\lambda$ we then have $f_\mu = f^a$. But $k[G]$ is integrally closed in $k(G)$ by Proposition 15.16, so $f \in k[G]$. By Proposition 2.4, f lies in a finite-dimensional subspace of $k[G]$ which is invariant under the action of G by right multiplication. We claim that f generates a submodule of $k[G]$ of highest weight λ. Indeed, for $xy \in U^- B$, $z \in B$ we have

$$\rho_z(f)(xy) = f(xyz) = \lambda(yz) = \lambda(z)\lambda(y) = \lambda(z)f(xy),$$

so $\rho_z(f), \lambda(z)f \in k[G]$ agree on the dense open set O, hence on all of G. Thus f is a maximal vector of weight λ. We obtain an irreducible quotient of highest weight λ from Proposition 15.6.

Now for G connected reductive, let $G' := [G, G]$, $Z := Z(G)^\circ$, so $G = ZG'$ by Corollary 8.22. If $\lambda \in X(T)$ is dominant, then $\lambda' := \lambda|_{T'} \in X(T')$, with

$T' := T \cap G'$ a maximal torus of G', is again dominant, so by the previous step there exists an irreducible highest weight module $V := L(\lambda')$ for G'. Note that elements z of the central subgroup $G' \cap Z \leq Z(G')$ act by scalars on the irreducible module V, hence necessarily by $\lambda(z)$. Thus we may define an action of Z on V by $z.v := \lambda(z)v$ for $z \in Z$, $v \in V$, to obtain a kG-module structure on V. Clearly a maximal vector of V with respect to G' is also a maximal vector with respect to G, of weight λ. $\qquad\square$

Definition 15.18 The irreducible kG-module with highest weight λ constructed in Theorem 15.17 is denoted $L(\lambda)$. (The notation $V(\lambda)$ is also used, for example in [32].)

Example 15.19 Let G be a simple algebraic group, $B \leq G$ a Borel subgroup with maximal torus T, Φ^+ the corresponding set of positive roots and $\alpha_0 \in \Phi^+$ the highest root (from Proposition 13.10). Let Ad be the adjoint representation of G. By definition the weights of T acting on Lie(G) are given by 0 and the roots of G. Now by Theorem 11.8, $[U_\beta, U_{\alpha_0}] = 1$ for all $\beta \in \Phi^+$. This implies that Ad (U_β) fixes Lie(U_{α_0}) $= \mathfrak{g}_{\alpha_0}$ pointwise, whence any $v_0 \in \mathfrak{g}_{\alpha_0} \setminus \{0\}$ is a maximal vector for Ad (B) of weight α_0.

Thus, if M is the unique maximal kG-submodule of $\langle Gv_0 \rangle$, then $\langle Gv_0 \rangle / M$ is an irreducible kG-module of highest weight α_0. In fact, in most cases, the submodule $\langle Gv_0 \rangle$ is already an irreducible kG-module (see [31] or [60, 1.9]; the case of SL_n is Exercise 20.17):

Theorem 15.20 *Let G be a simple algebraic group over an algebraically closed field k of characteristic p. Then* Lie(G) *is an irreducible module under the adjoint representation of G, except in the following cases, where $0 \neq M \subset$* Lie(G) *is a proper kG-submodule of* Lie(G)*:*

(1) *$M \subseteq Z($Lie$(G))$ and the root system of G has type A_n, respectively D_n, E_6, E_7 and $p \mid n+1$, respectively $p = 2$, $p = 3$, $p = 2$.*

(2) *The root system of G has type B_n, C_n, F_4, G_2 and $p = 2, 2, 2, 3$ respectively and M contains all root subspaces belonging to short roots in Φ.*

In the cases where Lie(G) is reducible, one has precise information about the highest weights of the remaining composition factors.

Example 15.21 (1) Let $G = G_2$ (see Example 9.5 for its root system). The highest root is $3\alpha + 2\beta$, with $\langle 3\alpha + 2\beta, \alpha^\vee \rangle = 0$, $\langle 3\alpha + 2\beta, \beta^\vee \rangle = 1$. So $3\alpha + 2\beta$ is in fact one of the fundamental dominant weights of G_2.

When $p = 3$, Lie(G) has a seven-dimensional ideal containing all root vectors corresponding to short roots. The highest such (with respect to the partial order on weights) is the highest short root $2\alpha + \beta$; here we have $\langle 2\alpha + \beta, \alpha^\vee \rangle = 1$ and $\langle 2\alpha + \beta, \beta^\vee \rangle = 0$. So $\text{Lie}(G)$ has a composition factor with the second fundamental dominant weight as highest weight. These are in fact the only composition factors of $\text{Lie}(G)$ in this case.

(2) Let $G = \text{SL}_2$ acting on its natural module V with fixed basis $\{x, y\}$. Let $T = D_2 \cap \text{SL}_2$, $B = T_2 \cap \text{SL}_2$ and $\Phi = \{\pm\alpha\}$. Then $\langle x \rangle$ is invariant under the Borel subgroup B and so x is a maximal vector of V with respect to B. Now $X(T) = \mathbb{Z}\lambda$, where $\lambda(\text{diag}(c, c^{-1})) = c$. As

$$\text{diag}(c, c^{-1})x = cx = \lambda(\text{diag}(c, c^{-1}))x,$$

x is of weight λ. Moreover,

$$\text{diag}(c, c^{-1})y = c^{-1}y = \lambda(\text{diag}(c, c^{-1}))^{-1}y$$

shows that the other weight of V is $-\lambda$. As $\alpha(\text{diag}(c, c^{-1})) = c^2$ we have $\alpha = 2\lambda$, so $-\lambda = \lambda - \alpha$.

More generally, it is straightforward to show that the natural module for SL_n is irreducible with highest weight λ_1, when numbering the base of Φ such that $\alpha_i(\text{diag}(t_1, t_2, \ldots, t_n)) = t_i t_{i+1}^{-1}$. The remaining weights of the representation are $\lambda_1 - \alpha_1 - \cdots - \alpha_j$ for $1 \le j \le n - 1$. (See Example 9.8; we have taken $\alpha_i = \chi_{i,i+1}$.)

Over fields of characteristic 0, much more is known about the irreducible representations of semisimple groups. For example, there is the Weyl degree formula giving the dimension of an irreducible module of a given highest weight, and formulas which allow one to calculate the dimensions of individual weight spaces (Freudenthal's formula and Kostant's formula). These are precisely those coming from the representation theory of semisimple complex Lie algebras. See [33, §24] for precise statements.

When $\text{char}(k)$ is positive and large with respect to the weight λ (in some precise sense), these same formulas remain valid. Even when $\text{char}(k)$ is small, they give upper bounds for the dimensions of irreducible modules and their weight spaces.

Finally, we mention the Lusztig conjecture which predicts the dimensions of irreducible modules whose highest weights satisfy certain geometric conditions, and assuming $\text{char}(k)$ is not too small relative to Φ, in terms of the Kazhdan–Lusztig polynomials of the affine Weyl group associated to G. Andersen, Jantzen and Soergel have shown that for each root system Φ, there exists an (unknown) bound $n(\Phi)$, such that the conjecture holds for all groups with this root system if $\text{char}(k) > n(\Phi)$. We refer the reader to [36, §3] for an expository account of this theory.

16
Representation theory and maximal subgroups

In the previous chapter, we described the parametrization via highest weights of the irreducible representations of a semisimple linear algebraic group. There are many further questions which can be raised, for example, on extensions of simple modules and on the validity of dimension formulas in positive characteristic. In this section, we consider three aspects of the representation theory of a semisimple algebraic group G which play a particularly important role in the study of the maximal subgroups of the classical algebraic groups, namely, a criterion for the existence of a G-invariant bilinear form, the restriction of an irreducible representation to a Levi subgroup and Steinberg's tensor-product decomposition of certain irreducible highest weight modules.

16.1 Dual modules and restrictions to Levi subgroups

A faithful rational representation $\rho : G \to \mathrm{SL}_n$ identifies G with a subgroup of SL_n. When is this subgroup maximal? Certainly, if ρ is not an irreducible representation, then G is contained in the stabilizer of a proper non-trivial subspace and so if G is maximal it must be the full stabilizer. Now suppose ρ is irreducible and that G stabilizes some non-degenerate form on k^n, and thus lies in the isometry group of this form. Then if G is maximal it must be the full isometry group of the form. So we are led to ask when G stabilizes a non-degenerate form on a given irreducible representation. Now G stabilizes a non-degenerate form if and only if the corresponding kG-module is isomorphic to its dual. When this is the case, the corresponding representation $\rho : G \to \mathrm{SL}(V)$ has image lying in one of the classical groups $\mathrm{Sp}(V)$ or $\mathrm{SO}(V)$; in particular if $\rho(G) \neq \mathrm{Sp}(V), \mathrm{SO}(V)$ then $\rho(G)$ is not a maximal subgroup of $\mathrm{SL}(V)$. We will see in Chapter 18 that in the large majority of cases, $\rho(G)$ is a maximal subgroup of the smallest classical group containing it.

Note that for V a kG-module, the dual $V^* := \mathrm{Hom}(V, k)$, with the natural induced G-action

$$(gf)(v) := f(g^{-1}v) \qquad \text{for } g \in G, \ f \in \mathrm{Hom}(V, k), \ v \in V,$$

is again a rational kG-module, irreducible if V is irreducible.

Fix a Borel subgroup B of G and a maximal torus T of B. Let Φ^+ be the corresponding set of positive roots. We consider the case where V (and hence V^*) is an irreducible kG-module, so $V = L(\lambda)$ for some dominant $\lambda \in X(T)$ (see Theorem 15.17). In order to decide when $L(\lambda)^*$ is isomorphic to $L(\lambda)$, we will determine its highest weight and apply Theorem 15.17(a).

Proposition 16.1   *Let G be a connected reductive algebraic group and let $\lambda \in X(T)$ be dominant. Then $L(\lambda)^* \cong L(-w_0(\lambda))$, where $w_0 \in W$ is the longest element.*

Proof   We have $w_0(\Phi^+) = \Phi^-$ by Corollary 11.18, and if $\dot{w}_0 \in N_G(T)$ is a preimage of w_0, then $\dot{w}_0 B \dot{w}_0^{-1} = B^- = \langle T, U_\alpha \mid \alpha \in \Phi^- \rangle$. Let v^+ be a maximal vector (with respect to B) of the irreducible kG-module $L(\lambda)$. Write $V = \langle v^+ \rangle \oplus V'$, with V' the sum of the weight spaces for weights other than λ, and let $f \in L(\lambda)^*$ denote the linear form with $f(v^+) = 1$, $f|_{V'} = 0$, as in Section 15.2. We claim that f is a maximal vector of $L(\lambda)^*$ for B^- of weight $-\lambda$. Indeed, if $g \in B^-$ then $g = tu$ for some $t \in T$, $u \in \langle U_\alpha \mid \alpha \in \Phi^- \rangle$ by Theorem 11.1, and hence

$$(gf)(v^+) = f(g^{-1}v^+) = f(u^{-1}t^{-1}v^+) = f(u^{-1}\lambda(t^{-1})v^+)$$
$$= \lambda(t^{-1})f(u^{-1}v^+) = \lambda(t^{-1})f(v^+ + v') = \lambda(t^{-1})f(v^+)$$

for some $v' \in V'$ by Lemma 15.4. Furthermore, for $v \in V'$, $(gf)(v) = f(g^{-1}v) = f(u^{-1}t^{-1}v) = 0 = f(v)$ as $u^{-1}t^{-1}v$ is a sum of weight vectors of weight strictly less than λ. So $gf = \lambda(t^{-1})f$, i.e., f is a maximal vector for B^- of weight $-\lambda$.

But then $\dot{w}_0^{-1}f$ is a maximal vector for B:

$$B\dot{w}_0^{-1}f = \dot{w}_0^{-1}B^-\dot{w}_0(\dot{w}_0^{-1}f) = \dot{w}_0^{-1}B^-f \in \langle \dot{w}_0^{-1}f \rangle.$$

Moreover, $\dot{w}_0^{-1}f$ is of weight $w_0^{-1}(-\lambda) = -w_0(\lambda)$. The result now follows from Corollary 15.10.   □

Remark 16.2   In view of the previous result, it is interesting to know for which Weyl groups $-w_0 = \mathrm{id}$, as in this case all irreducible kG-modules are self-dual.

It is known that $w_0 = -\mathrm{id}$ (on $X(T)$) when Φ is of type A_1, B_n, C_n, D_n (n even), E_7, E_8, F_4, G_2, see Corollary B.23. (These are precisely the irreducible Weyl groups which contain $-\mathrm{id}$.) In the remaining irreducible cases, $-w_0$ induces a non-trivial graph automorphism of the Dynkin diagram as indicated below:

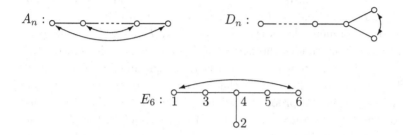

For example, for type E_6 and $\lambda = \lambda_1 + \lambda_6$, $L(\lambda) \cong L(\lambda)^* \cong L(-w_0(\lambda))$.

For the purposes of studying embeddings of reductive groups $H < G$, in particular when G is a classical group and H acts irreducibly on the natural module for G, it will also be useful to know something about the restriction of irreducible representations to Levi complements of parabolic subgroups. The result we state will enable us to apply inductive arguments in the study of irreducible representations.

Let Δ denote the base of Φ with respect to B. For some subset $I \subseteq S = \{s_\alpha \mid \alpha \in \Delta\}$, let P denote the corresponding parabolic subgroup of G with Levi decomposition $P = QL$, where $Q = R_u(P)$ and $L = L_I$, as in Section 12.2. We consider the restriction of an irreducible representation of G to the Levi subgroup L.

Proposition 16.3 *Let G be connected reductive, $\lambda \in X(T)$ a dominant weight, and write $L(\lambda)^Q$ for the fixed point space of Q on the irreducible kG-module $L(\lambda)$. Then:*

(a) $L(\lambda)^Q = \oplus_{\gamma \in \mathbb{N}_0 \Delta_I} L(\lambda)_{\lambda-\gamma}$.

(b) $L(\lambda)^Q$ *is the irreducible kL-module with highest weight λ.*

Proof Recall from Proposition 12.6 that $Q = \langle U_\alpha \mid \alpha \in \Phi^+ \setminus \Phi_I \rangle$. Set $V = L(\lambda)^Q$. For $\gamma \in \mathbb{Z}\Delta_I$ and $\alpha \in \Phi^+ \setminus \Phi_I$, Corollary 15.10 implies that $\lambda - \gamma + \alpha$ is not a weight of the module $L(\lambda)$ and hence by Lemma 15.4, U_α fixes pointwise the weight space $L(\lambda)_{\lambda-\gamma}$. Thus, $L(\lambda)_{\lambda-\gamma} \subseteq V$.

Now L normalizes Q and hence stabilizes the subspace V; in particular T stabilizes V and so V is a sum of T-weight spaces. For each weight

$\mu \in X(T)$ with $V_\mu \neq 0$, $\oplus_{\gamma \in \mathbb{Z}\Delta_I} V_{\mu+\gamma}$ is a non-trivial L-invariant submodule of V (again by Corollary 15.10). Let $U = R_u(B)$. By the Lie–Kolchin Theorem 4.1, the unipotent subgroup $L \cap U$ must have a non-trivial fixed point v on $\oplus_{\gamma \in \mathbb{Z}\Delta_I} V_{\mu+\gamma}$. But then v is fixed by the whole of $U = (L \cap U)Q$ (see Proposition 12.6), hence a maximal vector, of weight λ. That is, for all μ with $V_\mu \neq 0$, there exists $\gamma \in \mathbb{Z}\Delta_I$ with $\mu + \gamma = \lambda$, so $\mu = \lambda - \gamma$. Thus $V \subseteq \oplus_{\gamma \in \mathbb{Z}\Delta_I} L(\lambda)_{\lambda-\gamma}$ and (a) follows, using Corollary 15.10.

The above argument shows that $V = \oplus_{\gamma \in \mathbb{N}_0\Delta_I} L(\lambda)_{\lambda-\gamma}$ has a unique 1-space of fixed points under the action of $L \cap U$, namely the space $L(\lambda)_\lambda$. But $L \cap U$ is the unipotent radical of a Borel subgroup of L. Thus the sum of all simple kL-submodules of V, that is the *socle* of the restriction $V|_L$, is irreducible. As in the proof of Corollary 15.10, we see that the socle is the kL-submodule generated by any $v^+ \in V_\lambda \setminus \{0\}$: $\mathrm{soc}(V|_L) = \langle Lv^+ \rangle$. Now as $L(\lambda)$ is an irreducible kG-module, we have as in the proof of Proposition 15.6 $\langle Gv^+ \rangle = \langle U^- T v^+ \rangle = L(\lambda)$, where $U^- = \dot{w}_0 U \dot{w}_0^{-1}$ with w_0 the longest element of the Weyl group. Writing $U^- = Q^-(U^- \cap L)$, where $Q^- = \langle U_\alpha \mid \alpha \in \Phi^- \setminus \Phi_I \rangle$, we see that for all $\gamma \in \mathbb{N}_0\Delta_I$, the weight space $L(\lambda)_{\lambda-\gamma}$ is contained in $\langle (U^- \cap L)v^+ \rangle \subseteq \langle Lv^+ \rangle = \mathrm{soc}(V|_L)$. Hence $V = \oplus_{\gamma \in \mathbb{N}_0\Delta_I} L(\lambda)_{\lambda-\gamma} \subseteq \mathrm{soc}(V|_L)$, which is irreducible, establishing (b). $\square$

Example 16.4 Let G be semisimple, with simple roots Δ and fundamental dominant weights $\{\lambda_\alpha \mid \alpha \in \Delta\}$. For $I \subset S$, let Δ_I be the corresponding base of Φ_I. Then $\{\lambda_\alpha \mid \alpha \in \Delta_I\}$ is the dual basis of $\Delta_I^\vee$ in $\mathbb{Z}\Phi_I \otimes_{\mathbb{Z}} \mathbb{R}$, so these are the fundamental dominant weights of $T' := T \cap L_I'$ with respect to Δ_I, where $L_I' = [L_I, L_I]$. Thus if $\lambda = \sum_{\alpha \in \Delta} a_\alpha \lambda_\alpha \in X(T)$ is dominant, then its restriction to T' equals $\lambda|_{T'} = \sum_{\alpha \in \Delta_I} a_\alpha \lambda_\alpha$. So the highest weight of $L(\lambda)^Q$ as a kL_I'-module in Proposition 16.3 is just $\sum_{\alpha \in \Delta_I} a_\alpha \lambda_\alpha$.

16.2 Steinberg's tensor product theorem

Let $\rho : G \to \mathrm{GL}(V)$ be a rational representation which embeds G as a maximal subgroup into $\mathrm{GL}(V)$. If G leaves invariant a tensor product decomposition $V = V_1 \otimes V_2$, then it lies in the stabilizer of this decomposition, so is equal to it by maximality. Thus, it is important for the study of maximal subgroups to know which irreducible representations are tensor decomposable. A partial answer in positive characteristic is given by Steinberg's tensor product theorem, which is a major result in the representation theory of semisimple groups.

Let G be a semisimple algebraic group with root system Φ with respect to the maximal torus $T \leq G$. As already mentioned in Section 11.1 Steinberg has given a presentation of G in terms of the root subgroups $U_\alpha = \{u_\alpha(t) \mid t \in k\}$, $\alpha \in \Phi$; a description can be found in [73, §6]. Let τ be an automorphism of the field k. Then it is immediate from this so-called Steinberg presentation that it induces an automorphism of the abstract group G such that $u_\alpha(t) \mapsto u_\alpha(t)^\tau := u_\alpha(\tau(t))$, for all $\alpha \in \Phi$, $t \in k$.

For the classical matrix groups as defined in Section 1.2, such an automorphism can be obtained by componentwise application of τ to all matrix entries, as can be seen from an explicit determination of the root subgroups with respect to the maximal torus consisting of diagonal matrices, see Examples 11.4 and 11.7.

Here, we will be interested in the case of positive characteristic only. So throughout this section k will be an algebraically closed field of characteristic $\mathrm{char}(k) = p > 0$. Then we obtain the following special case:

Theorem 16.5 (Chevalley, Steinberg) *Let G be a semisimple algebraic group over k with $\mathrm{char}(k) = p$, with root system Φ with respect to a maximal torus T. Then the Frobenius automorphism $F_p : k \to k$, $c \mapsto c^p$, of k induces an endomorphism $F : G \to G$ of algebraic groups by*

$$u_\alpha(c) \mapsto F(u_\alpha(c)) := u_\alpha(F_p(c)) = u_\alpha(c^p) \qquad \text{for all } \alpha \in \Phi, c \in k.$$

Moreover, $F(t) = t^p$ for all $t \in T$.

For the classical groups constructed in Section 1.2 the desired map can be obtained by raising all matrix entries to their pth power. In general, the claim follows from the form of the relations in the Steinberg presentation in [73, p.66], see [66, Thm. 9.4.3]. Such endomorphisms play a crucial role in the construction of the finite groups of Lie type and will be discussed in more detail in Chapter 3.

Now given a rational representation $\rho : G \to \mathrm{GL}(V)$ and $i \geq 1$, we can define a new rational representation on V by $\rho^{(p^i)} := \rho \circ F^i$, with F from Theorem 16.5. The corresponding kG-module is denoted $V^{(p^i)}$, the action being given by

$$\rho^{(p^i)}(g)v = \rho(F^i(g))v \qquad \text{for } g \in G, v \in V.$$

The behavior of highest weights under this construction is easily described:

Proposition 16.6 *Let G be semisimple, $\lambda \in X(T)$ a dominant weight. Then $L(\lambda)^{(p)} \cong L(p\lambda)$ as kG-modules.*

Proof Since B is F-invariant, a maximal vector v of $L(\lambda)$ is also maximal for $L(\lambda)^{(p)}$. Moreover, $F(t).v = t^p.v = \lambda(t)^p.v$ for $t \in T$ by Theorem 16.5, so v is of weight $p\lambda$. Clearly, if V is an irreducible kG-module, then so is $V^{(p)}$. Thus, $L(\lambda)^{(p)}$ is the irreducible highest weight module of highest weight $p\lambda$.   □

We now relate the representations of G to those of its Lie algebra $\mathfrak{g} :=$ Lie(G) as follows. If $\rho : G \to \mathrm{GL}(V)$ is a rational representation, then its differential $d\rho : \mathfrak{g} \to \mathfrak{gl}(V)$ is a *Lie algebra representation* of $\mathfrak{g}$, i.e., a Lie algebra homomorphism to $\mathfrak{gl}(V)$. The irreducible constituents of $\mathfrak{g}$-representations obtained in this way are called *restricted*.

Note that $\mathfrak{g}$ acts trivially on $V^{(p)}$: indeed, we have $\rho^{(p)}|_T = \rho \circ F$, where $F : T \to T$, $t \mapsto t^p$, is the p-power map by Theorem 16.5. Thus $d\rho^{(p)}|_{\mathrm{Lie}(T)} = d\rho \circ dF|_{\mathrm{Lie}(T)}$, but clearly $dF = 0$, so Lie(T) lies in the kernel of $d\rho^{(p)}$. Similarly, since the action of F on the root group U_α, for $\alpha \in \Phi$, is simply the composition of a fixed isomorphism $\mathbf{G_a} \to U_\alpha$ with the p-power map on $\mathbf{G_a}$, we see that Lie(U_α) also lies in $\ker d\rho^{(p)}$, for all $\alpha \in \Phi$. Then Theorem 8.17(a) shows that $d\rho^{(p)} = 0$.

Clearly, if $V_1 \le V$ is a G-invariant subspace, then it is also $\mathfrak{g}$-invariant, hence if $d\rho$ is irreducible, then so was ρ. The converse does not hold in general. But there is the following crucial lifting result:

Theorem 16.7 (Cline–Parshall–Scott) *Let G be semisimple of simply connected type. Then every irreducible restricted $\mathfrak{g}$-module has a (unique) extension to a rational irreducible kG-module.*

Here $\rho : G \to \mathrm{GL}(V)$ is an *extension* of $\varphi : \mathfrak{g} \to \mathfrak{gl}(V)$ if $d\rho = \varphi$. The proof of this result is not very difficult, see [16, Thm. 1], but it needs the notion of restricted universal enveloping algebra, which we did not introduce. This allows one to construct a first tensor decomposition:

Lemma 16.8 *Let V be an irreducible rational kG-module, $V_1 \le V$ an irreducible $\mathfrak{g}$-submodule. Then $V \cong V_1 \otimes V_2$ as kG-modules, with V_1, V_2 irreducible as kG-modules and with trivial action of $\mathfrak{g}$ on V_2.*

Proof By Theorem 16.7, V_1 has a structure of irreducible kG-module. Then the space $V_2 := \mathrm{Hom}_\mathfrak{g}(V_1, V)$ of $\mathfrak{g}$-invariant homomorphisms from V_1 to V can be made into a kG-module by $(g.\varphi)(v) := g.\varphi(g^{-1}v)$ for $\varphi \in V_2$, $g \in G$ and $v \in V_1$. The k-linear map $V_1 \otimes V_2 \to V$, $(v, \varphi) \mapsto \varphi(v)$, is then a kG-homomorphism, non-zero by assumption and thus surjective by the irreducibility of V. As $\dim V_2 \le \dim V / \dim V_1$ we see that in fact it must be an isomorphism.

Now clearly, if V_2 has a G-invariant proper non-zero subspace, then so has the tensor product $V_1 \otimes V_2$, which is not the case, thus V_2 is irreducible for

G. Differentiating the action of G on V_2 shows that $X \in \mathfrak{g}$ acts on V_2 by $(X.\varphi)(v) = X\varphi(v) - \varphi(Xv)$. But V_2 consists of $\mathfrak{g}$-homomorphisms, so it is a trivial $\mathfrak{g}$-module. □

To continue, we need one further piece of notation. Let Φ be the root system relative to T with positive system Φ^+ and base $\Delta \subset \Phi$ determined by B.

Definition 16.9 A dominant weight $\lambda \in X(T)$ is said to be *p-restricted* if $\langle \lambda, \alpha^\vee \rangle < p$ for all $\alpha \in \Delta$. An irreducible kG-module with highest weight λ is called *restricted* if λ is p-restricted, and the corresponding representation is called a *restricted representation*.

Note that there are only finitely many p-restricted weights for a fixed semisimple group G and a fixed prime p.

Lemma 16.10 *Let G be semisimple of simply connected type with maximal torus T. If $\lambda \in X(T)$ is p-restricted and $d\lambda = 0$ then $\lambda = 0$.*

Proof Let's write $T = \mathbf{G}_{\mathrm{m}} \times \cdots \times \mathbf{G}_{\mathrm{m}}$ (l factors). Clearly the differentials of the coordinate functions form a basis of $\mathrm{Hom}(\mathrm{Lie}(T), k) = \mathrm{Hom}(k^l, k)$. Since G is simply connected, we have $\Omega = X(T)$, so the fundamental dominant weights $\{\lambda_i\}$ form another $\mathbb{Z}$-basis of $X(T)$, whence their differentials also form a basis of $\mathrm{Hom}(\mathrm{Lie}(T), k)$.

Now write $\lambda = \sum_i a_i\lambda_i$, so $0 = d\lambda = \sum_i a_i d\lambda_i$ by the formula for the differential of multiplication in Example 7.8(1), which shows that $a_i \equiv 0$ (mod p) for all i by the above. As λ is p-restricted, this implies that $a_i = 0$ for all i. □

The following result shows that restricted kG-modules behave particularly well with respect to the differential:

Corollary 16.11 (Curtis) *Let $\lambda \in X(T)$ be p-restricted. Then $L(\lambda)$ is an irreducible $\mathfrak{g}$-module.*

Proof Let V_1 be an irreducible $\mathfrak{g}$-submodule of $L(\lambda)$. By Lemma 16.8 there is a tensor decomposition of $L(\lambda)$ into irreducible kG-modules, of highest weights λ', μ say: $L(\lambda) \cong L(\lambda') \otimes L(\mu)$ with $V_1 \cong L(\lambda')$. But then $\lambda = \lambda' + \mu$ by Proposition 15.12, so μ is necessarily also p-restricted. Now $\mathrm{Lie}(T)$ acts by $d\mu$ on a maximal vector of $V_2 = L(\mu)$, but $\mathfrak{g}$ acts trivially by Lemma 16.8, so $d\mu = 0$. By Lemma 16.10 this forces $\mu = 0$, and so $L(\lambda) \cong L(\lambda') \cong V_1$ is an irreducible $\mathfrak{g}$-module. □

Now each dominant weight $\lambda \in X(T)$ can be uniquely expressed as

$$\lambda = \mu_0 + p\mu_1 + \cdots + p^a\mu_a,$$

where $a \in \mathbb{N}_0$ and $\mu_i \in X(T)$ is a p-restricted (dominant) weight for each $0 \le i \le a$. The main result of this section shows that there is an analogous decomposition of the corresponding irreducible kG-module $L(\lambda)$:

Theorem 16.12 (Steinberg) *Let G be semisimple of simply connected type, let $\lambda \in X(T)$ be dominant, and write $\lambda = \mu_0 + p\mu_1 + \cdots + p^a\mu_a$, with $a \in \mathbb{N}_0$ and p-restricted weights $\mu_i \in X(T)$ for each $0 \le i \le a$. Then*

$$L(\lambda) \cong L(\mu_0) \otimes L(\mu_1)^{(p)} \otimes \cdots \otimes L(\mu_a)^{(p^a)}$$

as kG-modules.

Proof Write $\lambda = \mu_0 + p\mu'$ (so $\mu' = \mu_1 + \cdots + p^{a-1}\mu_a$). Then $L(\lambda)$ is a kG-composition factor of $V := L(\mu_0) \otimes L(\mu')^{(p)} \cong L(\mu_0) \otimes L(p\mu')$ by Propositions 15.12 and 16.6. As $\mathfrak{g}$ acts trivially on $L(p\mu')$, all $\mathfrak{g}$-composition factors of V are isomorphic to the irreducible $\mathfrak{g}$-module $L(\mu_0)$ (see Corollary 16.11), so $L(\mu_0)$ is an irreducible $\mathfrak{g}$-submodule of $L(\lambda)$. By Lemma 16.8 we have $L(\lambda) \cong L(\mu_0) \otimes L(\nu)$ for some dominant weight ν, so $\lambda = \mu_0 + \nu$. Uniqueness of the highest weight shows that $\nu = p\mu'$, so $L(\lambda) \cong L(\mu_0) \otimes L(p\mu') \cong L(\mu_0) \otimes L(\mu')^{(p)}$. Now an easy induction, applied to μ', completes the proof. ☐

Thus, in positive characteristic, the investigation of all rational irreducible kG-modules is reduced to the study of the finitely many restricted ones.

Example 16.13 We use the tensor product theorem to illustrate that dimensions of irreducible highest weight modules may depend on the underlying characteristic.

(1) Let $G = \mathrm{SL}_n$ and let X be the natural n-dimensional module for G, of highest weight λ_1 (as mentioned in Example 15.21(2)). Then by Proposition 15.12 the highest weight in $V = X \otimes X$ is $2\lambda_1$, occurring with multiplicity 1. In particular V possesses a kG-composition factor with highest weight $2\lambda_1$. If $p > 2$, this weight is p-restricted and it can be shown that the corresponding highest weight module is precisely the symmetric square $S^2(X)$ of X and $V = S^2(X) \oplus \wedge^2(X)$. If however $p = 2$, the irreducible module with highest weight $2\lambda_1$ is simply the "twist" of the natural module: $L(2\lambda_1) \cong L(\lambda_1)^{(2)} = X^{(2)}$ by Proposition 16.6. Nevertheless, $\wedge^2(X)$ is an irreducible kG-module and V has three composition factors: $X^{(2)}$ and two factors isomorphic to $\wedge^2(X)$.

(2) Consider the kSL_3-module with highest weight $\lambda = 2\lambda_1 + 3\lambda_2$, where λ_i is the fundamental dominant weight corresponding to the simple root α_i. If $p > 3$, then λ is p-restricted and $\dim L(\lambda) = 39$ if $p = 5$ and $\dim L(\lambda) = 42$ if $p \neq 5$. (See [54].) If however $p = 3$, the p-adic expansion of λ is $\lambda = \mu_0 + 3\mu_1$ where $\mu_0 = 2\lambda_1$ and $\mu_1 = \lambda_2$, so $L(\lambda) \cong L(2\lambda_1) \otimes L(\lambda_2)^{(3)}$ in this case by Theorem 16.12. Thus, by the previous example and by Proposition 16.1 and Remark 16.2, $L(\lambda) \cong S^2(X) \otimes X^{*(3)}$, where X is the natural three-dimensional module for SL_3. Hence, $\dim L(\lambda) = 6 \cdot 3 = 18$. If $p = 2$, then Theorem 16.12 gives $L(\lambda) \cong L(\lambda_2) \otimes L(\lambda_1 + \lambda_2)^{(2)}$. Using Example 15.19, one can show that $L(\lambda_1 + \lambda_2)$ is isomorphic to the Lie algebra of SL_3, and so $\dim L(\lambda) = 3 \cdot 8 = 24$.

17

Structure of parabolic subgroups, II

We continue the investigation of the structure of a parabolic subgroup by now looking at its unipotent radical. We will see that the conjugation action of a Levi factor on the unipotent radical induces a family of representations of the Levi factor. Now that we know something about the representation theory of reductive groups, we will use this to obtain more information about the unipotent radical.

17.1 Internal modules

Let G be a connected reductive group with maximal torus T contained in a Borel subgroup B. Let Φ be the root system of G with respect to T, and $\Delta \subseteq \Phi$ the base determined by B. Let Φ^+ be the corresponding set of positive roots, with set of simple reflections S. So in particular, $B = \langle T, U_\alpha \mid \alpha \in \Phi^+ \rangle$ by Theorem 11.1(a).

For a subset $I \subseteq S$, recall the parabolic subsystem Φ_I of Φ. Let $P = BW_I B$ be the corresponding standard parabolic subgroup with unipotent radical $Q := R_u(P)$ and Levi decomposition

$$P = QL = \langle U_\alpha \mid \alpha \in \Phi^+ \setminus \Phi_I \rangle \cdot \langle T, U_\beta \mid \beta \in \Phi_I \rangle$$

from Proposition 12.6. Note that L is reductive by Proposition 12.6, so $[L, L]$ is semisimple by Corollary 8.22.

Definition 17.1 Let $I \subseteq S$ with corresponding set of simple roots Δ_I. For $\beta \in \Phi^+$, write

$$\beta = \sum_{\alpha \in \Delta_I} c_\alpha \alpha + \sum_{\gamma \in \Delta \setminus \Delta_I} d_\gamma \gamma.$$

We say that β is *of level* $\sum d_\gamma$, and we call $(d_\gamma)_{\gamma \in \Delta \setminus \Delta_I}$ the *shape* of β. For

$j \geq 1$, set

$$Q_j = \langle U_\beta \mid \beta \in \Phi^+ \text{ of level } \geq j \rangle \leq Q.$$

Example 17.2 Let $G = \mathrm{SL}_n$, the simply connected simple group of type A_{n-1}, with root system base $\{\alpha_1, \ldots, \alpha_{n-1}\}$ and corresponding set of positive roots Φ^+ (see Example 9.8).

Let P be the standard parabolic subgroup corresponding to the subset $I = \{s_{\alpha_1}\}$ of S with unipotent radical

$$Q = \Big\langle U_\beta \;\Big|\; \beta = \sum_{i=1}^{n-1} a_i \alpha_i \in \Phi^+, \; \sum_{i=2}^{n-1} a_i \neq 0 \Big\rangle$$

and Levi complement $L = \langle U_{\pm\alpha_1}, T \rangle$. Then the roots of level i for $1 \leq i \leq n-2$ are $\alpha_j + \cdots + \alpha_{j+i-1}$ for $2 \leq j \leq n-i$ and $\alpha_1 + \alpha_2 + \cdots + \alpha_{i+1}$.

We obtain the following filtration of unipotent radicals:

Proposition 17.3 *Let $P \leq G$ be a parabolic subgroup of the connected reductive group G, with unipotent radical Q. Then with the notation as in the definition above, we have for all $j \geq 1$:*

(a) $Q_j \trianglelefteq P$.
(b) $Q_j/Q_{j+1} \cong \prod_\beta \mathbf{G}_\mathrm{a}$, *where the product runs over $\beta \in \Phi^+$ of level j.*

Proof (a) Recall the commutator formula from Theorem 11.8:

$$[U_\alpha, U_\beta] \subseteq \prod_{m,n>0} U_{m\alpha+n\beta} \qquad \text{for all roots } \alpha \neq \pm\beta.$$

Let $\beta \in \Phi^+$ of level at least j and $\alpha \in \Phi_I \cup \Phi^+$. Then in particular, for $x \in U_\alpha$, $xU_\beta x^{-1}$ lies in the product of root subgroups for roots of level at least j. Since Q_j is a product of root subgroups for T, Q_j is also normalized by T and (a) follows.

For (b), let's choose a total order on the positive roots so that $\mathrm{ht}(\alpha) > \mathrm{ht}(\beta)$ implies that $\alpha > \beta$. For each $\alpha \in \Phi$, fix an isomorphism $u_\alpha : \mathbf{G}_\mathrm{a} \to U_\alpha$. Then by Theorem 11.1(a) any element u of Q_j may be written as a product $\prod_\beta u_\beta(c_\beta)y$ over $\beta \in \Phi^+$ of level j, with $y \in Q_{j+1}$ and the c_β uniquely determined by u. Thus, as Q_j/Q_{j+1} is abelian by the commutator formula, the map $(c_\beta)_\beta \mapsto \prod_\beta u_\beta(c_\beta)Q_{j+1}$ defines an isomorphism $\prod_\beta \mathbf{G}_\mathrm{a} \cong Q_j/Q_{j+1}$, with β as before. $\qquad\square$

As $Q_j \trianglelefteq P$ for all j, L acts on the quotient Q_j/Q_{j+1}. It is immediate to check that the rule $c u_\beta(t)Q_{j+1} := u_\beta(ct)Q_{j+1}$, for $\beta \in \Phi^+ \setminus \Phi_I$ of level j and $c \in k$, defines a k-vector space structure on Q_j/Q_{j+1} which commutes with the action of L.

Now for a shape S of level j let $V_S := (\prod_{\beta \text{ of shape } S} U_\beta) Q_{j+1}/Q_{j+1}$. From the commutator relations it is then easy to verify:

Proposition 17.4   *Let S be a shape of level $j \geq 1$.*

(a) *As k-vector space, V_S is isomorphic to the direct sum of the U_β, for β of shape S.*

(b) *V_S is normalized by L, hence a kL-submodule of Q_j/Q_{j+1}.*

The subgroups Q_j are closely related to the descending central series of Q. Recall (from Section 1.2) that $C^{i-1}Q$ denotes the ith term of the descending central series for Q. Then one has [3, Lemma 4]:

Proposition 17.5   *Let G be a simple algebraic group and assume that $(G, \text{char}(k)) \neq (B_n, 2), (C_n, 2), (F_4, 2), (G_2, 2), (G_2, 3)$. Then for any parabolic subgroup $P = QL$ of G we have $Q_j = C^{j-1}Q$ for all $j \geq 1$.*

About the proof   Essentially this follows easily from an inductive argument and the commutator relations, Theorem 11.8. The exceptions occur when certain structure constants $c_{\alpha\beta}^{mn}$ appearing in the commutator formula are divisible by char(k).   □

In fact one has very good control over the kL-modules which occur in the quotients Q_j/Q_{j+1}.

Theorem 17.6 (Azad–Barry–Seitz)   *Let G be a simple algebraic group and assume that $(G, \text{char}(k)) \neq (B_n, 2), (C_n, 2), (F_4, 2), (G_2, 2), (G_2, 3)$. Then for any parabolic subgroup $P = QL$ of G we have:*

(a) *There exists a unique root β_S of shape S having maximal height with respect to this property.*

(b) *For each shape S, V_S is an irreducible kL-module of highest weight β_S.*

(c) *If $S \neq S'$, then $V_S \not\cong V_{S'}$ as kL-modules.*

(d) *Q_j/Q_{j+1} is the direct sum of the V_S, for the shapes S of level j, hence a completely reducible kL-module.*

Sketch of proof   The assertion of (a) is just Exercise A.7. As all roots of shape S and of the same length are conjugate under the Weyl group W_I of L, by Lemma A.31, this already shows that V_S is irreducible when there is only one root length. By avoiding the small characteristic configurations excluded in the theorem, it is straightforward to show this in general. Part (b) then follows from the classification of highest weight modules given in Theorem 15.17(a) since clearly any non-identity element in the root group U_{β_S} is a maximal vector of weight β_S for the Levi factor L. Part (c) is now immediate by Theorem 15.17(a). We have already shown in Proposition 17.4(b) that Q_j/Q_{j+1} decomposes as a direct sum of the V_S, whence (d). □

Remark 17.7 Under the assumptions of Theorem 17.6, Azad, Barry and Seitz also show that the Levi complement L has finitely many orbits on each V_S. In the proof, they apply Richardson's result [59, Thm. E] to see that there exist a finite number of L-orbits on $Q = Q_1$ and so on Q_1/Q_2, and therefore on the V_S occurring in Q_1/Q_2 as well. For the general case, one realizes Q_j/Q_{j+1} as the commutator quotient of the unipotent radical of a parabolic subgroup of a certain reductive subgroup of G. See [3, Thm. 2(f)] for the details.

Definition 17.8 The kL-modules V_S appearing in the previous theorem are called *internal modules* for L.

Example 17.9 We illustrate some types of internal modules which may occur.

(1) Let $G = \mathrm{SL}_n$, the simply connected group of type A_{n-1}, with the parabolic subgroup P as in Example 17.2. The Levi complement has the form $L = \langle U_{\pm\alpha_1}, T \rangle = [L, L]Z(L)^\circ$, where $Z(L)^\circ$ is an $(n-2)$-dimensional torus and $[L, L]$ is a rank 1 semisimple group, so isomorphic to SL_2 or PGL_2 by Theorem 8.8 (in fact isomorphic to SL_2 by Proposition 12.14). Let's consider some of the levels worked out in Example 17.2.

At level 1, for shape $S_1 = (1, 0, \ldots, 0)$ we have $V_{S_1} = U_{\alpha_2} \cdot U_{\alpha_1+\alpha_2}$. This is easily seen to be the natural two-dimensional module for $[L, L]$. For each $i > 1$, $V_{S_i} = U_{\alpha_{i+1}}$ is the trivial module for the shape $S_i = (0, \ldots, 0, 1, 0, \ldots, 0)$.

At level i for $i \geq 2$, let $S = (\epsilon_1, \ldots, \epsilon_{n-2})$, $\epsilon_j = 0$ or 1 for all j and $\sum_{j=1}^{n-2} \epsilon_j = i$ be a shape of level i. Then if $\epsilon_1 = 1$, the corresponding simple module V_S is the natural two-dimensional module for $[L, L]$ and if $\epsilon_1 = 0$, then V_S is the trivial module.

(2) For a more elaborate example, let $G = \mathrm{SO}_{2n}$ (see Definition 1.15) and B the Borel subgroup of G consisting of upper triangular matrices exhibited in Example 6.7(4). Recall the basis $\{v_1, \ldots, v_{2n}\}$ of the orthogonal space on which G naturally acts, such that the quadratic form with respect to this basis is given by $f(x_1, \ldots, x_{2n}) = x_1 x_{2n} + x_2 x_{2n-1} + \cdots + x_n x_{n+1}$. Then set $V_1 = \langle v_1, \ldots, v_n \rangle$, a totally singular subspace of V, and let $P = \mathrm{Stab}_G(V_1)$, a parabolic subgroup (see Proposition 12.13). Then one

checks that P has Levi factor

$$L = \left\{ \begin{pmatrix} A & 0 \\ 0 & K_n A^{-\mathrm{tr}} K_n \end{pmatrix} \; \middle| \; A \in \mathrm{GL}_n \right\},$$

where K_n is as in Section 1.2, and unipotent radical

$$Q = \left\{ \begin{pmatrix} I_n & K_n S \\ 0 & I_n \end{pmatrix} \; \middle| \; S \in k^{n \times n} \text{ skew-symmetric} \right\}.$$

Let's consider the action of L on Q. We have

$$\begin{pmatrix} A & 0 \\ 0 & B \end{pmatrix} \begin{pmatrix} I_n & K_n S \\ 0 & I_n \end{pmatrix} \begin{pmatrix} A & 0 \\ 0 & B \end{pmatrix}^{-1} = \begin{pmatrix} I_n & A K_n S B^{-1} \\ 0 & I_n \end{pmatrix},$$

where $B = K_n A^{-\mathrm{tr}} K_n$. Hence the action of L on Q is equivalent to the natural action of GL_n on the space of skew-symmetric matrices: $S \mapsto A S A^{\mathrm{tr}}$, for $A \in \mathrm{GL}_n$ and $S \in k^{n \times n}$ skew-symmetric.

Let V be the natural n-dimensional $k\mathrm{GL}_n$-module (viewed as column vectors written with respect to a fixed basis). A skew-symmetric matrix A defines a bilinear map $\varphi_A : V \otimes V \to k$ via $v \otimes w \mapsto v^{\mathrm{tr}} A w$. As A is skew-symmetric, φ_A induces a linear map $\bar{\varphi}_A : \wedge^2 V \to k$. The map $A \mapsto \bar{\varphi}_A$ defines a $k\mathrm{GL}_n$-module isomorphism between the space of skew-symmetric matrices and $\mathrm{Hom}(\wedge^2 V, k) = (\wedge^2 V)^*$. It turns out that this is an irreducible module for GL_n. Hence Q is irreducible, $Q_j = 1$ for all $j > 1$ and $Q = V_{\mathcal{S}}$ for a unique shape $\mathcal{S} = (1)$, that is, coefficient 1 of the unique simple root lying outside of the root system of L.

(3) In this example we look at some internal modules for $G = \mathrm{SO}_8$, with root system of type D_4 and Dynkin diagram

Fix a Borel subgroup B and a maximal torus $T \leq B$, such that $\{\alpha_1, \alpha_2, \alpha_3, \alpha_4\}$ is the corresponding base of Φ.

Let P be the standard parabolic subgroup of G corresponding to the subset $\{\alpha_1, \alpha_3, \alpha_4\}$, with Levi decomposition $P = QL$, where $Q = R_u(P)$. Then $[L, L] = \langle U_{\pm\alpha_1}, U_{\pm\alpha_3}, U_{\pm\alpha_4} \rangle$. Here, $Q_1 = Q$ and $Q_2 = U_{\alpha_1 + 2\alpha_2 + \alpha_3 + \alpha_4}$.

Now Q_2 is one-dimensional and so must be the trivial module for $[L, L]$, with highest weight 0.

As there is a unique shape of roots of level 1, Theorem 17.6 implies that Q_1/Q_2 is irreducible. Set $B_L = \langle U_{\alpha_1}, U_{\alpha_3}, U_{\alpha_4}, T \rangle$, a Borel subgroup of L. By Theorem 11.8, $[U_{\alpha_i}, U_{\alpha_1+\alpha_2+\alpha_3+\alpha_4}] = 1$ for $i = 1, 3, 4$. So $U_{\alpha_1+\alpha_2+\alpha_3+\alpha_4} Q_2/Q_2$ is a B_L-stable line in Q_1/Q_2. Therefore xQ_2 is a maximal vector in the irreducible kL-module Q_1/Q_2 for any non-identity element $x \in U_{\alpha_1+\alpha_2+\alpha_3+\alpha_4}$.

Note that the weight $\alpha_1 + \alpha_2 + \alpha_3 + \alpha_4$ restricted to the maximal torus $\mathrm{im}(\alpha_i^\vee)$ of $\langle U_{\pm\alpha_i} \rangle$, for $i = 1, 3, 4$, is precisely the fundamental dominant weight λ (for the group SL_2). In particular, each of these three subgroups is isomorphic to SL_2 and $[L, L]$ is an image of the rank 3 group $\mathrm{SL}_2 \times \mathrm{SL}_2 \times \mathrm{SL}_2$. The quotient Q_1/Q_2 as module for $[L, L]$ is then $L(\lambda) \otimes L(\lambda) \otimes L(\lambda)$.

17.2 The theorem of Borel and Tits

The internal modules arising in Theorem 17.6 are important when considering embeddings of reductive groups. To illustrate this, we first need the following fundamental result on normalizers of unipotent subgroups (see [32, Prop. 30.3] or [29, Thm. 3.1.1]):

Theorem 17.10 (Borel–Tits) *Let G be a connected reductive linear algebraic group. Let $U \leq G$ be a closed unipotent subgroup of G, which lies in a Borel subgroup of G. Then there exists a parabolic subgroup P of G with $U \leq R_u(P)$ and $N_G(U) \leq P$.*

Sketch of proof Let $U \leq G$ be a closed unipotent subgroup lying in a Borel subgroup B of G. Set $U_0 := U$ and define inductively $N_{i+1} = N_G(U_i)$ and $U_{i+1} = U_i.R_u(N_{i+1})$ for all $i \geq 0$. Then $U = U_0 \leq U_1 \leq U_2 \leq \ldots$ and $N_1 \leq N_2 \leq \ldots$ are chains of closed subgroups of G. (By Exercise 10.18, normalizers are closed.)

We claim that each U_i lies in a Borel subgroup of G. This is already the case for U_0, so we assume it is true for U_i. Thus, the set of fixed points of U_i acting by left multiplication on G/B is non-empty and closed, and stabilized by the connected solvable group $R_u(N_{i+1}) \leq N_G(U_i)$. Hence by Theorem 6.1, there exists gB which is fixed by $R_u(N_{i+1})$ as well as by U_i. So $U_{i+1} = U_i.R_u(N_{i+1})$ lies in the Borel subgroup gBg^{-1} as claimed. Now, for all i, as

$$U_{i+1}/U_i = U_i.R_u(N_{i+1})/U_i \cong R_u(N_{i+1})/(U_i \cap R_u(N_{i+1}))$$

is connected, either $\dim(U_{i+1}/U_i) \geq 1$ or $U_{i+1}/U_i = 1$. Since $\dim(G)$ is finite, the sequence $U \leq U_1 \leq \ldots$ stabilizes at some point, say $U_l = U_{l+j}$ for all $j \geq 1$ and hence $N_{l+1+j} = N_{l+1}$ for $j \geq 1$.

Set $P(U) := N_{l+1}$ and $V := U_l$. Then, by our previous argument, V lies in a Borel subgroup and $V \supseteq R_u(N_G(V))$. The difficult step in the proof, which is given in [32, Prop. 30.3], shows that, under the established conditions, V is connected, and hence $U \leq V = R_u(N_G(V))$ and $P(U)$ is a parabolic subgroup of G, and the result follows. $\square$

Corollary 17.11   *Let $U \leq G$ be a closed connected unipotent subgroup. Then there exists a parabolic subgroup P of G such that $U \leq R_u(P)$ and $N_G(U) \leq P$.*

Proof   Since U is a closed connected unipotent, hence solvable subgroup of G, U lies in a Borel subgroup. The result then follows from Theorem 17.10.
                                                                        $\square$

We now state a consequence which is useful when considering inclusions of one reductive group in another.

Corollary 17.12   *Let H, G be connected reductive algebraic groups with H a closed subgroup of G. Let P_H be a parabolic subgroup of H. Then there exists a parabolic subgroup P_G of G with $R_u(P_H) \leq R_u(P_G)$ and $P_H \leq P_G$.*

Proof   The subgroup $R_u(P_H)$ is a closed connected unipotent group. So, by Corollary 17.11 there exists a parabolic subgroup P_G of G such that

$$P_H \leq N_G(R_u(P_H)) \leq P_G$$

and $R_u(P_H) \leq R_u(P_G)$. $\square$

This can be used to study embeddings of connected reductive algebraic groups $H \leq G$. Let P_H be a parabolic subgroup of H, and P the parabolic subgroup of G with $Q_H = R_u(P_H) \leq R_u(P) = Q$ and $P_H \leq P$, whose existence is asserted by Corollary 17.12. Let $P_H = Q_H L_H$ and $P = QL$ be Levi decompositions of P_H, respectively P. Then we obtain an embedding of reductive groups $L_H \hookrightarrow L$ via

$$L_H \cong QL_H/Q \leq P/Q \cong L.$$

Now L acts on each of the quotients Q_j/Q_{j+1} in Q (defined in Section 17.1), with L_H stabilizing the image of Q_H in this quotient. Using Theorem 17.6, we obtain a composition series for Q_H as kL_H-module and this must be compatible with the restrictions of the kL-modules Q_j/Q_{j+1} to L_H. This can be used to provide precise information about the embeddings $L_H \hookrightarrow L$ and $Q_H \leq Q$.

Example 17.13 Let us apply this discussion to a known embedding of reductive groups, namely $G_2 \leq G$, where G is semisimple with root system of type B_3 and $\operatorname{char}(k) \neq 2, 3$. (The group G_2 has an irreducible seven-dimensional representation on which it fixes a non-degenerate quadratic form.) Let T be a maximal torus of G_2 and $B \leq G_2$ a Borel subgroup containing T. Let Φ^+ be the corresponding set of positive roots with base $\{\alpha_1, \alpha_2\}$ (ordered as in the Dynkin diagram in Table 9.1; see Example 9.5(4) for the root system). Let $P \leq G_2$ be the standard parabolic subgroup containing B corresponding to the simple root α_1, so P has a Levi factor $L = \langle U_{\pm\alpha_1}, T \rangle$ and unipotent radical $Q = U_{\alpha_2} U_{\alpha_1+\alpha_2} U_{2\alpha_1+\alpha_2} U_{3\alpha_1+\alpha_2} U_{3\alpha_1+2\alpha_2}$. By Theorem 17.6, Q_1/Q_2 is a four-dimensional irreducible kL-module and Q_2 is a one-dimensional trivial kL-module.

Now let P_G be a parabolic subgroup of G with $P \leq P_G$ and $Q \leq R_u(P_G) :=$ Q_G as in Corollary 17.12. Let $B_G \leq P_G$ be a Borel subgroup of G and Φ_G^+ the corresponding set of positive roots with base $\{\beta_1, \beta_2, \beta_3\}$. As $Q \leq Q_G$, we know that $\dim Q_G \geq 5$ and that a Levi factor L_G of P_G acting on Q_G must have a composition factor of dimension at least 4 on some $(Q_G)_j/(Q_G)_{j+1}$. This restricts the possibilities for $P_G = P_I$ to one of the following: $\Delta_I = \{\beta_1, \beta_3\}$ or $\Delta_I = \{\beta_2, \beta_3\}$. If we further assume that P_G is minimal with respect to $P \leq P_G$, then using Proposition 12.13 and the representation theory of SL_2, one can see that in the second case, L would act irreducibly on the five-dimensional L_G-module Q_G, and therefore would not admit the required composition series of Q. Hence we conclude that the parabolic P_G corresponds to the subset $\Delta_I = \{\beta_1, \beta_3\}$.

The above corollary also has the following important consequence for the study of maximal subgroups of a reductive group.

Corollary 17.14 *Let G be a connected reductive algebraic group. Let $H \leq$ G be maximal among proper closed subgroups of G (with respect to inclusion). Then either H° is reductive or H is a parabolic subgroup of G.*

Proof Suppose H° is not reductive. Then $R_u(H^\circ)$ is a non-trivial closed connected unipotent subgroup of G. By Corollary 17.12, there exists a proper parabolic subgroup P of G with $R_u(H^\circ) \leq R_u(P)$ and $N_G(R_u(H^\circ)) \leq P$. So, $H \leq N_G(R_u(H^\circ)) \leq P$. But since H is maximal, we have $H = P$. $\square$

We conclude our discussion of the Borel–Tits Theorem with a further corollary, which shows that the hypotheses of Corollary 17.11 may be relaxed.

Corollary 17.15 *Let G be a connected reductive group and $U \leq G$ a unipotent subgroup of G. Then U lies in a Borel subgroup of G.*

Proof   The proof is by induction on $\dim(G)$. If $\dim(G) = 1$ then G is solvable by Theorem 3.2 so it is a Borel subgroup of G containing U.

Suppose $\dim(G) > 1$. Since U is unipotent, it is nilpotent by Corollary 2.10; in particular $Z(U) \neq 1$. Let $1 \neq u \in Z(U)$ and set $U_1 = \overline{\langle u \rangle}$, the closure of the subgroup generated by u. Then $U \leq N_G(\langle u \rangle) \leq N_G(U_1)$.

By Theorem 6.10, u lies in some Borel subgroup B of G. So, $\langle u \rangle \leq B$; hence $U_1 \leq B$ and in fact $U_1 \leq R_u(B)$, by Theorem 4.4. By Theorem 17.10 there exists a parabolic subgroup P of G with $U \leq N_G(U_1) \leq P$ and $U_1 \leq R_u(P)$.

Now $U_1 \neq 1$, so $R_u(P) \neq 1$, hence $P \neq G$. The connected reductive group $P/R_u(P)$ has dimension less than $\dim(G)$. By induction, $U.R_u(P)/R_u(P)$ lies in a Borel subgroup of $P/R_u(P)$, of the form $H/R_u(P)$ for some closed connected solvable subgroup H of G with $R_u(P) \leq H$. Now there exists a Borel subgroup B_1 of G with $B_1 \leq P$ and $R_u(P) \leq B_1$. So $B_1/R_u(P)$ is a closed connected solvable subgroup of $P/R_u(P)$ and hence lies in some conjugate of $H/R_u(P)$. But then $B_1 \leq H^x$ for some $x \in G$ whence H is a Borel subgroup of G and $U \leq H$ as required.   □

Remarks 17.16   (a) With U as in Corollary 17.15, Theorem 17.10 applies to produce a parabolic subgroup P of G with $U \leq R_u(P)$ and $N_G(U) \leq P$. In fact, as shown in [7, Cor. 3.9], P is canonical in the sense that any automorphism of G which stabilizes U must stabilize P.

(b) We now see that maximal unipotent subgroups of G are unipotent radicals of Borel subgroups.

18

Maximal subgroups of classical type simple algebraic groups

Let V be a finite-dimensional vector space over k equipped with the 0 form, a non-degenerate symplectic bilinear form (and $\dim V \geq 4$), or a non-degenerate symmetric bilinear form and associated quadratic form (and $\dim V \geq 7$). In particular, when $\mathrm{char}(k) = 2$ and V is equipped with a non-zero form, then $\dim V$ is even. Let $\mathrm{Isom}(V)$ denote the full group of isometries of V, so $\mathrm{GL}(V)$, $\mathrm{Sp}(V)$, $\mathrm{GO}(V)$, respectively. Let $\mathrm{Cl}(V)$ denote the group $\mathrm{SL}(V)$, $\mathrm{Sp}(V)$, $\mathrm{SO}(V)$, respectively. (Recall that by our definition $\mathrm{SO}(V) = \mathrm{GO}(V)^\circ$.) Set $G = \mathrm{Cl}(V)$ for the remainder of this section; so G is a simple algebraic group (see Table 9.2).

18.1 A reduction theorem

We first establish a reduction theorem for the study of the closed connected subgroups of simple classical groups.

Before stating this result, we recall the relationship between homogeneous representations, tensor products of representations and central products of groups, which will be used. For H a group, a kH-module V is said to be *homogeneous* if it is the direct sum of isomorphic simple kH-modules. For representations $\rho_i : H_i \to \mathrm{GL}(V_i)$ of groups H_i, $i = 1, 2$, we write $H_1 \otimes H_2 \leq \mathrm{GL}(V_1 \otimes V_2)$ for the central product generated by the images of the H_i, letting $\rho_i(H_i)$ act trivially on V_{3-i}.

Proposition 18.1 *Let K be a field, H a group and $V = V_1 \oplus \cdots \oplus V_1$ ($r > 1$ summands) a homogeneous finite-dimensional KH-module, with V_1 an absolutely irreducible KH-module. Then there is a tensor product decomposition $V = V_1 \otimes V_2$ with $\dim V_2 = r$, such that $H \leq \mathrm{GL}(V_1) \otimes 1$. Moreover,*

$$C_{\mathrm{GL}(V)}(H) = 1 \otimes \mathrm{GL}(V_2), \quad \text{and} \quad C_{\mathrm{GL}(V)}(C_{\mathrm{GL}(V)}(H)) = \mathrm{GL}(V_1) \otimes 1.$$

Proof   As V_1 is absolutely irreducible, $V_2 := \mathrm{Hom}_{KH}(V_1, V)$ is an r-dimensional K-vector space. Then the K-linear map

$$\psi : V_1 \otimes V_2 \to V, \qquad v \otimes \varphi \mapsto \varphi(v),$$

defines an isomorphism, since indeed both sides have the same dimension, and surjectivity is easily verified. Endowing V_2 with the trivial H-action, ψ even becomes an isomorphism of KH-modules, as

$$\psi(h(v \otimes \varphi)) = \psi(hv \otimes \varphi) = \varphi(hv) = h\varphi(v) = h\psi(v \otimes \varphi)$$

for all $h \in H$, $v \in V_1$, $\varphi \in V_2$.

Now clearly $1 \otimes \mathrm{GL}(V_2) \le C_{\mathrm{GL}(V)}(H)$. Conversely, if $g \in C_{\mathrm{GL}(V)}(H)$ then g acts on the embeddings of V_1 into V, so on V_2 via $g : V_2 \to V_2$, $(g\varphi)(v) = g\varphi(v)$. This action is faithful, since an element g in the kernel satisfies $g\varphi(v) = \varphi(v)$ for all $\varphi \in V_2$, $v \in V_1$. Thus $C_{\mathrm{GL}(V)}(H)$ embeds into $1 \otimes \mathrm{GL}(V_2)$. The last statement follows as before by replacing H acting on V_1 by $\mathrm{GL}(V_2)$ acting on V_2.   □

Corollary 18.2   *Let K be a field, $H = H_1 \cdots H_r$ a commuting product of normal subgroups $H_i \trianglelefteq H$, V a finite-dimensional absolutely irreducible KH-module.*

(a) *There exist absolutely irreducible KH_i-modules V_i such that $V = V_1 \otimes \cdots \otimes V_r$, with H_i acting trivially on V_j for $j \ne i$.*

(b) *If V is a self-dual KH-module, then V_i is self-dual for H_i, for $1 \le i \le r$.*

Proof   Clearly it suffices to consider the case when $r = 2$, so $H = H_1 H_2$. Then the socle of $V|_{H_1}$ is an H-invariant submodule of V, so all of V by irreducibility, whence $V|_{H_1}$ is completely reducible. Moreover, H_2 must permute the irreducible summands transitively, and since $[H_1, H_2] = 1$, they must all be H_1-isomorphic, that is, $V|_{H_1}$ is homogeneous. Then Proposition 18.1 applies to show that $V = V_1 \otimes V_2$ and $H_2 \le C_{\mathrm{GL}(V)}(H_1 \otimes 1) = 1 \otimes \mathrm{GL}(V_2)$.

For (b) let's write $V|_{H_1} = \bigoplus_{i=1}^{t} V_1$ for some absolutely irreducible KH_1-module V_1. Since V is self-dual, we also have $V|_{H_1} = \bigoplus_{i=1}^{t} V_1^*$, hence $V_1 \cong V_1^*$ as KH_1-modules, as claimed.   □

As a final preparation we state a result about forms on tensor products ([50, Prop. 2.2]) which gives a kind of converse to Corollary 18.2(b), see Exercise 20.24:

Proposition 18.3   *Let f_i be non-degenerate bilinear forms on the finite-dimensional k-vector spaces V_i, $i = 1, 2$, and set $V := V_1 \otimes V_2$. Then:*

(a) *There is a unique bilinear form f on V such that*

$$f(v_1 \otimes v_2, w_1 \otimes w_2) = f_1(v_1, w_1)f_2(v_2, w_2) \qquad \text{for all } v_i, w_i \in V_i.$$

(b) *The form f is symmetric if and only if f_1, f_2 are either both symmetric or both skew-symmetric.*

(c) *If* $\mathrm{char}(k) = 2$ *then there is a unique quadratic form Q on V, with associated bilinear form f, such that $Q(v_1 \otimes v_2) = 0$ for all $v_i \in V_i$, and Q is preserved by* $\mathrm{Sp}(V_1) \otimes \mathrm{Sp}(V_2)$.

The first weak reduction result for closed connected subgroups of classical groups is now the following:

Proposition 18.4 *Let H be a closed connected subgroup of $G = \mathrm{Cl}(V)$. Then one of the following holds:*

(1) $H \leq \mathrm{Stab}_G(X)$ *with $X \leq V$ a proper non-zero subspace which is either totally singular or non-degenerate, or $p = 2$ and X is non-singular of dimension 1 in an orthogonal space;*

(2) $V = V_1 \otimes V_2$ *and H lies in a subgroup of the form $\mathrm{Cl}(V_1).\mathrm{Cl}(V_2)$ acting naturally on $V_1 \otimes V_2$. Here V is equipped with the product form and* $\dim V_i \geq 2$ *for $i = 1, 2$; or*

(3) H *is a simple algebraic group acting irreducibly on V and $V|_H$ is tensor indecomposable.*

Proof If $V|_H$ is reducible then choose $0 \neq X \leq V$ minimal with respect to being H-invariant. Then $X \cap X^\perp$ is an H-invariant subspace, so the minimality of X implies that either $X \cap X^\perp = 0$ and X is non-degenerate, or $X \cap X^\perp = X$ and X is totally isotropic. Moreover in the latter case, if V is an orthogonal space and $\mathrm{char}(k) = 2$, then the set of singular vectors in X is again H-invariant and by minimality either X is totally singular or X is a one-dimensional space of non-singular vectors, as in (1).

So, we may and will assume now that $V|_H$ is irreducible. Then H is reductive (see Proposition 15.1). Thus we have $H = [H, H]Z(H)^\circ$ by Corollary 8.22. Now, H acts irreducibly, so by Schur's Lemma $Z(H)$ acts by scalars on V. Thus, $Z(H) \leq Z(\mathrm{Cl}(V))$ which is finite, whence H is semisimple and $H = [H, H]$.

By Theorem 8.21(d), $H = H_1 \cdots H_t$ is a central product of simple algebraic groups H_i, and so $V|_H = V_1 \otimes \cdots \otimes V_t$ with V_i an irreducible representation of H_i, by Corollary 18.2(a). Moreover, if V is self-dual, then so are the kH_i-modules V_i by Corollary 18.2(b), and thus $H_i \leq \mathrm{Cl}(V_i)$ for the corresponding classical group. Since V_i is a faithful module for the simple group

H_i, $\dim V_i \geq 2$. Hence if $t > 1$, H stabilizes a tensor product decomposition on V as in the statement of (2).

We now assume that $t = 1$, that is, H is simple. Again, if $V|_H$ is a tensor product of irreducible kH-modules, we have H as in (2); otherwise we have (3).   □

Note that the above result gives no information about finite subgroups of G.

In order to formulate a more powerful reduction theorem for arbitrary (not necessarily connected) closed subgroups H of a classical group $G = \mathrm{Cl}(V)$, we first need to define five classes $\mathcal{C}_1, \ldots, \mathcal{C}_5$ of natural subgroups:

Class $\mathcal{C}_1$ (Subspace stabilizers). Here $H \in \mathcal{C}_1$ if $H = \mathrm{Stab}_G(X)$ where X is a proper non-zero subspace of V, with X totally singular or non-degenerate or non-singular of dimension 1 if $G = \mathrm{SO}(V)$ with $p = 2$.

Class $\mathcal{C}_2$ (Stabilizers of orthogonal decompositions). Here $H \in \mathcal{C}_2$ if H is the stabilizer of an orthogonal decomposition $V = \bigoplus_{i=1}^{t} V_i$ with V_i mutually orthogonal and isometric subspaces of V. Thus $H = \mathrm{Stab}_G(V_1 \perp \ldots \perp V_t) \cong (\mathrm{Isom}(V_1) \wr \mathfrak{S}_t) \cap G$, where we write $\mathfrak{S}_t$ for the symmetric group on t letters.

Class $\mathcal{C}_3$ (Stabilizers of totally singular decompositions). Here $G = \mathrm{Sp}(V)$ or $\mathrm{SO}(V)$ and $H \in \mathcal{C}_3$ if H is the stabilizer of a direct sum decomposition $V = X \oplus X'$ with X, X' maximal totally isotropic (singular if V is equipped with a quadratic form) subspaces of V. In particular, $\dim V = 2m$ is even and $H \cong \mathrm{GL}_m.2$ or GL_m, the latter happens when m is odd and $G = \mathrm{SO}(V)$. (The element of order 2 exchanges the two spaces X and X'; the condition of lying in the connected component of $\mathrm{Isom}(V)$ shows that this element does not lie in G when m is odd and $G = \mathrm{SO}(V)$.)

Class $\mathcal{C}_4$ (Stabilizers of tensor product decompositions). Here, there are two cases:

(a) $V = V_1 \otimes V_2$ with V_1 not isometric to V_2 and $\dim(V_i) > 1$. The central product $\mathrm{Cl}(V_1).\mathrm{Cl}(V_2)$ acts naturally on V and $H = N_G(\mathrm{Cl}(V_1).\mathrm{Cl}(V_2))$.

(b) $V = \bigotimes_{i=1}^{t} V_i$ with $t > 1$ where V_i are mutually isometric with $\dim(V_i) \geq 2$ and

$$H = N_G\left(\prod_{i=1}^{t} \mathrm{Cl}(V_i)\right) = \left(\left(\prod_{i=1}^{t} \mathrm{Cl}(V_i)\right).\mathfrak{S}_t\right) \cap G,$$

where $\prod_i \mathrm{Cl}(V_i)$ acts naturally on $\otimes_i V_i$ and the $\mathrm{Cl}(V_i)$ are simple.

Class $\mathcal{C}_5$ (Finite local subgroups). Recall that a subgroup of a finite group is called *local* if it normalizes a non-trivial r-subgroup for some prime r. Let r be a prime with $r \neq \text{char}(k)$. Let R be an extra-special r-group of order r^{2m+1} (and we will write $R = r^{2m+1}$), or when $r = 2$ we may take R to be 2^{2m+1} or a central product of such a group with a cyclic group of order 4, i.e., $Z_4 \circ 2^{2m+1}$.

Every faithful irreducible k-representation of R has dimension r^m; moreover, these representations are uniquely determined by the action of $Z(R)$. (See [28, 5.4, 5.5].) Take $\rho : R \longrightarrow \text{GL}_{r^m}(k)$ to be such a representation and V the corresponding kR-module. Now $Z(R)$ is cyclic of order r or of order 4, if $R = r^{2m+1}$, respectively $R = Z_4 \circ 2^{2m+1}$; a generator of $Z(R)$ acts as a scalar ω on V and as ω^{-1} on the dual V^*. So V can be self-dual only if $Z(R)$ has order 2. In particular, if r is odd or if $r = 2$ and $R = Z_4 \circ 2^{2m+1}$, then R fixes no non-degenerate form on V, while in fact if $R = 2^{2m+1}$, then $\rho(R) \leq \text{SO}_{r^m}(k)$ or $\rho(R) \leq \text{Sp}_{r^m}(k)$. We consider the representation $\rho : R \longrightarrow \text{Cl}(V)$. Then $H \in \mathcal{C}_5$ if $H = N_{\text{Cl}(V)}(\rho(R))$. Here the structure of $H/Z(G)$ is described in Table 27.2 in Part III.

Definition 18.5 For a classical group $G = \text{Cl}(V)$, write $\mathcal{C}(G) := \mathcal{C}_1 \cup \mathcal{C}_2 \cup \mathcal{C}_3 \cup \mathcal{C}_4 \cup \mathcal{C}_5$. For $H \leq G$ a closed subgroup, write H^∞ for H° if H is infinite and for the last term of the derived series of H if H is finite.

Recall that an abstract group H is said to be *almost simple* if there is a simple group S such that $S \leq H \leq \text{Aut}(S)$. A finite group $H \neq 1$ is called *quasi-simple*, if H is perfect and $H/Z(H)$ is simple. The subnormal quasi-simple subgroups of a finite group H are called the *components* of H (see [2, §31], for example).

We can now state the reduction theorem ([50, Thm. 1]):

Theorem 18.6 (Liebeck–Seitz) *Let $G = \text{Cl}(V)$ be a classical algebraic group and H a closed subgroup of G. Then one of the following holds:*

(1) *there exists $M \in \mathcal{C}(G)$ such that $H \leq M$, or*
(2) *$H.Z(G)/Z(G)$ is almost simple and H^∞ acts irreducibly on V. Moreover, if H is infinite, H° acts tensor indecomposably on V.*

Sketch of proof Let $H \leq G$ be a closed subgroup not satisfying the conclusion. We consider the action of H on the natural kG-module V. If $V|_H$ is reducible, then we get $H \leq M \in \mathcal{C}_1$ by the first paragraph in the proof of Proposition 18.4, so $V|_H$ is irreducible.

Step 1: $V|_{H^\circ}$ is homogeneous.
Since H stabilizes the socle of $V|_{H^\circ}$, this socle is V. Suppose that $V|_{H^\circ}$ is

not homogeneous, that is $V|_{H^\circ} \cong V_1 \oplus \cdots \oplus V_t$ and $V_j \cong Y_j \oplus \cdots \oplus Y_j$, for some set of pairwise non-isomorphic irreducible kH°-modules $Y_1, \ldots, Y_t$, with $t > 1$. The irreducibility of the kH-module V forces H to permute the $V_1, \ldots, V_t$ transitively. If $G = \mathrm{SL}(V)$, then $H \leq \mathrm{Stab}_G(V_1 \perp \ldots \perp V_t)$, the latter subgroup lying in $\mathcal{C}_2$, contradicting our assumption on H.

So we have $G = \mathrm{Sp}(V)$ or $\mathrm{SO}(V)$. Consider $V_1 \cong Y_1 \oplus \cdots \oplus Y_1$ as kH°-module, and assume for the moment that Y_1 is not a self-dual kH°-module. The non-degenerate form induces an isomorphism $V \cong V^*$ and, via this isomorphism, we obtain $V/Y_1^\perp \cong Y_1^*$. As $Y_1^* \not\cong Y_1$, we have $V_1 \subseteq Y_1^\perp$, which in turn implies that V_1 is totally isotropic. Moreover, if $G = \mathrm{SO}(V)$ and $p = 2$, then the irreducibility of Y_1 implies that Y_1 is either totally singular or a non-singular 1-space. In the latter case, we have Y_1 the trivial module for H° and hence $Y_1 \cong Y_1^*$, contradicting our assumption. So in all cases we have Y_1, and hence V_1, totally singular, and by the transitivity of H on the V_j, all V_j are totally singular. Finally, we have $V/V_1^\perp \cong V_1^* \not\cong V_1$, so, relabeling if necessary, we get $V = V_1^\perp \oplus V_2$ and $V_1 \oplus V_2$ is a non-degenerate subspace of V. Working inside $(V_1 \oplus V_2)^\perp$ we find (again after relabeling) that $V|_{H^\circ} = (V_1 \oplus V_2) \perp \ldots \perp (V_{t-1} \oplus V_t)$. If $t = 2$, then $V|_{H^\circ} = V_1 \oplus V_2$ is a decomposition into two totally isotropic subspaces, so H lies in a subgroup $M \in \mathcal{C}_3$ and if $t > 2$, H lies in a subgroup $M \in \mathcal{C}_2$, contradicting our assumption on H.

Now we turn to the case where $Y_1 \cong Y_1^*$ and so $V_j \cong V_j^*$ for all j. Then $V/V_1^\perp \cong V_1^* \cong V_1$ and so $V_1 \subseteq (\bigoplus_{j>1} V_j)^\perp$. Then H preserves the decomposition $V_1 \perp \ldots \perp V_t$, with each V_j non-degenerate and so $H \leq M \in \mathcal{C}_2$, again contradicting our assumption on H.

Step 2: $V|_{H^\circ}$ is irreducible or H is finite.

Suppose that $V|_{H^\circ}$ is reducible (and homogeneous by Step 1), that is, $V|_{H^\circ} \cong Y \oplus \cdots \oplus Y$ ($r > 1$ summands), for some simple H°-module Y. If $\dim Y > 1$, we apply Proposition 18.1 to find that $V = V_1 \otimes V_2$ with

$$H^\circ \leq \mathrm{GL}(V_1) \otimes 1 = C_{\mathrm{GL}(V)}(C_{\mathrm{GL}(V)}(H^\circ)).$$

Moreover, H normalizes the product of derived subgroups

$$(\mathrm{GL}(V_1) \otimes 1)'(1 \otimes \mathrm{GL}(V_2))' = \mathrm{SL}(V_1) \otimes \mathrm{SL}(V_2)$$

and hence in the case $G = \mathrm{SL}(V)$, we have $H \leq M \in \mathcal{C}_4$.

When $G = \mathrm{Sp}(V)$ or $\mathrm{SO}(V)$, a long argument (to be found in the proof of [50, Lemma 3.3]) shows that $H \leq M \in \mathcal{C}_4$ as before. So in all cases, we have $H \leq M \in \mathcal{C}(G)$, contradicting our assumption. So $\dim Y = 1$, and $H^\circ \leq Z(G)$ and so H is finite.

Step 3: H is finite.

Assume that H is infinite, so $H° \neq 1$. By Step 2 we have that $V|_{H°}$ is irreducible. As in the proof of Proposition 18.4, we see that $H°$ is reductive, indeed semisimple as $Z(H°)$ must lie in $Z(G)$. Thus $H° = H_1 \cdots H_r$, a commuting product of simple algebraic groups. If $r > 1$, then $V|_{H°}$ is a tensor product of irreducible H_i-modules by Corollary 18.2 and $H \leq M \in \mathcal{C}_4$. So we must have that $r = 1$ and $H° = H_1$ acts irreducibly on V, and indeed tensor indecomposably (else $H \leq M \in \mathcal{C}_4$). So finally we have H as in (2) of the theorem, contradicting our assumption on H.

Step 4: H is finite with no components and $H \not\leq Z(G)$.

By Step 3, H is finite. As H is not contained in a subgroup lying in $\mathcal{C}(G)$ and does not satisfy the conclusion (2) of the result, the above arguments show that any non-trivial normal subgroup of H acts irreducibly on V, and so $H \not\leq Z(G)$. Let $E := E_1 \circ \cdots \circ E_t$ be the product of the components of H, a characteristic subgroup of H. Suppose $E \neq 1$. Arguing as in Step 3 above, replacing $H°$ by E, we see that either H lies in a member of $\mathcal{C}_4$, or $t = 1$ and $E = E_1$ acts irreducibly on V. In the latter case $C_G(E) \leq Z(G)$, so $HZ(G)/Z(G) \leq \mathrm{Aut}(EZ(G)/Z(G))$ and so $HZ(G)/Z(G)$ is almost simple and $H^\infty \geq E$ acts irreducibly on V as in conclusion (2). Thus we have $E = 1$.

Step 5: Final contradiction.

So finally we have that H is finite and has no components. Set $\bar{H} := HZ(G)/Z(G) \neq 1$. Then all minimal normal subgroups of $\bar{H}$ are elementary abelian, that is, direct products of cyclic groups of equal prime order. Fix one such prime r, and let Q be a Sylow r-subgroup of the preimage in H of the product of all minimal normal r-subgroups of $\bar{H}$, a characteristic subgroup of H. Let R denote its subgroup generated by all elements of order r, respectively of order 4 if $r = 2$. Then R is a normal r-subgroup of H.

Then as above, $V|_R$ is irreducible. In particular, $r \neq p = \mathrm{char}(k)$ by Proposition 2.9. In this case the irreducibility of $V|_R$ forces every characteristic abelian subgroup of R to be cyclic. Then R has precisely the structure of the r-groups described in $\mathcal{C}_5$, by [2, 23.9]. Hence $H \leq M \in \mathcal{C}_5$, a final contradiction. □

18.2 Maximal subgroups of the classical algebraic groups

We turn now to the determination of the maximal positive-dimensional subgroups of the classical algebraic groups; let $G = \mathrm{Cl}(V)$ be as defined in the previous section. We must discuss the maximality of the subgroups in the various families in Theorem 18.6(1), excluding however the finite groups in the family $\mathcal{C}_5$, as well as of those subgroups satisfying condition (2) of loc. cit. It will be convenient to define a subclass of the family of subgroups satisfying Theorem 18.6(2):

Class $\mathcal{C}_6$ (Normalizers of classical groups). Here $H \in \mathcal{C}_6$ if $H = N_G(\mathrm{Sp}(V))$ or $H = N_G(\mathrm{SO}(V))$ in $G = \mathrm{SL}(V)$, and $H = N_G(\mathrm{SO}(V))$ in $G = \mathrm{Sp}(V)$ if $p = 2$.

We let $\mathcal{S}^k$ denote the remaining positive-dimensional subgroups, that is:

Class $\mathcal{S}^k$. Here $H \in \mathcal{S}^k$ if $HZ(G)/Z(G)$ is almost simple, H° acts irreducibly and tensor-indecomposably on V, and if H is of classical type, V is not the natural module for H.

Let $H_1, H_2 \in \mathcal{C}_1 \cup \mathcal{C}_2 \cup \mathcal{C}_3 \cup \mathcal{C}_4 \cup \mathcal{C}_6 \cup \mathcal{S}^k$ and suppose $H_1 < H_2$. It turns out that the most elusive configurations are those where $H_2 \in \mathcal{S}^k$ and $H_1 \in \mathcal{C}_4 \cup \mathcal{S}^k$. In that case H_1° is semisimple and acts irreducibly on V, and if H_2 is of classical type, V is not the natural module for H_2. Moreover if $H_1^\circ = H_2^\circ$, then $H_1^\circ \leq H_1 \leq N_G(H_2^\circ)$ and so H_1 corresponds to some subgroup of the finite (by Theorem 11.11) group $N_G(H_2^\circ)/H_2^\circ$. So finally we may assume $H_1^\circ < H_2^\circ$.

Thus in order to complete the classification of the maximal positive-dimensional closed subgroups of the classical algebraic groups, we must determine all triples (X, Y, V) satisfying the following conditions:

  (i) Y is a simple algebraic group,
  (ii) X is a semisimple closed proper subgroup of Y,
  (iii) V is a non-trivial tensor indecomposable kY-module,
  (iv) X acts irreducibly on V, and
  (v)   if Y is a classical group then V is not the natural module for Y.

If V is a non-restricted kY-module, then by Theorem 16.12, $V \cong M^{(p^i)}$ for some restricted irreducible kY-module M. As X acts irreducibly on V, it must also act irreducibly on M, and hence we may assume that

  (iii)* V is a non-trivial tensor indecomposable restricted kY-module.

Also, as our aim is to obtain a complete list of triples (X, Y, V), we will assume further that

  (i)* Y is a simple algebraic group of simply connected type.

The condition (i)* implies that the character group of a maximal torus of Y coincides with the lattice of abstract weights Ω, and thus in particular, by Theorem 15.17(b), there exist irreducible highest weight modules V for any dominant weight from Ω. Finally, the condition (v) corresponds to the fact that we are interested in the maximality of $N_G(X)$ in the smallest classical group containing it.

The main results of Dynkin [23, 24] (which cover the case when $\mathrm{char}(k) = 0$) and Seitz [60] and Testerman [80] (which cover the case when $\mathrm{char}(k) = p > 0$) are summarized in the following theorem. For the purposes of this theorem, we will take $p = \infty$ when $\mathrm{char}(k) = 0$. Before stating the result, let us introduce some additional notation. Let $\{\omega_i \mid 1 \leq i \leq \mathrm{rk}(X)\}$, respectively $\{\lambda_j \mid 1 \leq j \leq \mathrm{rk}(Y)\}$, be a set of fundamental dominant weights with respect to a fixed choice of maximal torus and Borel subgroup of X, respectively Y, where we label Dynkin diagrams as in Table 9.1.

Theorem 18.7 (Dynkin, Seitz, Testerman) *Let Y and V satisfy conditions (i)*, (iii)* and (v), and let $\rho : Y \rightarrow \mathrm{SL}(V)$ be the corresponding rational representation of Y. Let X be a subgroup of Y satisfying (ii). Then $V|_X$ is irreducible if and only if one of the following occurs:*

(1) *the triple $(\rho(X), \rho(Y), V)$ (or $(\rho(X), \rho(Y), V^*)$) belongs to one of the infinite series given in Table 18.1; or*

(2) *the triple is among a specified finite list of additional triples, each corresponding to a fixed embedding of simple algebraic groups and a kY-module V of fixed p-restricted highest weight.*

Remarks 18.8 (a) Recall that V is a restricted kY-module and hence all of the parameters a, b, a_i are non-negative integers, strictly less than p. Moreover, when there is a congruence relation modulo p, it is understood that $\mathrm{char}(k) > 0$.

(b) The majority of the examples of Table 18.1 above correspond to natural embeddings of X in the group Y. For example, the first two configurations $C_n < A_{2n-1}$ and $B_n < A_{2n}$ arise from the natural embeddings of a symplectic or orthogonal group in the linear group. The embeddings $A_n < A_{(n^2+n-2)/2}$ and $A_n < A_{(n^2+3n)/2}$ correspond to the action of SL_{n+1} on the exterior square or symmetric square of its natural module.

The embeddings in groups of type D_m all correspond to a usual decomposition of an orthogonal space into a direct sum of non-degenerate subspaces, each corresponding to the natural module of one of the factors of the subgroup X, with one exception: the embedding $B_n A_1 < D_{n+3}$ arises from a

Table 18.1 *Irreducible triples*

$X < Y$	$V\vert_X$	$V\vert_Y$	conditions
$C_n < A_{2n-1}$, $n \geq 2$	$a\omega_1$	$a\lambda_1$	$a \geq 2$
$C_n < A_{2n-1}$, $n \geq 2$	$a\omega_j + b\omega_{j+1}$, $j < n$	$a\lambda_j + b\lambda_{j+1}$	$a + b = p - 1 > 1$ $a \neq 0$ if $j = n - 1$
$B_n < A_{2n}$, $n \geq 3$	ω_j, $2 \leq j < n$	λ_j	$p \neq 2$
$B_n < A_{2n}$, $n \geq 2$	$2\omega_n$	λ_n	$p \neq 2$
$D_n < A_{2n-1}$, $n \geq 4$	ω_j, $2 \leq j < n - 1$	λ_j	$p \neq 2$
$D_n < A_{2n-1}$, $n \geq 4$	$\omega_{n-1} + \omega_n$	λ_{n-1}	$p \neq 2$
$A_n < A_{(n^2+n-2)/2}$, $n \geq 3$	$\omega_1 + \omega_3$	λ_2	$p \neq 2$
$A_n < A_{(n^2+3n)/2}$, $n \geq 2$	$2\omega_1 + \omega_2$	λ_2	$p \neq 2$
$B_n < D_{n+1}$, $n \geq 3$	$a\omega_n$	$a\lambda_n, a\lambda_{n+1}$	$a > 0$
$B_n < D_{n+1}$, $n \geq 3$	$a\omega_i + b\omega_n$, $1 \leq i < n$	$a\lambda_i + b\lambda_j$, $j = n, n+1$	$a + b + n \equiv i$ (p) $a \neq 0 \neq b$
$B_{n-j}B_j < D_{n+1}$, $n \geq 2$	$\omega_{n-j} + \omega_n$	λ_n, λ_{n+1}	
$X \to' B_{n-j}B_j < D_{n+1}$, $n \geq 2$	$\omega_{n-j} + \omega_n$	λ_n, λ_{n+1}	$p \geq 5$ if some $\pi_i(X) = A_1$
$B_n A_1 < D_{n+3}$, $n \geq 2$	$\omega_n + 3\omega_{n+1}$	$\lambda_{n+2}, \lambda_{n+3}$	$p \geq 5$
$X \to' B_{n_1} \cdots B_{n_j} < D_m$, $m = n_1 + \ldots + n_j + 1$	$\sum_{i=1}^{j} \omega_{n_1 + \ldots + n_i}$	λ_{m-1}, λ_m	$p = 2$
$D_n < C_n$	$\sum_{i=1}^{n-1} a_i\omega_i + a_{n-1}\omega_n$	$\sum_{i=1}^{n-1} a_i\lambda_i$	$p = 2$, $\sum_{i=2}^{n-1} a_i \neq 0$
$X \to' B_{n_1} \cdots B_{n_j} < B_m$, $m = n_1 + \ldots + n_j$	$\sum_{i=1}^{j} \omega_{n_1 + \ldots + n_i}$	λ_m	$p = 2$

decomposition of the $2(n+3)$-dimensional space into a direct sum of the natural module for the B_n and the irreducible module with p-restricted highest weight $4\omega_{n+1}$ (the five-dimensional irreducible p-restricted module for A_1).

Finally, the last two embeddings are maximal rank configurations corresponding to maximal rank subsystem subgroups generated by short roots. See Theorems 13.12 and 13.14, as well as Proposition 13.15.

(c) The symbol $X \to' M$ occurs when M is a central product and the notation means that either X projects surjectively to each of the simple factors of M or some factor is of type B_2 and the projection is an A_1 acting irreducibly on the spin module for the B_2. Moreover, in order to ensure that $V\vert_X$ is irreducible, it may be necessary for the projections to involve field twists.

(d) The specified list of additional triples comprises approximately 45 examples, depending on how one counts the examples arising from graph automorphisms of Y.

(e) In [80], there were three question marks in the final result, due to the absence of an existence proof for particular embeddings in the exceptional groups F_4 and E_6. The existence was later established in [81] and hence there are no remaining question marks in the above result.

(f) The above result in case Y is of type A_m and char$(k) > 0$ was established independently by Suprunenko in [77].

(g) Note that there are essentially three types of examples:

(I) Positive characteristic analogues of characteristic 0 configurations. For example, in characteristic 0, the naturally embedded $X = \mathrm{Sp}_{2n}$ in $Y = \mathrm{SL}_{2n} = \mathrm{SL}(W)$ acts irreducibly on the symmetric powers $S^a(W)$ for all a; in characteristic $p > 0$, X acts irreducibly on $S^a(W)$ for all $1 < a < p$. This is the first family of examples in the table.

(II) For a fixed embedding $X < Y$, families which exist only in positive characteristic, for each positive characteristic p. For example, taking X and Y as in (I), the second family of examples in the table has no characteristic 0 analogue.

(III) Examples which occur for a fixed prime characteristic. For example, the adjoint representation of the group SL_3 induces a seven-dimensional irreducible representation when $p = 3$. (See Theorem 15.20(1).) Moreover, SL_3 preserves a non-degenerate orthogonal form on the corresponding module (see [11, Table 2]). Hence, in characteristic 3, we have an embedding $X := \mathrm{PGL}_3 < Y := \mathrm{SO}_7 = \mathrm{SO}(W)$. It turns out that the symmetric square $S^2(W)$ is irreducible for Y when char$(k) = 0$, but has a 21-dimensional quotient when char$(k) = 3$. One can show that the restriction of this quotient module to the subgroup X is an irreducible kX-module, hence giving an example. This particular example, which is unique with respect to one further condition (see Proposition 18.9 below) is among the isolated examples mentioned in the statement of the theorem.

The relatively short length of this list of triples justifies the following statement: If $\rho : H \to \mathrm{SL}(V)$ is a rational irreducible, tensor indecomposable representation of a simple algebraic group H, then, most of the time, $\rho(H)$ is maximal among closed connected subgroups of $G = \mathrm{Cl}(V)$, the smallest classical group containing $\rho(H)$. In particular, the following result is deduced directly from the classification given by Theorem 18.7 (see [60, Cor.4]).

Corollary 18.9 *Let H, ρ and G be as above. If Y is a closed connected subgroup of G with $\rho(H) < Y < G$, then with just two exceptions, Y is maximal among closed connected subgroups of G. The exceptions are as follows:*

(1) *when $p = 3$, there is an embedding of irreducible subgroups $A_2 < G_2 < B_3 < \mathrm{SO}_{27}$, and*

(2) *when $p = 2$, there is an embedding of irreducible subgroups $D_4 < C_4 < F_4 < \mathrm{Cl}(V)$, where $\dim V = 26$.*

Remark 18.10 Note that the exception of (2) above is not listed in [60, Cor. 4]. Moreover, the precise statement concerning the type of the classical group $\mathrm{Cl}(V)$ is not at this time a published result. The three groups do fix a symplectic form on V; see Remark 16.2.

We conclude this section with a discussion of how the results of this chapter are applied to determine the triples (X, Y, V). As a first step we classify the triples where X is the group SL_2 or PGL_2. We will apply the following lemma.

Lemma 18.11 *Let $H = \mathrm{SL}_2$, with maximal torus T, and let V be a rational irreducible kH-module. Then the T-weight spaces of V are all one-dimensional.*

Proof Let α be a generator of the root lattice of H with respect to T and $\lambda \in X(T)$ the corresponding fundamental dominant weight. Let $\mu \in X(T)$ be the highest weight of V. With the p-adic expansion $\mu = a_1 q_1 \lambda + \cdots + a_t q_t \lambda$, for some $1 \le a_i < p$ and for distinct p-powers $q_1, \cdots, q_t$, we have

$$V = L(a_1 \lambda)^{(q_1)} \otimes \cdots \otimes L(a_t \lambda)^{(q_t)}$$

by Steinberg's tensor product theorem (Theorem 16.12). Moreover, Exercise 20.18(d) shows that the weights of the $(a_i + 1)$-dimensional module $L(a_i \lambda)$ are $a_i \lambda, a_i \lambda - \alpha, \ldots, a_i \lambda - a_i \alpha$, and hence the weights of $L(a_i \lambda)^{(q_i)}$ are

$$a_i q_i \lambda, q_i(a_i \lambda - \alpha), \ldots, q_i(a_i \lambda - a_i \alpha),$$

and each is of multiplicity 1.

Now a basis of V consisting of weight vectors is obtained by tensoring weight vectors of the $L(a_i \lambda)^{(q_i)}$. The weight vectors thus obtained have weights of the form

$$\sum_{i=1}^{t} q_i(a_i \lambda - m_i \alpha) = \mu - \sum_{i=1}^{t} m_i q_i \alpha,$$

for some $0 \le m_i \le a_i$. Two such weights are equal only if the coefficients of $q_1, \ldots, q_t$ are the same. Thus, there is a one-dimensional space of vectors of weight $\mu - (m_1 q_1 + \cdots + m_t q_t)\alpha$, for each choice of m_i with $0 \le m_i \le a_i$, which establishes the result. $\square$

For the remainder of the section we assume the triple (X, Y, V) satisfies the conditions (i)*, (ii), (iii)*, (iv) and (v) above. Fix a maximal torus T_X of X and a maximal torus T_Y of Y with $T_X \leq T_Y$. Thus, when $X = \mathrm{SL}_2$ or PGL_2, Lemma 18.11 shows that V must have one-dimensional weight spaces with respect to T_Y as well. This leads to a short list of possibilities for λ, the highest weight of V; for example, if Y has type B_n, $\lambda = \lambda_1$ or λ_n, or if Y has type E_6, then $\lambda = \lambda_i$ for $i = 1$ or 6. At this point, case-by-case considerations lead to two isolated examples, one in G_2 and another in D_5, as well as the configurations in Table 18.1 where X is diagonally embedded in a central product of a certain number of copies of A_1, when $p = 2$. See [60, Thm. 7.1] for the proof.

Now one is in a position to proceed by induction on $\mathrm{rk}(X)$. In order to do so, we will want to consider the action of a Levi subgroup of X on the irreducible module V. Choose a parabolic subgroup P_X of X; by Corollary 17.12, there exists a parabolic subgroup P_Y of Y, with $P_X \leq P_Y$ and $Q_X := R_u(P_X) \leq Q_Y := R_u(P_Y)$. Let L_X, respectively L_Y, be Levi subgroups of P_X, respectively P_Y. The following result plays a crucial role in applying the induction hypothesis:

Proposition 18.12   *With notation as above, we have that the fixed point spaces V^{Q_X} and V^{Q_Y} are equal.*

Proof   As $L_X \leq P_Y$, L_X stabilizes $V^{Q_Y} \leq V^{Q_X}$. But by Proposition 16.3, we know that V^{Q_X} is an irreducible kL_X-module, and since $V^{Q_Y} \neq 0$ by Proposition 2.9, we get the result.   □

With P_X and P_Y as above, write $[L_Y, L_Y] = L_1 \cdots L_t$, a central product of simple algebraic groups and let π_i be the projection of $[L_Y, L_Y]$ onto L_i. Then by Corollary 18.2 the irreducible L_Y-module V^{Q_Y} (which is equal to the irreducible L_X-module V^{Q_X}) is a tensor product $M_1 \otimes \cdots \otimes M_t$, where M_i is an irreducible kL_i-module. By Proposition 16.3, M_i has a restricted highest weight. Moreover, the closed subgroup $\pi_i([L_X, L_X])$ must act irreducibly on M_i for all i. Choose P_X such that $[L_X, L_X]$ is simple. Then for each i such that M_i is non-trivial and $\pi_i([L_X, L_X]) < L_i$, we obtain a triple $(\pi([L_X, L_X]), L_i, M_i)$ which satisfies the hypotheses (i), (ii), (iii)*. In fact, Proposition 12.14 shows that the condition (i)* is satisfied, since Y, and so $[L_Y, L_Y]$, is simply connected. As $\mathrm{rk}([L_X, L_X]) < \mathrm{rk}(X)$, our inductive hypothesis applies and we will have a list of these smaller rank configurations. If we choose P_X to be a maximal parabolic subgroup of X, the induction will determine the highest weight of $V|_X$, in terms of the highest weight of the kY-module V, up to one missing coefficient of the fundamental dominant weight corresponding to the unique simple root of the root system of X which is not in the root system of L_X.

In case Y is a classical type algebraic group, one also considers the action of X on W, the natural module for Y. Here one shows that either $W|_X$ is again an irreducible kX-module or the highest weight of the kY-module V is explicitly known (see [60, 5.1]). In the latter case, it is fairly straightforward to then determine the groups X which can act irreducibly on the known module V.

Various further techniques come into play to determine explicitly the highest weight of the kY-module V. Dynkin's methods differ significantly from those used in [60] and [80]. In characteristic 0, representations of reductive groups are completely reducible and the dimensions of irreducible modules and of the weight spaces in these modules can be calculated (see the remarks at the end of Section 15.2), while in characteristic $p > 0$, representations are not completely reducible, the extension theory of the simple modules is not completely understood and the dimensions of the irreducible modules can at best be bounded above and below.

At this point, we have determined which of the members of the class $\mathcal{S}^k$ are in fact maximal in the smallest classical group $\mathrm{Cl}(V)$ containing them. For if $H \in \mathcal{S}^k$ is not maximal, then $H < Y$ for some $Y \in \mathcal{C}_6$ or $Y \in \mathcal{S}^k$. The first possibility is ruled out by the minimality of $\mathrm{Cl}(V)$ and the second possibility occurs on the list of triples. To decide the type of the minimal classical group containing H, if $p \neq 2$, we can apply Proposition 16.1 and [73, Lemma 79].

What remains now is comparatively easy; we must decide which of the subgroups H in the families $\mathcal{C}_1, \ldots, \mathcal{C}_4, \mathcal{C}_6$ are indeed maximal. This is straightforward when one applies the structure theory of G and the known action of H on V. It turns out that the majority of the subgroups in $\mathcal{C}_1 \cup \mathcal{C}_2 \cup \mathcal{C}_3 \cup \mathcal{C}_4 \cup \mathcal{C}_6$ are indeed maximal. We give the result in the cases $\mathrm{SL}(V)$ and $\mathrm{Sp}(V)$, the orthogonal case being slightly more technical; we refer the reader to [60, Thm. 3] for a discussion of the connected case.

Proposition 18.13   *For* $\mathrm{Cl}(V) \in \{\mathrm{SL}(V), \mathrm{Sp}(V)\}$, *the positive-dimensional members in* $\mathcal{C}_1 \cup \ldots \cup \mathcal{C}_6$ *which are maximal inside* $\mathrm{Cl}(V)$ *are as given in Tables 18.2 and 18.3.*

Here, P_m denotes the parabolic subgroup corresponding to the set $I = S \setminus \{s_m\}$, where the simple reflections are numbered as in the Dynkin diagrams in Table 9.1. The subgroup H of $G = \mathrm{Cl}(V)$ is obtained as the intersection with G of the normalizer in $\mathrm{GL}(V)$ of the group in the second column.

Table 18.2 *Maximal subgroups in* $\mathcal{C}_1 \cup \ldots \cup \mathcal{C}_4 \cup \mathcal{C}_6$ *in* SL_n

class of H	structure	conditions	$\mathrm{rk_{ss}}(H)$
$\mathcal{C}_1$	P_m	$1 \leq m \leq n-1$	$n-2$
$\mathcal{C}_2$	$\mathrm{GL}_m \wr \mathfrak{S}_t$	$n = mt, t \geq 2$	$t(m-1)$
$\mathcal{C}_4$	$\mathrm{GL}_{n_1} \otimes \mathrm{GL}_{n_2}$	$n = n_1 n_2,\ 2 \leq n_1 < n_2$	$n_1 + n_2 - 2$
	$(\otimes_{i=1}^{t} \mathrm{GL}_m).\mathfrak{S}_t$	$n = m^t, m \geq 3, t \geq 2$	$t(m-1)$
$\mathcal{C}_6$	Sp_n	n even	$n/2$
	SO_n	$p \neq 2$	$\lfloor n/2 \rfloor$

Table 18.3 *Maximal subgroups in* $\mathcal{C}_1 \cup \ldots \cup \mathcal{C}_4 \cup \mathcal{C}_6$ *in* Sp_{2n}, $n \geq 2$

class of H	structure	conditions	$\mathrm{rk_{ss}}(H)$
$\mathcal{C}_1$	P_m	$1 \leq m \leq n$	$n-1$
	$\mathrm{Sp}_{2m} \times \mathrm{Sp}_{2(n-m)}$	$1 \leq m < n/2$	n
$\mathcal{C}_2$	$\mathrm{Sp}_{2m} \wr \mathfrak{S}_t$	$n = mt, t \geq 2$	n
$\mathcal{C}_3$	$\mathrm{GL}_n.2$	$p \neq 2$	$n-1$
$\mathcal{C}_4$	$\mathrm{Sp}_{2n_1} \otimes \mathrm{SO}_{n_2}$	$p \neq 2,\ n = n_1 n_2,\ n_2 \geq 3$	$n_1 + \lfloor n_2/2 \rfloor$
	$(\otimes_{i=1}^{t} \mathrm{Sp}_{2m}).\mathfrak{S}_t$	$2n = (2m)^t,\ t \geq 3,\ pt$ odd	tm
$\mathcal{C}_6$	GO_{2n}	$p = 2$	n

Proof First consider the subgroups $H = \mathrm{Stab}(X)$ of the family $\mathcal{C}_1$. If $X \leq V$ is a totally isotropic subspace, then by Proposition 12.13, H is a parabolic subgroup of G. By Proposition 12.2 any subgroup $K \leq G$ with $H \leq K$ is also a parabolic subgroup and the maximal proper parabolic subgroups are precisely those corresponding to subsets of Δ of the form $\Delta \setminus \{\alpha\}$, for some simple root α. Conversely, again by Proposition 12.13 these subgroups are precisely stabilizers of totally isotropic subspaces. For $G = \mathrm{Sp}_{2n}$, the subgroups $H = \mathrm{Sp}_{2m} \times \mathrm{Sp}_{2(n-m)}$ have maximal semisimple rank, so by Theorem 13.14 and Proposition 13.15, we see that the only possible proper containment of H in a subgroup of $\mathrm{Sp}(V)$ would occur when $p = 2$ and the subgroup is of type D_n. But again by Theorem 13.14, we see that there is no such subgroup of D_n. Thus the listed members of $\mathcal{C}_1$ are certainly maximal. On the other hand, since the members of all other families act irreducibly on the natural module, none of them can be contained in any member of $\mathcal{C}_1$.

Secondly, the members H of $\mathcal{C}_2$ are maximal rank subgroups and as above one determines the possible maximal rank overgroups. Then one argues that none of these contain the full normalizer of H° and so H is maximal. The members of $\mathcal{C}_3$ have a simple component whose rank is larger than the rank of all simple components of any member of C_2. The members K of the remaining classes are such that K° acts irreducibly on V, hence none of these can lie in a member of C_2.

For the members of C_3, one notes again that these are of maximal rank. This time however, there is a maximal rank overgroup, namely the maximal rank D_n subgroup of $\mathrm{Sp}(V)$, by Proposition 13.15. Hence we must exclude $p = 2$ in order to obtain a maximal subgroup. Now since the connected component of a C_3 subgroup acts reducibly on V, no member of the remaining classes can be contained in such a subgroup.

The groups $H \in C_4$ fix a non-degenerate bilinear form of the same type as G, by Proposition 18.3. So they aren't contained in members of class C_6, except in the following cases: when $G = \mathrm{SL}_n$ and $m = 2$, then $\mathrm{SL}_2 = \mathrm{Sp}_2$ fixes a skew-symmetric form, so H lies in a symplectic or orthogonal subgroup of G, by Proposition 18.3; similarly, when $G = \mathrm{Sp}_{2n}$ and pt is even, then H fixes a quadratic form; and when G is symplectic and $p = 2$, then $\mathrm{Sp}_{2n_1} \cdot \mathrm{SO}_{n_2}$ is contained in the group SO_{2n} by Proposition 18.3(c). We must now determine whether H is contained in a member Y of $\mathcal{S}^k$. If this is the case, then the triple (H°, Y, V) is covered by Theorem 18.7. In fact, by inspection one sees that the only candidate is the embedding $B_n A_1 < D_{n+3}$ of Table 18.1. But clearly in that case, the subgroup H° does not act via the tensor product of the natural modules of the two factors on V.

Finally, the groups in C_6 are such that H° acts tensor indecomposably, hence do not lie in any member of C_4.   □

Remarks 18.14   (a) The above considerations, based upon the reduction theorem 18.6, do not treat the odd-dimensional orthogonal groups defined over fields of characteristic 2. Indeed, while these are isomorphic as abstract groups to the symplectic groups acting on a space of dimension one less, Theorem 9.13 and the identification of the different simple algebraic groups with certain classical groups (Table 9.2) shows that the groups SO_{2n+1} and Sp_{2n} are not isomorphic as algebraic groups. (See Exercise 20.25 for further details.) Nevertheless the natural map (homomorphism of abstract groups) $\varphi : \mathrm{SO}_{2n+1} \to \mathrm{Sp}_{2n}$ is indeed a bijective morphism of algebraic groups. Then the fact that φ is continuous, together with Proposition 1.5, shows that it suffices to determine the maximal closed positive-dimensional subgroups of Sp_{2n} in order to determine those of SO_{2n+1}.

(b) The classification of the maximal positive-dimensional closed subgroups of the simple algebraic groups of types A_n, B_n, C_n and D_n follows from the above considerations. Let G_{sc} be a simply connected simple algebraic group with the given root system. Then there exists an isogeny

$\pi : G_{\mathrm{sc}} \to \mathrm{Cl}(V)$, for one of the classical groups $\mathrm{Cl}(V)$. The maximal closed positive-dimensional subgroups of G_{sc} are then inverse images of those of $\mathrm{Cl}(V)$. Moreover, the maximal closed positive-dimensional subgroups of any homomorphic image $\varphi(G_{\mathrm{sc}})$ of G_{sc} are images of the corresponding subgroups of G_{sc}, since by Proposition 1.5 images of closed subgroups under morphisms are closed.

19

Maximal subgroups of exceptional type algebraic groups

We now consider the case of the exceptional type algebraic groups. The classification of the maximal closed connected subgroups was obtained by Dynkin [23] in the case where char$(k) = 0$. The case of positive characteristic is covered by three lengthy articles of Seitz [61] and Liebeck–Seitz [48, 53].

19.1 Statement of the result

We require some further notation before stating the results. In what follows we write AGL_n for the affine general linear group (semi-direct product of GL_n with the group of translations). By abuse of notation, we will write Φ for a semisimple algebraic group with root system of type Φ; if the subgroup is a subsystem subgroup corresponding to the p-closed subset $\Psi \subset \Phi$, we will write Ψ for those subsystem subgroups generated by long root subgroups and $\tilde{\Psi}$ for subsystem subgroups generated by short root subgroups. We also write $W(\Psi)$ for the Weyl group of Ψ. Finally, we will continue to adopt the convention $p = \infty$ when char$(k) = 0$.

The first result classifies the subgroups which are maximal among proper closed connected subgroups.

Theorem 19.1 *Let G be an exceptional algebraic group defined over an algebraically closed field of characteristic p. Let $X < G$ be a closed subgroup. Then X is maximal among proper closed connected subgroups of G if and only if one of the following holds:*

(1) *X is a maximal parabolic subgroup.*

(2) *X is a maximal rank subsystem subgroup as in the table below:*

G	X
G_2	$A_1\tilde{A}_1$, A_2, $\tilde{A}_2$ $(p=3)$
F_4	B_4, A_1C_3 $(p \neq 2)$, C_4 $(p=2)$, $A_2\tilde{A}_2$
E_6	A_1A_5, $A_2A_2A_2$
E_7	A_1D_6, A_7, A_2A_5
E_8	D_8, A_1E_7, A_8, A_2E_6, A_4A_4

(3) *X and G are as in the following tables:*

G	X simple
G_2	A_1 $(p \geq 7)$
F_4	A_1 $(p \geq 13)$, G_2 $(p=7)$
E_6	A_2 $(p \neq 2,3)$, G_2 $(p \neq 7)$, C_4 $(p \neq 2)$, F_4
E_7	A_1 (2 classes, $p \geq 17, 19$ respectively), A_2 $(p \geq 5)$
E_8	A_1 (3 classes, $p \geq 23, 29, 31$ respectively), B_2 $(p \geq 5)$

G	X semisimple, non-simple
F_4	A_1G_2 $(p \neq 2)$
E_6	A_2G_2
E_7	A_1A_1 $(p \neq 2,3)$, A_1G_2 $(p \neq 2)$, A_1F_4, G_2C_3
E_8	A_1A_2 $(p \neq 2,3)$, G_2F_4

The second result classifies the maximal closed positive-dimensional subgroups.

Theorem 19.2 *Let G be an exceptional algebraic group of adjoint type, defined over an algebraically closed field of characteristic p. Let $M < G$ be a closed subgroup. Then M is maximal among positive-dimensional closed subgroups of G if and only if one of the following holds:*

(1) *M is a maximal parabolic subgroup.*

(2) *$M = N_G(X)$ where X is connected reductive of maximal rank and the pair $(X, M/X)$ is as in the table below, where T_i denotes an i-dimensional subtorus of G:*

G	X	M/X
G_2	$A_1\tilde{A}_1$, A_2, $\tilde{A}_2$ $(p=3)$	1, Z_2, Z_2
F_4 $(p \neq 2)$	B_4, D_4, A_1C_3, $A_2\tilde{A}_2$	1, $\mathfrak{S}_3$, 1, Z_2
F_4 $(p = 2)$	B_4, C_4, D_4, $\tilde{D}_4$, $A_2\tilde{A}_2$	1, 1, $\mathfrak{S}_3$, $\mathfrak{S}_3$, Z_2
E_6	A_1A_5, $(A_2)^3$, D_4T_2, T_6	1, $\mathfrak{S}_3$, $\mathfrak{S}_3$, $W(E_6)$
E_7	A_1D_6, A_7, A_2A_5, $(A_1)^3D_4$,	1, Z_2, Z_2, $\mathfrak{S}_3$,
	$(A_1)^7$, E_6T_1, T_7	$PSL_3(2)$, Z_2, $W(E_7)$
E_8	D_8, A_1E_7, A_8, A_2E_6,	1, 1, Z_2, Z_2,
	$(A_4)^2$, $(D_4)^2$,	Z_4, $\mathfrak{S}_3 \times Z_2$,
	$(A_2)^4$, $(A_1)^8$, T_8	$GL_2(3)$, $AGL_3(2)$, $W(E_8)$

(3) $G = E_7$, $p \neq 2$ and $M = (Z_2^2 \times D_4).\mathfrak{S}_3$.

(4) $G = E_8$, $p \neq 2,3,5$ and $M = A_1 \times \mathfrak{S}_5$.

(5) $M = N_G(X)$ for X as in (3) of Theorem 19.1.

(6) $G = E_8$, $p \neq 2$, $M = N_G(X)$, where $X = A_1G_2G_2$, with A_1G_2 maximal closed connected in F_4, and $M/X = Z_2$.

Remark 19.3 When M is as in (5) of the above theorem, the index $|M : X|$ is 1 or 2. In all cases where X has a factor of type A_2, M induces a non-trivial graph automorphism of this factor.

19.2 Indications on the proof

The proofs of the above theorems in positive characteristic were established in two phases: the first phase (in [61] and [48]) proved Theorems 19.1 and 19.2 under some mild assumptions on char(k), while the second phase (in [53]) completed the proof of these theorems. The bulk of the work in the second phase was in showing that no "new" examples occur in small characteristics. We will not go into the second phase here, but rather outline the strategy used in the first phase under the additional assumption that X is connected.

We assume the exceptional group G to be simply connected, since any maximal subgroup of G must contain $Z(G)$ and hence under the natural isogeny of Proposition 9.15, we obtain an image which is a maximal subgroup. (Clearly, any maximal subgroup of an image of G is the image of a maximal subgroup of G.)

We establish a first easy reduction result:

Proposition 19.4 *In the above setting, let X be a proper closed connected* *subgroup of G which is maximal among such subgroups. Then either X is semisimple or X is a maximal proper parabolic subgroup of G.*

Proof If X is not reductive, then by Corollary 17.14, X is a parabolic subgroup. So we may assume X is reductive. If $Z(X)° \neq 1$, it is a torus by Proposition 6.20, and Proposition 12.10 then implies that $C_G(Z(X)°)$ is a proper Levi subgroup and hence X lies in a proper parabolic subgroup of G. The result then follows from the maximality assumption. □

Now by Proposition 12.2 the maximal proper parabolic subgroups of G correspond to maximal proper subsets of a base of the root system of G. Thus we will assume from now on that X is a semisimple group. In addition, we have the following information about $Z(X)$:

Lemma 19.5 *Let X be a semisimple maximal proper closed connected subgroup of G. Then either $Z(X) \leq Z(G)$ or X is a maximal rank subgroup of G.*

Proof As $Z(X) \leq C_X(T) = T$ for some maximal torus T of X, $Z(X)$ consists of semisimple elements. Now suppose $Z(X) \not\leq Z(G)$ and let $s \in Z(X) \setminus Z(G)$. Then by maximality $X = C_G(s)°$, which by Proposition 14.1 is a proper maximal rank connected subgroup of G. □

As maximal rank semisimple subgroups of G are well-understood (see Chapter 13), it is straightforward to determine which of these subgroups are maximal, leading to the cases in Theorem 19.1(2). We shall therefore assume henceforth that X is semisimple of rank strictly less than $\mathrm{rk}(G)$. We can then show the following:

Lemma 19.6 *Consider* $\mathrm{Lie}(X) \leq \mathrm{Lie}(G)$. *Then:*

(a) $X = \mathrm{Stab}_G(\mathrm{Lie}(X))°$, *and*
(b) $C_{\mathrm{Lie}(G)}(X) := \{v \in \mathrm{Lie}(G) \mid \mathrm{Ad}\,(x)v = v \text{ for all } x \in X\} \leq Z(\mathrm{Lie}(G))$.

Proof of (a) In the adjoint action of X on $\mathrm{Lie}(G)$, X stabilizes $\mathrm{Lie}(X)$ and so $X \leq \mathrm{Stab}_G(\mathrm{Lie}(X))°$. By maximality either $X = \mathrm{Stab}_G(\mathrm{Lie}(X))°$ or $\mathrm{Stab}_G(\mathrm{Lie}(X))° = G$. In the latter case, we apply Theorem 15.20 to see that either $\mathrm{Lie}(X) \subseteq Z(\mathrm{Lie}(G))$ or $\mathrm{Lie}(X)$ contains all root subspaces for short roots of G. In the first case, $\dim(\mathrm{Lie}(X)) = 1$, which is impossible as X is semisimple. In the second case, we are reduced to two possible configurations: G is of type G_2, $p = 3$ and $\dim(\mathrm{Lie}(X)) = 7$ (see Example 15.21(1)) or G is of type F_4, $p = 2$ and $\dim(\mathrm{Lie}(X)) = 26$. But there is no semisimple group of dimension 7, respectively 26 and of rank less than 2, respectively 4. Hence $X = \mathrm{Stab}_G(\mathrm{Lie}(X))°$ as required for (a). □

For the proof of (b), one must use the fact that the centralizer in G of a semisimple, or nilpotent, element in $\mathrm{Lie}(G)$ contains a maximal torus of G, or has a non-trivial unipotent radical. See [61, 1.3] for details.

With these lemmas in place, and after fixing some additional notation, we introduce the main tool used in the study of the embedding $X < G$. Let T be a maximal torus of G, T_X a maximal torus of X with $T_X \leq T$ and Φ_X the corresponding root system of X. Let B_X be a Borel subgroup of X containing T_X and let Δ_X be the corresponding base of Φ_X. For $\alpha \in \Phi_X$, let $\alpha^\vee : \mathbf{G}_m \to T_X$ be the coroot as in Lemma 8.19. For $c \in k^\times$, set $\gamma(c) = \prod \alpha^\vee(c)$, the product taken over all roots $\alpha \in \Phi_X^+$. Then $\mathrm{im}(\gamma)$ is a one-dimensional torus in X and we shall first consider the action of this torus on $\mathrm{Lie}(X)$. Let $\mathrm{Lie}(X) = \mathrm{Lie}(T_X) \oplus \bigoplus_{\alpha \in \Phi_X} \mathrm{Lie}(X)_\alpha$ and choose $v_\alpha \in \mathrm{Lie}(X)_\alpha \setminus \{0\}$.

Lemma 19.7   *With the above notations we have:*

(a) $\gamma(c)h = h$ for all $h \in \mathrm{Lie}(T_X)$;
(b) $\gamma(c)v_\beta = c^2 v_\beta$ for all $\beta \in \Delta_X$ and $c \in k^\times$.

Proof   For (a), we simply note that $\mathrm{im}(\gamma) \leq T_X$ and T_X acts trivially on itself via conjugation, hence T_X fixes pointwise its Lie algebra. For (b), let $\beta \in \Delta_X$, $c \in k^\times$. Then we note that $\gamma(c)v_\beta = (\prod \alpha^\vee(c))v_\beta = c^r v_\beta$ where $r = \sum_{\alpha \in \Phi_X^+} \langle \beta, \alpha^\vee \rangle$. Setting $\rho = \frac{1}{2} \sum_{\alpha \in \Phi_X^+} \alpha^\vee$, we have $r = \langle \beta, 2\rho \rangle$. By Exercise A.1 $\{\alpha^\vee \mid \alpha \in \Phi_X^+\}$ forms a positive system in the root system $\Phi_X^\vee$. Now by Exercise 20.26 we know that ρ is the sum of the fundamental dominant weights relative to the abstract root system $\Phi_X^\vee$ and $\Phi_X = \Phi_X^{\vee\vee}$. So using the isomorphism i of Exercise 10.35, we see that $r = 2\langle \beta, \rho \rangle = 2$ as claimed. $\square$

Finally we indicate one of the key tools in the analysis which follows:

Proposition 19.8   *There exists a base Δ of the root system of G with respect to T such that $\gamma(c)$ acts as c^0 or c^2 on each of the root spaces $\mathrm{Lie}(G)_\alpha$, for $\alpha \in \Delta$.*

About the proof   For $\beta \in \Phi$, the root system of G with respect to T, let $r_\beta \in \mathbb{Z}$ such that $\gamma(c)v = c^{r_\beta}v$ for all $v \in \mathrm{Lie}(G)_\beta$; that is, $r_\beta = \langle \beta, \gamma \rangle$. One first shows that all T_X-weights on $\mathrm{Lie}(G)$ are integral linear combinations of elements of Φ_X. See [61, 2.3] for this proof. Then apply Lemma 19.7 to see that the r_β must be even integers. Choosing an appropriate ordering on $X(T)$ one finds a base Δ of Φ such that $r_\beta \geq 0$ for all $\beta \in \Delta$ and so for all

$\beta \in \Phi^+$ (see Section A.1). Now set

$$J := \langle T, U_\beta, U_{-\beta} \mid \beta \in \Delta \text{ and } r_\beta = 0 \text{ or } 2\rangle.$$

Then J is the Levi factor of a suitable parabolic subgroup of G and it is clear that

$$J = \langle T, U_\beta, U_{-\beta} \mid \beta \in \Phi \text{ and } r_\beta = 0 \text{ or } 2\rangle.$$

Lemma 19.7 and Theorem 8.17 imply that $\mathrm{Lie}(X) \le \mathrm{Lie}(J)$; so

$$Z(J)^\circ \le C_G(\mathrm{Lie}(X))^\circ \le \mathrm{Stab}_G(\mathrm{Lie}(X))^\circ = X$$

by Lemma 19.6. In particular $Z(J)^\circ \le C_X(\mathrm{Lie}(X))^\circ$. Now $C_X(\mathrm{Lie}(X)) = \ker(\mathrm{Ad}\,_X)$ and as we are assuming X to be semisimple, Exercise 10.32 implies that $\ker(\mathrm{Ad}\,_X) = Z(X)$. Hence $C_X(\mathrm{Lie}(X))^\circ = 1$, so $Z(J)^\circ = 1$ and $J = G$ by Proposition 12.6 as required. $\square$

We can now explain the strategy of the proof of Theorem 19.1. By Proposition 19.8, there are $2^{|\Delta_G|}$ possibilities for the $|\Delta_G|$-tuple of integers $(r_\beta)_{\beta \in \Delta_G}$ as defined in the proof of loc. cit. For each of these possibilities, one can list the precise weights of $\mathrm{im}(\gamma)$ on $\mathrm{Lie}(G)$. Now there will be a maximum such weight, indeed given by $\sum_{\beta \in \Delta_G} n_\beta r_\beta$, where the highest root in Φ_G is $\sum_{\beta \in \Delta_G} n_\beta \beta$. As there are a finite number of irreducible kX-modules for which the $\mathrm{im}(\gamma)$ weights are less than or equal to this highest weight, one obtains a collection of possible X-composition factors of $\mathrm{Lie}(G)|_X$. Moreover, in many cases the precise dimensions of these irreducible modules are known, or at least upper and lower bounds for their dimensions. Thus, one has a finite list of possible composition series (up to permutation of the factors) for the module $\mathrm{Lie}(G)|_X$. For each of these composition series, one computes the precise set of weights of $\mathrm{im}(\gamma)$ on $\mathrm{Lie}(G)$ and compares this set of weights with the given set of weights obtained from the choice of $(r_\beta)_{\beta \in \Delta_G}$. For a fixed pair of groups X, G, one thus reduces to a finite number of choices for

(i) the tuples $(r_\beta)_{\beta \in \Delta_G}$, and
(ii) the corresponding set of composition factors of $\mathrm{Lie}(G)|_X$.

The next part of the proof is that which required the assumptions on $\mathrm{char}(k)$ in [61] and [48]. Many of the configurations remaining after the above analysis correspond to a composition series containing a large number of trivial composition factors. If $\mathrm{char}(k)$ is large enough, this often contradicts the conclusion of Lemma 19.6(b). The consideration of the small characteristic configurations required a detailed analysis of the remaining possible X-composition series of $\mathrm{Lie}(G)$.

20

Exercises for Part II

Let k be an algebraically closed field of characteristic $p \geq 0$. We will take the numbering of Dynkin diagrams of irreducible root systems as given in Table 9.1.

Exercise 20.1 (Existence of graph automorphisms)

(a) Show how to reduce the proof of Theorem 11.12 on the existence of graph automorphisms to the case of simple groups of simply connected type.
(b) Verify the details of the proof for type SL_n, $n \geq 3$.
(c) Show that a suitable element of GO_{2n} induces a non-trivial graph automorphism of SO_{2n}, $n \geq 2$.

[**Hint**: For (c) consider the element given in Example 22.9(2).]

Exercise 20.2 Let G be a group with a BN-pair, with $W = N/(B \cap N)$ generated by a set of involutions S. For $w \in W$ write $\ell(w)$ for the length of a shortest expression $w = s_1 \cdots s_r$ with $s_i \in S$. Show the following:

(a) If $s \in S$, $w \in W$ with $\ell(ws) \geq \ell(w)$ then $B\dot{w}B \cdot B\dot{s}B \subseteq B\dot{w}\dot{s}B$.
(b) If $s \in S$, $w \in W$ with $\ell(ws) \leq \ell(w)$ then $B\dot{w}B \cdot B\dot{s}B$ has non-empty intersection with $B\dot{w}B$.
(c) If $\ell(ws) < \ell(w)$, then $\dot{s} \in B\dot{w}^{-1}B\dot{w}B$.

[**Hint**: For (a), reason by induction on $\ell(w)$ and apply Theorem 11.17; for (b) apply (a). The result of (c) is a corollary of (a) and (b).]

Exercise 20.3 We continue using the notation of Exercise 20.2. For $w \in W$ let $w = s_1 \cdots s_r$, with $s_i \in S$, be a shortest expression, so that $r = \ell(w)$, and let $J := \{s_1, \ldots, s_r\} \subseteq S$. Show the following:

(a) We have $\langle B, \dot{w} \rangle = \langle B, \dot{w}^{-1} B \dot{w} \rangle = P_J$.
(b) All overgroups of B in G are of the form P_I for some $I \subseteq S$.
(c) Conclude that all P_I are self-normalizing and that distinct P_I cannot be conjugate.

[**Hint**: For (a), apply the preceding exercise. For (b) let P be a subgroup of G containing B. Any element in P can be written in the form $g = b\dot{w}b'$, with $b, b' \in B$, $w \in W$, by Theorem 11.17. We conclude that $\dot{w} \in P$. Thus, P is a union of double cosets $P = \bigcup_{w \in M} B\dot{w}B$ for some $M \subseteq W$. Now apply (a).]

Exercise 20.4 Let G be connected reductive, $I \subseteq S$ and L_I the corresponding standard Levi subgroup of G. Then there is a natural isomorphism $N_G(L_I)/L_I \xrightarrow{\sim} N_W(W_I)/W_I$.
[**Hint**: Let $g \in N_G(L_I)$, then T^g is a maximal torus in L_I, hence L_I-conjugate to T. Thus $gl \in N_G(T)$ for some $l \in L_I$. Show that this defines the required isomorphism.]

Exercise 20.5 Let G be a connected reductive algebraic group with root datum $(X, \Phi, Y, \Phi^\vee)$ with respect to the maximal torus T.

(a) Let $T' \leq T$ be a subtorus with root datum $(X', \emptyset, Y', \emptyset)$ (see Example 9.12). Show that the Levi subgroup $C_G(T')$ has corresponding root datum $(X, \Psi, Y, \Psi^\vee)$ with $\Psi := \Phi \cap \mathrm{Ann}(Y')$.
(b) Let L be a standard Levi subgroup of G with root datum $(X, \Psi, Y, \Psi^\vee)$. Show that the torus $Z(L)^\circ$ has the root datum $(X', \emptyset, Y', \emptyset)$, where $Y' = \mathrm{Ann}(\langle \Psi \rangle)$, $X' = X/\mathrm{Ann}(Y')$.

Here, for $Z \leq X$ we set $\mathrm{Ann}(Z) := \{ \gamma \in Y \mid \langle \chi, \gamma \rangle = 0 \text{ for all } \chi \in Z \}$.

Exercise 20.6 Let G be a simple algebraic group with simply laced Dynkin diagram. Let Δ denote a base of its root system. Show that the isomorphisms $u_\alpha : \mathbf{G}_a \to U_\alpha$, $\alpha \in \Delta$, may be chosen such that $[u_\alpha(t), u_\beta(u)] = u_{\alpha+\beta}(\pm tu)$ whenever $\alpha, \beta \in \Delta$ with $\alpha + \beta \in \Phi$.
[**Hint**: The statement holds for SL_3 by Example 11.9. Now use Example 12.9 to show it for SL_4 and then induction for the general case.]

Exercise 20.7 Prove the following generalization of Proposition 12.14: Let G be semisimple, $L_I \leq G$ the standard Levi subgroup corresponding to $I \subseteq S$. If $\gcd(\Lambda(G), \Lambda(\Phi_I)) = 1$ then $[L, L]$ is of simply connected type.

Exercise 20.8 Let G be semisimple with a maximal torus T with Weyl group W, $H \leq G$ a subsystem subgroup normalized by T with Weyl group W_H. Then there is a natural isomorphism $N_G(HT)/HT \cong N_W(W_H)/W_H$.
[**Hint**: Apply the arguments from Exercise 20.4 to the subgroup HT.]

Exercise 20.9 Let Φ be one of the indecomposable root systems of type A_n, B_n, C_n or D_n.

(a) Determine the proper closed subsystems of Φ of rank equal to the rank of Φ.

(b) Let $G = \mathrm{Sp}_{2n}$, a simple algebraic group of type C_n, acting on the natural $2n$-dimensional module V preserving a symplectic bilinear form. For $n_1, \ldots, n_t \in \mathbb{N}$ with $\sum n_i = n$, there exist subspaces $V_i \le V$ with $\dim V_i = 2n_i$ and $V = V_1 \perp \cdots \perp V_t$. Hence the group $\mathrm{Sp}_{2n_1} \times \cdots \times \mathrm{Sp}_{2n_t}$ is naturally embedded in G; let H denote the corresponding closed subgroup of G.

   (i) Show that H is a maximal rank subgroup of G, for all characteristics $\mathrm{char}(k)$.
   (ii) Find a closed subsystem Ψ of the root system of G such that H is the subsystem subgroup associated to this subsystem.

[**Hint:** (a) You should find no example for A_n, $n-1$ for type B_n, $\lfloor (n-1)/2 \rfloor$ for type C_n and $\lfloor (n-2)/2 \rfloor$ for type D_n.]

Exercise 20.10 Show that the unipotent element

$$u := \begin{pmatrix} 1 & 1 & 0 & 0 \\ 0 & 1 & 0 & 0 \\ 0 & 0 & 1 & 1 \\ 0 & 0 & 0 & 1 \end{pmatrix} \in \mathrm{Sp}_4$$

of $G = \mathrm{Sp}_4$ over a field of characteristic 2 is not contained in the connected component $C_G(u)^\circ$ of its centralizer.

Exercise 20.11 Let G be connected, $S \le G$ a torus. Show that the set of $s \in S$ such that $C_G(s) = C_G(S)$ is dense open in S.
[**Hint:** Embed $G \le \mathrm{GL}(V)$ and decompose V into weight spaces for S. Have a look at the proof of Corollary 14.10.]

Exercise 20.12 Let G be connected reductive with maximal torus T.

(a) Show that $N_G(T)$ controls G-fusion in T, that is, if $s, t \in T$ are G-conjugate, they are conjugate by an element of $N_G(T)$.
(b) Deduce that there is a natural bijection between the set of semisimple classes of G and the set T/W of W-orbits on T.

[**Hint:** For (a), use the uniqueness of expression in the Bruhat decomposition.]

Exercise 20.13 Let G be a simple algebraic group, with maximal torus T, root system Φ and Weyl group W over the algebraic closure k of a finite field $\mathbb{F}_p$. Show the following:

(a) There is a (non-canonical) isomorphism $e : \mathbb{Q}_{p'}/\mathbb{Z} \to k^\times$, where $\mathbb{Q}_{p'}$ denote the additive group of rational numbers with denominator prime to p.

(b) There is an isomorphism of groups

$$\tau : (Y \otimes \mathbb{Q}_{p'})/Y \longrightarrow Y \otimes \mathbb{Q}_{p'}/\mathbb{Z} \longrightarrow Y \otimes k^\times \longrightarrow T,$$

defined by $\gamma \otimes r \mapsto \gamma \otimes e(r) \mapsto \gamma(e(r))$, which is W-equivariant with respect to the natural W-actions on Y and T.

(c) The semisimple conjugacy classes of G are in bijection with W-orbits on $(Y \otimes \mathbb{Q}_{p'})/Y$, that is, with elements of $(Y \otimes \mathbb{Q}_{p'})/W_a$, where $W_a = Y.W$ denotes the affine Weyl group (of W acting on Y), see Section B.2.

(d) Conclude that any semisimple element of G is conjugate to some $s \in T$ such that the root system Ψ of $C_G(s)^\circ$ (as in Theorem 14.2) has a basis $\Delta_1 \subseteq \Delta \cup \{-\alpha_0\}$.

[**Hint:** (c) Use Proposition 14.6. (d) By Proposition B.12 every W_a-orbit on $Y \otimes \mathbb{Q}_{p'} \subset Y \otimes \mathbb{R}$ contains a point x from the fundamental alcove A. Then for $s = \tau(x)$, for $\alpha \in \Phi$ we have $\alpha(s) = 1$ if and only if $\langle s_\alpha, x \rangle = 0$, that is, if and only if $s_{\alpha^\vee}$ fixes x, that is, x lies on the wall of A corresponding to $\alpha^\vee$.]

Exercise 20.14 Assume $\mathrm{char}(k) \neq 2$ and let s be the image in $G = \mathrm{PGL}_2$ of $\mathrm{diag}(1, -1) \in \mathrm{GL}_2$. Show that $C_G(s)$ has two connected components. In particular, centralizers of semisimple elements in connected groups are not necessarily connected.

Exercise 20.15 Show that the torsion primes of a reductive group G are precisely the torsion primes of its root system Φ together with the prime divisors of the order of its fundamental group.
[**Hint:** First show that $\Lambda(G) \cong Y/\mathbb{Z}\Phi^\vee$ by applying $\mathrm{Hom}(-, \mathbb{Z})$ to the exact sequence defining $\Lambda(G)$. Now, if $G' \leq G$ is a subsystem group with root system Ψ, the torsion of $Y/\mathbb{Z}\Psi^\vee$ is the torsion of $Y/\mathbb{Z}\Phi^\vee$ together with that of $\mathbb{Z}\Phi^\vee/\mathbb{Z}\Psi^\vee$.]

Exercise 20.16 Let G be a semisimple algebraic group and $s \in G$ semisimple.

(a) Show that $C_G(s)/C_G(s)^\circ$ is isomorphic to a subgroup of $\ker(\pi)$, where $\pi : G_{\mathrm{sc}} \to G$ denotes the natural isogeny from a simply connected group with the same root system as G.

(b) If s is of finite order, the exponent of $C_G(s)/C_G(s)^\circ$ divides the order of s.

(c) If s is of finite order prime to $|\ker(\pi)|$ then $C_G(s)$ is connected.

[**Hint**: Fix a preimage $\hat{s} \in G_{sc}$ of s. For $g \in C_G(s)$ with preimage $\hat{g}$, $\hat{g}\hat{s}\hat{g}^{-1} = z\hat{s}$ for some $z \in \ker(\pi)$. Then $g \mapsto z = [\hat{g}, \hat{s}]$ is a well-defined group homomorphism with the required properties. For (b), use that $[\hat{g}, \hat{s}]$ is central, so $[\hat{g}, \hat{s}]^n = [\hat{g}, \hat{s}^n]$ for all n.]

Exercise 20.17   Show that the adjoint representation of SL_n is irreducible if and only if $\mathrm{char}(k)$ does not divide n.
[**Hint**: Think about the weights of the adjoint representation and the possible highest weights of composition factors. This establishes Theorem 15.20(1) in case $G = SL_n$.]

Exercise 20.18   Let $G = SL_2(K)$, the group of 2×2, determinant 1 matrices over an arbitrary (not necessarily algebraically closed) field K. Then G acts naturally on the polynomial ring $K[X, Y]$ in two indeterminates X, Y via

$$\begin{pmatrix} a & b \\ c & d \end{pmatrix} (X^i Y^j) = (aX + cY)^i (bX + dY)^j,$$

extended by linearity. Let $V_d \subset K[X, Y]$ be the subspace of homogeneous polynomials of degree d. We thus have a representation $\rho : G \to GL_{d+1}(K)$ of G.

(a) Verify that for K of characteristic 0 or for $\mathrm{char}(K) > d$, V_d is irreducible.
(b) Verify that if $\mathrm{char}(K) = p > 0$, $\langle X^p, Y^p \rangle$ is a G-invariant subspace of V_p.
(c) Show that if K is of characteristic $p > 0$ and $K \neq \mathbb{F}_2$, then V_p is not completely reducible as a KG-module.
(d) Now take $K = k$, an algebraically closed field of characteristic $p \geq 0$. As weights for the maximal torus consisting of diagonal matrices in G are all multiples of the weight $\lambda = \frac{1}{2}\alpha$ (where α is the unique positive root in the root system for G), the irreducible modules for G are parametrized by positive integers. Show that $V_d \simeq L(d\lambda)$ and that if $\mathrm{char}(k) = p > 0$, all restricted irreducible rational kSL_2-modules are of the form V_d, for $d < p$. In particular, SL_2 has a unique restricted irreducible rational representation of dimension $m + 1$ for all $m < p$, and the weights of the corresponding module are $m, m - 2, \ldots, -(m - 2), -m$.

Exercise 20.19   Let G be a connected reductive algebraic group with maximal torus T and Weyl group W. Let B be a Borel subgroup containing T and

Δ the corresponding set of simple roots in the root system Φ. Let $\lambda \in X(T)$ be a dominant weight.

(a) Show that $W_\lambda := \mathrm{Stab}_W(\lambda)$ is the subgroup generated by the s_α, $\alpha \in \Delta$, such that $\langle \lambda, \alpha^\vee \rangle = 0$; that is W_λ is the Weyl group of the Levi subgroup corresponding to the subset $\{\alpha \in \Delta \mid \langle \lambda, \alpha^\vee \rangle = 0\}$.

(b) Let W_λ be as in (a). Show that $\dim(L(\lambda)) \geq \sum_\mu |W : W_\mu|$, where the sum is taken over all dominant weights μ for which the multiplicity of μ in $L(\lambda)$ is non-zero.

[**Hint**: For (a), let $v^+ \in L(\lambda)$ be a maximal vector with respect to the Borel subgroup B. Then $B \leq \mathrm{Stab}_G(\langle v^+ \rangle)$, so $\mathrm{Stab}_G(\langle v^+ \rangle)$ is a parabolic subgroup of G. Now use the structure of parabolic subgroups (or see Corollary A.29). For (b), use Lemma 15.3 and Proposition 15.8.]

Exercise 20.20   Let $\mathrm{char}(k) = 2$ and let $G = \mathrm{SL}_2$. Fix a maximal torus T of G, a Borel subgroup B containing T and a simple root corresponding to the choice of B. Let $\lambda \in X(T)$ be the unique fundamental dominant weight. Let $F : k \to k$ be the 2-power map.

(a) Show that the adjoint representation of G equips $\mathrm{Lie}(G)$ with the structure of an indecomposable kG-module, having a composition series $0 \subset V_1 \subset \mathrm{Lie}(G) = V$, with $V_1 \cong L(0)$ and $V/V_1 \cong L(\lambda)^{(2)}$.

(b) Show that V_2, as in Exercise 20.18, is an indecomposable kG-module with a composition series $0 \subset W_1 \subset V_2$, such that $W_1 \cong L(\lambda)^{(2)}$ and $V_2/W_1 \cong L(0)$.

Exercise 20.21   Let G be a simple algebraic group of type G_2 defined over the field k and assume $\mathrm{char}(k) = 0$. Let $V = \mathrm{Lie}(G)$, the irreducible kG-module of highest weight λ_2, the second fundamental dominant weight with respect to a choice of maximal torus and positive system of roots. (See Example 15.21(1).) Show that the kG-module $V \wedge V \cong \mathrm{Lie}(G) \oplus L(3\lambda_1)$.

[**Hint**: You may use the fact that kG-modules are completely reducible when $\mathrm{char}(k) = 0$ (see [33, 14.3]) and that $\dim(V(a_1\lambda_1 + a_2\lambda_2)) = \frac{1}{5!}(a_1 + 1)(a_2 + 1)(a_1 + a_2 + 2)(a_1 + 2a_2 + 3)(a_1 + 3a_2 + 4)(2a_1 + 3a_2 + 5)$, which follows from the Weyl degree formula, see [33, §24].]

Exercise 20.22   Let G be a simple algebraic group of type D_4. For each of the maximal parabolic subgroups P of G, with unipotent radical Q, show that each of the internal modules Q_j/Q_{j+1}, for $j \geq 1$ is irreducible and determine their highest weights as modules for the Levi factor. (See also Example 17.9(3).)

Exercise 20.23 Let $G = SL_3$ and let W be the three-dimensional vector space on which G naturally acts with fixed ordered basis $\{e_1, e_2, e_3\}$. Let $B_G = \mathrm{Stab}_G(ke_1 \subset ke_1 + ke_2 \subset V)$, a Borel subgroup of G. Set $V = \wedge^2 W$, a six-dimensional vector space equipped with the natural G-action inherited from the action of G on $W \otimes W$. Let $\rho : G \to GL_6$ be the corresponding rational representation.

(a) Show that $\mathrm{im}\,\rho < SL_6$.
(b) Let $P_G = \mathrm{Stab}_G(ke_1)$, a proper parabolic subgroup of G, with $B_G < P_G$. Find $P \leq SL_6$, a parabolic subgroup such that $\rho(P_G) \leq P$ and $\rho(R_u(P_G)) \leq R_u(P)$.

Exercise 20.24 Let f_i be non-degenerate bilinear forms on the finite-dimensional k-vector spaces V_i, $i = 1, 2$, and set $V := V_1 \otimes V_2$. Show the following:

(a) There is a unique bilinear form $f = f_1 \otimes f_2$ on V defined by

$$f(v_1 \otimes v_2, w_1 \otimes w_2) := f_1(v_1, w_1)f_2(v_2, w_2) \qquad \text{for } v_i, w_i \in V_i.$$

(b) The form f is symmetric if and only if f_1, f_2 are either both symmetric or both alternating.
(c) If $\mathrm{char}(k) = 2$ then there is a unique quadratic form Q on V, with associated bilinear form f, such that $Q(v_1 \otimes v_2) = 0$ for all $v_i \in V_i$, and Q is preserved by $Sp(V_1) \otimes Sp(V_2)$.

[**Hint:** See [44, §4.4].]

Exercise 20.25 Let $\mathrm{char}(k) = 2$ and let V be a three-dimensional vector space over k, with fixed basis $\{e, w, f\}$. Define a quadratic form on V by $Q(ae + bw + cf) := b^2 + ac$ and let SO_3 be the corresponding orthogonal group, as in Section 1.2. Show that $\langle w \rangle$ is the radical of the associated bilinear form and that $\varphi : SO_3 \to SL(V/\langle w \rangle)$ defines a morphism of algebraic groups whose image is the group $Sp(V/\langle w \rangle)$. Show that $\ker(\varphi) = 1$ and so $\varphi : SO_3 \to Sp(V/\langle w \rangle)$ is an isomorphism of abstract groups. Now use Proposition 7.7 to see that φ is not an isomorphism of algebraic groups. (One can also see directly that the inverse is not a polynomial map.)

Exercise 20.26 Let Φ be a root system in a Euclidean space E, with base $\Delta \subset \Phi$ and corresponding set of positive roots Φ^+. Set $\rho = \frac{1}{2}\sum_{\alpha \in \Phi^+} \alpha$. Show that ρ is the sum of the fundamental dominant weights with respect to the base Δ.
[**Hint:** Use Lemma A.8 to deduce that $s_\beta(\rho) = \rho - \beta$ and hence $\langle \rho, \beta^\vee \rangle = 1$ for all $\beta \in \Delta$.]

PART III

FINITE GROUPS OF LIE TYPE

In this part we introduce and study finite analogues of the simple linear algebraic groups which were the subject of the first two parts. The construction of the various classical groups over algebraically closed fields in Section 1.2 generalizes in a straightforward way to give versions over arbitrary base fields. It is much less obvious how to obtain versions of the simple exceptional groups. This was first achieved by Chevalley [15] who showed how to construct analogues of all simple algebraic groups as automorphism groups of simple Lie algebras over any base field. Still this approach falls short of producing all versions in which we will be interested. For example, over an algebraically closed field there is just one class of non-degenerate orthogonal forms up to similarity in any dimension, while over arbitrary fields there may be many, with corresponding non-isomorphic isometry groups. Also, the isometry groups of unitary forms do not arise in Chevalley's setup.

Shortly after Chevalley's construction, Steinberg [68] presented a variation of this by considering fixed points of field automorphisms composed with algebraic group automorphisms. This allows one to recover the unitary groups, for example. But this wasn't yet the end of the story. In 1960 M. Suzuki discovered an infinite series of finite simple permutation groups which at that time did not seem to have any relation with algebraic groups. It was recognized in the same year by Steinberg how his construction could be generalized to yield these as subgroups of the four-dimensional symplectic group over an algebraically closed field of characteristic 2.

We will follow this approach of Steinberg, which now seems to be the best way to recover via a general construction all of the versions described above: given a field K, the various analogues of linear algebraic groups are obtained from algebraic groups over the algebraic closure k of K by means of a Galois descent. The situation becomes particularly simple when $K = \mathbb{F}_q$ is a finite field, since then the absolute Galois group $\mathrm{Gal}(\overline{\mathbb{F}_q}/\mathbb{F}_q)$ is profinite cyclic, topologically generated by the Frobenius automorphism F with respect to the base field $\mathbb{F}_q$. The finite groups of Lie type then occur as the fixed point groups of suitable variants of this Frobenius automorphism. More precisely, they arise as fixed point groups of endomorphisms some power of which is the standard Frobenius. In most but not all cases, such endomorphisms are obtained by composing the Frobenius with an automorphism of the underlying algebraic group.

This point of view allows one to answer many questions on the families of finite groups of Lie type by interpreting them as questions on the F-fixed points of corresponding structures in algebraic groups and transferring the results from there.

In Chapter 21 we describe the setting for this method of Galois descent over finite fields by introducing the notion of Steinberg endomorphism (which is sometimes also called Frobenius endomorphism or generalized Frobenius map) and formulate the important theorem of Lang–Steinberg. Chapter 22 is devoted to the study of properties of Steinberg endomorphisms and their classification. The fixed point groups under such endomorphisms are what we call the finite groups of Lie type, the central topic of this part.

We begin the structural investigation in Chapter 23 by exhibiting a root system and root subgroups for these groups. In Chapter 24 we first derive a Bruhat decomposition from which, using some results from the invariant theory of finite reflection groups, we compute the orders of the finite groups of Lie type. We then show that these groups also possess a BN-pair, as do their algebraic counterparts, which enables us to study simplicity and automorphism groups. The structure of tori, Sylow subgroups, their centralizers and normalizers is the subject of Chapter 25. In Chapter 26 we investigate subgroups of maximal rank such as parabolic subgroups, Levi subgroups and centralizers of semisimple elements.

The last three chapters are devoted to the study of maximal subgroups of finite groups of Lie type, using the corresponding results for simple algebraic groups in the previous part. We present the reduction theorems of Aschbacher for the classical groups and of Liebeck–Seitz for the exceptional groups and comment on the current status of the determination of all maximal subgroups.

21

Steinberg endomorphisms

In this section, we introduce the notion of Steinberg endomorphisms on linear algebraic groups over the algebraic closure of a finite field and define the finite groups of Lie type. We present the crucial theorem of Lang–Steinberg and several of its consequences, which will be essential for most of the results to come. Throughout, the base field k is the algebraic closure of a finite field of characteristic $\operatorname{char}(k) = p$.

21.1 Endomorphisms of linear algebraic groups

We start by looking at certain endomorphisms of the algebraic group GL_n:

Example 21.1 Let $k = \overline{\mathbb{F}_q}$ where $q = p^f$. The map $F_q : k \to k$, $t \mapsto t^q$, is a field automorphism of k which fixes $\mathbb{F}_q$ pointwise. In fact the Galois group $\operatorname{Gal}(k/\mathbb{F}_q)$ is generated (as a profinite group) by this Frobenius automorphism.

Letting F_q act on the matrix entries, this induces a group homomorphism

$$F_q : \operatorname{GL}_n \longrightarrow \operatorname{GL}_n, \qquad (a_{ij}) \longmapsto (a_{ij}^q),$$

with fixed point group

$$(\operatorname{GL}_n)^{F_q} := \{g \in \operatorname{GL}_n \mid F_q(g) = g\} = \operatorname{GL}_n(\mathbb{F}_q),$$

the general linear group over $\mathbb{F}_q$ which we will henceforth denote by $\operatorname{GL}_n(q)$. We call F_q the *standard Frobenius of* GL_n with respect to $\mathbb{F}_q$.

This example may be generalized as follows: Let V be an affine algebraic variety over $k = \overline{\mathbb{F}_q}$ defined by a set of polynomials $I \subseteq \mathbb{F}_q[T_1, \ldots, T_n]$ with coefficients in $\mathbb{F}_q$. We then say that V is *defined over* $\mathbb{F}_q$. The Frobenius automorphism $F_q : k \to k$, $t \mapsto t^q$, of k which acts naturally on $k[T_1, \ldots, T_n]$

via the coefficients, then leaves I invariant. Thus F_q also acts on the set V of common zeros of I in k^n, by

$$F_q : V \to V, \qquad (v_1, \ldots, v_n) \longmapsto (v_1^q, \ldots, v_n^q).$$

This induced map F_q is called the *Frobenius morphism of V with respect to the $\mathbb{F}_q$-structure given by I.* As in the example above, we write

$$V^{F_q} := V(\mathbb{F}_q) := \{v \in V \mid F_q(v) = v\}$$

for the F_q-fixed points of V. Note that, as F_q is induced by an element of the Galois group of $k/\mathbb{F}_q$, it is a bijective map.

Thus, if a closed subgroup $G \leq \mathrm{GL}_n$ is defined by equations over $\mathbb{F}_q$, this gives rise to a Frobenius morphism $F_q : G \to G$, $(a_{ij}) \mapsto (a_{ij}^q)$, with respect to this $\mathbb{F}_q$-structure, which clearly is a morphism of algebraic groups, with finite fixed point group $G(\mathbb{F}_q) = G^{F_q} \leq \mathrm{GL}_n^{F_q} = \mathrm{GL}_n(q)$. Note, however, that this bijective group homomorphism is not an isomorphism of algebraic groups.

Still, this approach for the construction of finite fixed point groups is not yet general enough to cover all situations we want to treat.

Example 21.2 Let's consider again $G = \mathrm{GL}_n$, but now with the endomorphism

$$F : \mathrm{GL}_n \longrightarrow \mathrm{GL}_n, \qquad (a_{ij}) \longmapsto (a_{ij}^q)^{-\mathrm{tr}}.$$

Thus F is the composite of the previous F_q with the map sending a matrix to the transpose of its inverse. Note that these two maps commute. Then certainly $F^2 : \mathrm{GL}_n \to \mathrm{GL}_n$, $(a_{ij}) \mapsto (a_{ij}^{q^2})$, is the standard Frobenius map F_{q^2} with respect to $\mathbb{F}_{q^2}$. So, the fixed points under F satisfy

$$G^F \leq G^{F^2} = \mathrm{GL}_n^{F_{q^2}} = \mathrm{GL}_n(q^2).$$

Here, the fixed point group $\mathrm{GU}_n(q) := G^F$ is the so-called *general unitary group* over $\mathbb{F}_{q^2}$. We also set $\mathrm{SU}_n(q) := \mathrm{SL}_n^F = \mathrm{GU}_n(q) \cap \mathrm{SL}_n(q^2)$, the *special unitary group*. (Note that, despite the suggestive notation, $\mathrm{GU}_n(q)$ *cannot* be obtained as a subgroup of $\mathrm{GL}_n(q)$. Some justification for this notation will be given in Example 22.11.) The above definition shows that it is the group of invertible $n \times n$-matrices over $\mathbb{F}_{q^2}$ leaving invariant the non-degenerate sesquilinear form

$$\langle \, , \, \rangle : \mathbb{F}_{q^2}^n \times \mathbb{F}_{q^2}^n \to \mathbb{F}_{q^2}, \qquad \left\langle \sum_{i=1}^n x_i e_i, \sum_{i=1}^n y_i e_i \right\rangle = \sum_{i=1}^n x_i y_i^q,$$

where $\{e_i \mid 1 \leq i \leq n\}$ denotes the standard basis of $\mathbb{F}_{q^2}^n$.

Definition 21.3 An endomorphism $F : G \to G$ of a linear algebraic group G such that for some $m \geq 1$ the power $F^m : G \to G$ is the Frobenius morphism with respect to some $\mathbb{F}_{p^a}$-structure of G is called a *Steinberg endomorphism* of G. We write G^F for the group of fixed points of F on G.

Let's note that since some power of a Steinberg endomorphism is induced by a Galois automorphism of k, Steinberg endomorphisms are always automorphisms of abstract groups, but not of algebraic groups in general. See Exercise 30.1 for a statement on arbitrary endomorphisms of linear algebraic groups.

Example 21.4 Steinberg endomorphisms on $\mathbf{G}_m$ and $\mathbf{G}_a$.

(1) By Exercise 10.11 the only algebraic group endomorphisms of $\mathbf{G}_m$ are of the form $\sigma_m : c \mapsto c^m$, for some $m \in \mathbb{Z}$. The standard Frobenius on $\mathbf{G}_m$ maps all $c \in \mathbf{G}_m$ to some fixed power which is a power of $p = \operatorname{char}(k)$. Clearly, some power of σ_m is of this form if and only if m itself is a power of p. Thus, the only Steinberg endomorphisms on $\mathbf{G}_m$ are of the form $F : \mathbf{G}_m \to \mathbf{G}_m$, $c \mapsto c^q$, for some $q = p^f$, with fixed point group $\mathbf{G}_m^F \cong \mathbb{F}_q^\times$, the multiplicative group of a finite field.

(2) Similarly, it is easily seen that in characteristic $p > 0$ the only endomorphisms of $\mathbf{G}_a$ are of the form $\sigma_h : c \mapsto h(c)$, where $h = \sum_{i=0}^{n} a_i T^{p^i} \in k[T]$, that is, h is a so-called *additive polynomial*. Some power of σ_h is the standard Frobenius endomorphism on $\mathbf{G}_a$ if and only if $h = T^{p^i}$. So, the only Steinberg endomorphisms on $\mathbf{G}_a$ are of the form $F : \mathbf{G}_a \to \mathbf{G}_a$, $c \mapsto c^q$, with $\mathbf{G}_a^F = \mathbb{F}_q^+$.

In particular, for $\mathbf{G}_a$ and $\mathbf{G}_m$ all Steinberg endomorphisms are Frobenius maps.

Examples 21.1 and 21.2 have already shown two instances of Steinberg endomorphisms of GL_n. Both of them fall into the second case of the fundamental dichotomy for endomorphisms of simple algebraic groups proved by Steinberg ([72, Thm. 10.13]):

Theorem 21.5 (Steinberg) *Let G be a simple linear algebraic group, $\sigma : G \to G$ an endomorphism of G. Then precisely one of the following holds:*

(1) *σ is an automorphism of algebraic groups, or*

(2) *the group $G^\sigma := \{g \in G \mid \sigma(g) = g\}$ is finite.*

The second case occurs if and only if σ is a Steinberg endomorphism.

(Clearly the statement does not generalize to semisimple groups!) The automorphisms occurring in (1) have been studied in Theorem 11.11: up to inner automorphisms, these come from the graph automorphisms on the Dynkin diagram of G. From now on, we will be interested exclusively in the endomorphisms of the second type.

Definition 21.6   Let G be a semisimple algebraic group, $F : G \to G$ a Steinberg endomorphism. Then the finite group of fixed points G^F is called a *finite group of Lie type*.

21.2 The theorem of Lang–Steinberg

The crucial tool for transferring results from algebraic groups G to finite groups G^F of fixed points under a Steinberg endomorphism F is the theorem of Lang–Steinberg (see [72, Thm. 10.1] or [76, Thm. J]):

Theorem 21.7 (Lang–Steinberg)   *Let G be a connected linear algebraic group over $\overline{\mathbb{F}}_p$ with a Steinberg endomorphism $F : G \to G$. Then the morphism*

$$L : G \to G, \qquad g \mapsto F(g)g^{-1},$$

is surjective.

Proof for F a standard Frobenius map   Since F is standard there exists an embedding $G \hookrightarrow \mathrm{GL}_n$ as a closed subgroup such that F is given by $F : (a_{ij}) \mapsto (a_{ij}^q)$ for some power q of the characteristic of k.

The differential dF of F is 0 (see Example 7.8(3)), so by Proposition 7.7(a) and Example 7.8(1),(2), $d_1L = dF - 1 = -1$ is an isomorphism. Now for any $x \in G$, the morphism $L' := L \circ \rho_x$ satisfies

$$L'(g) = L \circ \rho_x(g) = L(gx) = F(gx)x^{-1}g^{-1} = F'(g)g^{-1} \qquad \text{for } g \in G,$$

with $F' = \rho_{L(x)} \circ F$. By Proposition 7.7(a) its differential equals

$$d_xL \circ d_1\rho_x = d_1L' = dF' - 1 = -1,$$

hence also is an isomorphism. This shows that d_xL is an isomorphism for all $x \in G$. Now by [66, Thm. 4.3.3(ii)] there exists $x \in G$ such that

$$\dim \overline{L(G)} = \dim(d_xL)(\mathrm{Lie}(G)) = \dim \mathrm{Lie}(G) = \dim(G),$$

and thus $\overline{L(G)} = G$ as G is connected. Hence $L(G)$ contains a dense open subset of G by Proposition 1.6.

Now, for arbitrary $x \in G$ consider the map $L_x : G \to G$, $g \mapsto F(g)xg^{-1}$.

As above, we can conclude that $L_x(G)$ contains a dense open subset of G. Since G is connected, by Proposition 1.9 there exists $y \in L(G) \cap L_x(G)$, that is, $y = F(g_1)g_1^{-1} = F(g_2)xg_2^{-1}$ for some $g_1, g_2 \in G$. But then $x = L(g_2^{-1}g_1)$ lies in the image of L. □

Note that the assumption of G being connected is crucial here, otherwise the conclusion fails. For example, take $1 \neq G \leq \mathrm{GL}_n(p)$, a finite, hence disconnected algebraic group, and F the standard Frobenius map, then clearly L is a constant map.

In the remainder of this chapter we derive various direct consequences of this central result. The first concerns the semidirect product $G \rtimes \langle F \rangle$ of an algebraic group with the cyclic subgroup generated by a Steinberg endomorphism.

Corollary 21.8 *Let G be connected with a Steinberg endomorphism F : $G \to G$. Then, in the semidirect product $G \rtimes \langle F \rangle$, the coset $G.F$ of F consists of a single conjugacy class, that is, $G.F = F^G$. In particular, G^{gF} and G^F are G-conjugate for any $g \in G$.*

Proof This is Exercise 30.4. □

Definition 21.9 Let H be a group, σ an (abstract group) automorphism of H. We say that h_1, h_2 are σ-*conjugate* if there exists an $x \in H$ with $h_2 = \sigma(x)h_1x^{-1}$. The equivalence classes for this relation are called σ-*conjugacy classes* of H.

Note that $h_1, h_2 \in H$ are σ-conjugate if and only if the elements $h_1\sigma, h_2\sigma \in H\sigma$ are conjugate (by an element of H) in the semidirect product $H \rtimes \langle \sigma \rangle$ of H with σ. So indeed σ-conjugacy is an equivalence relation. Clearly, for σ acting trivially we recover the usual conjugacy classes of H.

Lemma 21.10 *Let G be a linear algebraic group with Steinberg endomorphism $F : G \to G$, H a closed connected normal F-stable subgroup. Then the quotient map induces a natural bijection from F-conjugacy classes of G to F-conjugacy classes of G/H.*

Proof Clearly the natural map $\pi : G \to G/H$ induces a surjection from the set of F-conjugacy classes of G to those of G/H. Now let $g_1, g_2 \in G$ have F-conjugate images, so $g_2H = F(x)g_1x^{-1}H$ for some $x \in G$. Then $g_2 = F(x)g_1hx^{-1}$ for some $h \in H$. Now the Lang–Steinberg Theorem, applied to the endomorphism $F' = g_1^{-1}F : H \to H$, $g \mapsto g_1^{-1}F(g)g_1$, of the connected group H yields that $h = F'(y)y^{-1} = g_1^{-1}F(y)g_1y^{-1}$ for some $y \in H$, whence $g_2 = F(x)F(y)g_1y^{-1}x^{-1} = F(xy)g_1(xy)^{-1}$ are F-conjugate, as claimed. □

Let's prove a further useful consequence of the Lang–Steinberg Theorem, which has many applications.

Theorem 21.11   *Let G be a connected linear algebraic group with a Steinberg endomorphism $F : G \to G$, acting transitively on a non-empty set V with a compatible F-action $F : V \to V$ (i.e., $F(g.v) = F(g).F(v)$ for all $g \in G$, $v \in V$). Then:*

(a) *F has fixed points on V, i.e., $V^F \neq \emptyset$.*
(b) *If the stabilizer G_v is closed for some $v \in V$, then for any $v \in V^F$ there is a natural 1–1 correspondence:*

$$\{G^F\text{-orbits on } V^F\} \longleftrightarrow \{F\text{-classes in } G_v/G_v^\circ\}.$$

Proof   (a) Let $v \in V$. By transitivity of the action there exists some $g \in G$ with $g.F(v) = v$. By the theorem of Lang–Steinberg we may write $g = x^{-1}F(x)$ for some $x \in G$. Hence, $F(x.v) = F(x).F(v) = x.v$, so $x.v \in V^F$.

(b) We have $G_{g.v} = gG_vg^{-1}$ for all $g \in G$, hence since conjugation is an isomorphism of algebraic groups, by transitivity all stabilizers G_v of $v \in V$ are closed. By part (a), there exists $v \in V^F$. Its stabilizer G_v is a closed subgroup, containing G_v° with finite index, and both are F-stable.

Now, if $w \in V^F$, then by transitivity there exists $g \in G$ such that $w = g.v$. Since $v, w \in V^F$ we have $g.v = w = F(w) = F(g).F(v) = F(g).v$, so $g^{-1}F(g) \in G_v$. Now if $g.v = h.v = w$ for $h \in G$ then $h^{-1}g \in G_v$, whence $g^{-1}F(g)$ and $h^{-1}F(h)$ are F-conjugate in G_v. By Exercise 30.5 this map $V^F/G^F \to G_v$, $g \mapsto g^{-1}F(g)$, induces a bijection from G^F-orbits in V^F to F-classes in G_v. By Lemma 21.10 the latter are in natural bijection with F-classes in G_v/G_v°.   □

Let's apply this result to the action of G on itself by conjugation:

Corollary 21.12   *Let G be connected reductive, $F : G \to G$ a Steinberg endomorphism. Then there exists a pair $T \leq B$ consisting of an F-stable maximal torus T contained in an F-stable Borel subgroup B of G. All such pairs $T \leq B$ are G^F-conjugate.*

Proof   By Theorems 6.4 and 4.4 the group G acts transitively by conjugation on

$$V := \{(B, T) \mid T \text{ a maximal torus in a Borel subgroup } B \text{ of } G\},$$

compatibly with F. So by Theorem 21.11(a) there exists an F-stable pair $v = (B, T)$ in V. Moreover, the stabilizer of v is just $G_v = N_G(B, T) := N_G(B) \cap N_G(T)$. We claim that this is a closed, connected subgroup of G.

As $N_G(B) = B$ by Theorem 6.12, we have $G_v = N_B(T)$. By Theorem 4.4(b), $N_B(T) = C_B(T) \leq C_G(T) = T$, the last equality by Corollary 8.13(b), as G is connected. So $G_v = T$ is closed connected, and hence all F-stable pairs in V are G^F-conjugate by Theorem 21.11(b). □

Definition 21.13 A maximal torus of G as in Corollary 21.12 is called *maximally split* with respect to F.

For the standard Frobenius map F on $G = \mathrm{GL}_n$, the F-fixed points of maximally split tori of G are again isomorphic to a product of copies of the multiplicative group (now of the finite field), but this need not be true for general Steinberg endomorphisms, as the next example shows.

Example 21.14 The F-structure of maximally split tori in GL_n.

(1) Let $G = \mathrm{GL}_n$ with $F_q : G \to G$, $(a_{ij}) \mapsto (a_{ij}^q)$, the standard Frobenius map. Then $\mathrm{T}_n \leq G$ is an F_q-stable Borel subgroup, $\mathrm{D}_n \leq \mathrm{T}_n$ is an F_q-stable maximal torus, so D_n is a maximally split torus of GL_n with respect to F_q. Clearly, we have $\mathrm{D}_n^{F_q} \cong \mathbb{F}_q^\times \times \cdots \times \mathbb{F}_q^\times = (\mathbb{F}_q^\times)^n$ of order $|\mathrm{D}_n^{F_q}| = (q-1)^n$.

(2) Let $G = \mathrm{GL}_n$, $F = F_q\sigma$ with F_q as before and $\sigma : (a_{ij}) \mapsto ((a_{ij})^{-\mathrm{tr}})^w$ with $w = \begin{pmatrix} 0 & & .1 \\ & \cdot\cdot\cdot & \\ 1 & & 0 \end{pmatrix}$. Then again T_n is an F-stable Borel subgroup, and $\mathrm{D}_n \leq \mathrm{T}_n$ is an F-stable maximal torus, hence maximally split in GL_n with respect to F. Let's compute its structure: For $t = \mathrm{diag}(t_1, \ldots, t_n)$ we have

$$F(t) = F(\mathrm{diag}(t_1, \ldots, t_n)) = \mathrm{diag}(t_n^{-q}, \ldots, t_1^{-q}).$$

So, $F(t) = t$ if and only if $t_1 = t_n^{-q}, t_2 = t_{n-1}^{-q}, \ldots, t_n = t_1^{-q}$. In particular, $t_i^{q^2} = t_i$ for all $1 \leq i \leq n$, and $t_{m+1}^{q+1} = 1$ if $n = 2m+1$ is odd. Hence,

$$\mathrm{D}_n^F \cong \begin{cases} (\mathbb{F}_{q^2}^\times)^m \\ (\mathbb{F}_{q^2}^\times)^m \times \mathbb{F}_{q^2}^\times/\mathbb{F}_q^\times \end{cases} \text{ of order } \begin{cases} (q^2-1)^m & \text{if } n = 2m, \\ (q^2-1)^m(q+1) & \text{if } n = 2m+1. \end{cases}$$

Note that here $G^F \cong \mathrm{GU}_n(q)$ is again isomorphic to the general unitary group by Corollary 21.8, since we have changed the Frobenius map from Example 21.2 only by the inner element w.

Classification of finite groups of Lie type

We study the Steinberg endomorphisms more closely via their action on the character group and on the Weyl group and present their classification which in turn gives a classification of the finite groups of Lie type. As for semisimple algebraic groups, this classification can be given in combinatorial terms via so-called complete root data. We then comment on the construction of the various finite groups of Lie type.

22.1 Steinberg endomorphisms

Let G be a connected reductive linear algebraic group, $F : G \to G$ a Steinberg endomorphism. Recall from Corollary 21.12 that there exists an F-stable maximal torus T contained in an F-stable Borel subgroup B of G. As T is F-stable, so is $N_G(T)$, so F naturally acts on the Weyl group $W = N_G(T)/T$ of G. Similarly, F also acts on the character group $X := X(T)$ and the cocharacter group $Y := Y(T)$ via

$$F(\chi)(t) := \chi(F(t)) \quad \text{for } \chi \in X, \ t \in T,$$

$$F(\gamma)(c) := F(\gamma(c)) \quad \text{for } \gamma \in Y, \ c \in k^\times.$$

Let's consider these actions in some more detail:

Lemma 22.1 *Let T be a torus with character group X, and $F : T \to T$ a Steinberg endomorphism. Then:*

(a) *There exists $\delta \in \mathbb{N}$ and a power r of $p = \text{char}(k)$ such that $F^\delta|_X = r \, \text{id}_X$ on X.*

(b) *We have $T^F \cong X/(F-1)X$.*

Proof Let's first consider the case where $T = \mathrm{D}_n \leq \mathrm{GL}_n$ and $F = F_q$ is the standard Frobenius map on GL_n. Then

$$F(\chi)(t) = \chi(F(t)) = \chi(t^q) = \chi(t)^q = (q\chi)(t) \qquad \text{for all } \chi \in X, \ t \in T,$$

so $F|_X = q\,\mathrm{id}_X$ as claimed. Moreover, $T^F \cong (\mathbb{F}_q^\times)^n$ (see Example 21.14(1)).

In the general situation, as F is a Steinberg endomorphism, there exists an embedding $T \hookrightarrow \mathrm{GL}_n$ such that some power F' of F is the restriction to T of a standard Frobenius on GL_n. Let T_1 be a maximal torus of GL_n containing the torus T. This is conjugate to the maximal torus D_n, so $T_1 = \mathrm{D}_n^g$ for some $g \in \mathrm{GL}_n$. Now g has entries in some finite subfield of k, so there is a positive power F^δ of F' which fixes g. Then conjugation by g defines an isomorphism $T_1 \cong \mathrm{D}_n$ which commutes with F^δ. By the first part, F^δ acts by multiplication with some power r of p on $X(\mathrm{D}_n)$, so on $X(T_1)$. Restriction of characters from T_1 to T is F-equivariant, with F-stable kernel $T^\perp$ (see Proposition 3.8(a)), so F^δ also acts by the scalar r on $X = X(T) \cong X(T_1)/T^\perp$, proving (a).

For (b), again apply Proposition 3.8(a) to the closed subgroup $T^F \leq T$ to obtain that $X(T^F) \cong X/(T^F)^\perp$. Now we claim that

$$t \in T^F \quad \Longleftrightarrow \quad F(t) = t \quad \Longleftrightarrow \quad (F-1)\chi(t) = 1 \text{ for all } \chi \in X.$$

The implication from left to right is clear. Conversely, if $F(t) \neq t$ then clearly there is $\chi \in X$ with $\chi(F(t)t^{-1}) \neq 1$. So we have $T^F = ((F-1)X)^\perp$. As $(F-1)X$ contains $(F^\delta - 1)X = (r-1)X$, it is of index prime to p in X, so we have $(T^F)^\perp = (F-1)X$ by Proposition 3.8(b). Finally, as $|T^F|$ is prime to p and k contains all p'-roots of unity, $X(T^F) = \mathrm{Hom}(T^F, k^\times) \cong \mathrm{Hom}(T^F, \mathbb{C}^\times)$, and the latter is isomorphic to T^F. Thus indeed

$$T^F \cong X(T^F) \cong X/(T^F)^\perp \cong X/(F-1)X. \qquad \square$$

Write $\Phi \subset X$ for the root system of G, with positive system Φ^+ with respect to T and B. For $\alpha \in \Phi$ let's choose isomorphisms $u_\alpha : \mathbf{G}_a \to U_\alpha$ onto the root subgroups (see Theorem 8.17(c)). We set $X_\mathbb{R} := X \otimes_\mathbb{Z} \mathbb{R}$.

Proposition 22.2 *Let G be a connected reductive algebraic group with Steinberg endomorphism $F : G \to G$, and T, B, X, Φ as above.*

(a) *There exists a permutation ρ of Φ^+ and, for each $\alpha \in \Phi^+$, a positive integral power $q_\alpha > 1$ of $p = \mathrm{char}(k)$ and $a_\alpha \in k^\times$ such that $F(\rho(\alpha)) = q_\alpha \alpha$ and $F(u_\alpha(c)) = u_{\rho(\alpha)}(a_\alpha c^{q_\alpha})$ for all $c \in k$.*

(b) *There exists $\delta \geq 1$ such that $F^\delta|_X = q^\delta \mathrm{id}_X$ and $F = q\phi$ on $X_\mathbb{R}$ for some positive fractional power q of p and some $\phi \in \mathrm{Aut}(X_\mathbb{R})$ of order δ inducing ρ^{-1} on Φ^+.*

In particular both F and ρ induce graph automorphisms of the Coxeter diagram of Φ.

Proof Part (a) is just Lemma 11.10 applied to the endomorphism F. For part (b), by Lemma 22.1 there exists some minimal $\delta \in \mathbb{N}$ such that $F^\delta|_X = r \operatorname{id}_X$. Thus all eigenvalues of F on $X \otimes_{\mathbb{Z}} \mathbb{C}$ have absolute value $q := r^{\frac{1}{\delta}} > 1$, a fractional power of p. Then with $\phi := \frac{1}{q} F : X_{\mathbb{R}} \to X_{\mathbb{R}}$ we get $\phi^\delta = q^{-\delta} F^\delta = \operatorname{id}_{X_{\mathbb{R}}}$, so ϕ has finite order and permutes the roots in the same way as ρ^{-1} by (a).

By Lemma 11.10, ρ stabilizes the set of simple roots $\Delta \subseteq \Phi^+$. Hence, in its action on the Weyl group F stabilizes the set of simple reflections of W and thus both F and ρ induce graph automorphisms of the Coxeter diagram of Φ. $\square$

Note that F^δ fixes all the one-dimensional submodules of X generated by the roots, so it also fixes all simple reflections in W, hence acts trivially on W.

In order to obtain further information on the q_α we need to assume that F permutes the simple components of the derived group $[G, G]$ transitively, which is the case for example if G is simple.

Lemma 22.3 *In the situation of Proposition 22.2 assume that G is simple. Then q_α is constant on roots of the same length.*

Moreover, either all $q_\alpha = q$, in which case ρ preserves root lengths, or else G is of type B_2, G_2 or F_4 with $p = 2, 3, 2$, respectively, ρ interchanges long and short roots, $q_\alpha q_\beta = q^2$ and $q_\beta / q_\alpha = p$ for all long roots α and all short roots β.

Proof We first show that q_α is constant on roots of a given length. As G is simple its root system Φ is indecomposable, see Exercise 10.33. Then any two roots $\alpha, \beta \in \Phi$ of the same length lie in the same W-orbit by Corollary A.18, so $\beta = w\alpha$ for some $w \in W$. Let $w_1 \in W$ such that $F(w_1) = w$, then by Proposition 22.2(a)

$$q_\alpha F(\rho(\beta)) = q_\alpha q_\beta \beta = q_\alpha q_\beta w\alpha = q_\beta w F(\rho(\alpha)) = q_\beta F(w_1 \rho(\alpha)).$$

Thus $\rho(\beta)$ and $w_1\rho(\alpha)$ are proportional roots with positive proportionality factor q_α / q_β, so they are equal by axiom (R2) whence $q_\alpha = q_\beta$ is constant on roots of the same length.

Since G is simple, W is irreducible, so there is a unique W-invariant scalar product on $X_{\mathbb{R}}$ up to scalar multiples, which must hence be fixed by ϕ, up to a scalar. If q_α is constant on Φ^+, then by Proposition 22.2 we necessarily have $q_\alpha = q$, and $\phi \circ \rho = \mathrm{id}$ on Φ^+. Thus ϕ is a permutation of the roots, and it preserves quotients of root lengths, so the same holds for ρ.

Since ρ induces a graph automorphism of the Coxeter diagram, G must be of type B_2, G_2 or F_4 if ρ does not preserve root lengths. This case is discussed in Exercise 30.6. □

Definition 22.4 We'll say that a Steinberg endomorphism $F : G \to G$ of a connected reductive group G is $\mathbb{F}_q$-*split* if there exists an F-stable maximal torus $T \le G$ such that $F(t) = t^q$ for all $t \in T$, or equivalently, if F acts as $q\,\mathrm{id}$ on $X(T)$.

It is called *twisted* if it is not split, and the product of an $\mathbb{F}_q$-split endomorphism with an (algebraic group) automorphism of G (see Theorem 11.11 for a description of these). Finally, we call F *very twisted* if some power of F induces the non-trivial graph automorphism on the Coxeter diagram of an irreducible component of G of type B_2, G_2 or F_4.

It turns out that a Steinberg endomorphism of a simple algebraic group is already essentially determined by $q > 0$ and $\phi \in \mathrm{Aut}(X_\mathbb{R})$ as follows (see [72, 11.7]):

Theorem 22.5 (Classification of Steinberg endomorphisms) *Let G be a simple simply connected algebraic group, $F : G \to G$ a Steinberg endomorphism. Then F is uniquely determined, up to inner automorphisms of G, by q and the Coxeter diagram automorphism $\rho|_\Delta$ in Proposition 22.2.*

Conversely, for every pair (q, ρ), with q an integral power of $\mathrm{char}(k)$ *and ρ a Dynkin diagram automorphism of Δ, there exists a Steinberg endomorphism F of G with parameters q, ρ. For ρ the non-trivial Coxeter diagram automorphism of Δ of type B_2, G_2, respectively F_4, there exists a Steinberg endomorphism with associated graph automorphism ρ if and only if $q^2 = 2^{2f+1}, 3^{2f+1}$, respectively 2^{2f+1}, for some $f \in \mathbb{N}_0$.*

The uniqueness essentially follows from the classification of simple algebraic groups via their root data, see Theorem 9.13. Let's discuss the question of existence of the various Steinberg endomorphisms.

Example 22.6 Let G be simple of simply connected type and let $F_p : G \to G$ be the endomorphism from Theorem 16.5 which acts as p-powering on a maximal torus and on the parameters of the associated root subgroups. It can be shown that this is an $\mathbb{F}_p$-split Steinberg endomorphism, so the same is true for all of its powers $F_q = F_p^f$.

Now let ρ be a permutation of the set of simple roots which induces a symmetry of the Dynkin diagram and σ the graph automorphism of G induced by ρ whose existence was asserted in Theorem 11.12, which permutes the root subgroups for simple roots according to ρ. Then F_p commutes with σ, so the composition $F = F_q\sigma$ has the property that $F^\delta = F_p^{f\delta}$ is $\mathbb{F}_{q^\delta}$-split. Thus, as F_q acts trivially on the root system, F is a Steinberg endomorphism which induces ρ on Δ.

More explicitly, for the groups of type A_n, $n \geq 2$, we have already exhibited in Example 21.2 a twisted Steinberg endomorphism inducing the non-trivial graph automorphism, with fixed point group the unitary group, and we will construct the two different types of Steinberg endomorphisms for SO_{2n} (of type D_n) in Example 22.9 below.

The following lifting result allows one to pass to simple groups of arbitrary isogeny type:

Proposition 22.7 *Let G be semisimple, $\pi : G_{sc} \to G$ the natural isogeny from a simply connected group of the same type. Then every isogeny $\sigma : G \to G$ can be lifted to an isogeny $\sigma_{sc} : G_{sc} \to G_{sc}$ such that $\pi \circ \sigma_{sc} = \sigma \circ \pi$.*

Proof Take $H_1 = G_{sc}$, $H_2 = G$ and $\varphi = \sigma \circ \pi$ in Proposition 9.18. $\square$

Thus, all Steinberg endomorphisms of semisimple groups are induced by Steinberg endomorphisms of simply connected groups.

Example 22.8 Conversely, a Steinberg endomorphism $F : G_{sc} \to G_{sc}$ of a simply connected group descends to an epimorphic image $G = \pi(G_{sc})$ under an isogeny π if $\ker(\pi)$ is F-stable. Since $\ker(\pi)$ is a subgroup of the fundamental group, in the case of simple groups the latter condition can only possibly fail when G is of type D_{2n}, $\ker(\pi)$ has order 2 and F acts non-trivially on the Dynkin diagram (see Table 9.2). Indeed, the graph automorphism of order 2 of D_{2n} interchanges the two subgroups of order 2 of the fundamental group which belong to the half-spin groups (see Example 9.16(3)). So the Steinberg endomorphism of G_{sc} inducing a non-trivial graph automorphism does not descend to the half-spin groups.

Also, the triality graph automorphism of D_4 of order 3 (see Example 12.12) acts non-trivially on the fundamental group, hence there do not exist Steinberg endomorphisms associated to triality graph automorphisms on SO_8 or on $HSpin_8$.

Thus, finally, by Examples 22.6 and 22.8 the existence question in Theorem 22.5 is only about the very twisted automorphisms of groups of exceptional type, which we will not discuss here. Note that there necessarily q is a non-integral power of 2 or 3. See [73, §11] for this case.

22.2 The finite groups G^F

Let G be a simple linear algebraic group with Steinberg endomorphism $F : G \to G$ inducing a graph automorphism of order δ on the Dynkin diagram such that $F^\delta = q^\delta \mathrm{id}$ on $X_{\mathbb{R}}$. If the root system of G has type R, the corresponding finite group of fixed points G^F is sometimes denoted $^\delta R(q)$ (an abuse of notation since it does not specify the isogeny type of G). The groups of type $^2B_2$ are called *Suzuki groups* since they were discovered by M. Suzuki [78] as a particular class of Zassenhaus groups; the groups of types $^2G_2$ and $^2F_4$ were first constructed by R. Ree [57, 58] and are hence called *Ree groups*. In the case where F induces a non-trivial diagram automorphism, the group G^F is sometimes also called a *twisted group of Lie type*.

Table 22.1 shows the various possibilities for the finite groups of Lie type G^F according to Theorem 22.5 and Proposition 22.7, and the identification of some of them with classical groups over finite fields.

Table 22.1   *Finite groups of Lie type*

Φ	δ	G_{sc}^F	G_{ad}^F	other isogeny types
A_{n-1}	1	$\mathrm{SL}_n(q)$	$\mathrm{PGL}_n(q)$	$Z_{d/e}.\mathrm{PSL}_n(q).Z_e$ $e\|d := (n, q-1)$
B_n	1	$\mathrm{Spin}_{2n+1}(q)$	$\mathrm{SO}_{2n+1}(q)$	$-$
C_n	1	$\mathrm{Sp}_{2n}(q)$	$\mathrm{PCSp}_{2n}(q)$	$-$
D_n	1	$\mathrm{Spin}_{2n}^+(q)$	$(\mathrm{PCO}_{2n}^\circ)^+(q)$	$\mathrm{SO}_{2n}^+(q), \mathrm{HSpin}_{2n}^+(q)$
G_2	1	$G_2(q)$		$-$
F_4	1	$F_4(q)$		$-$
E_6	1	$(E_6)_{\mathrm{sc}}(q)$	$(E_6)_{\mathrm{ad}}(q)$	$-$
E_7	1	$(E_7)_{\mathrm{sc}}(q)$	$(E_7)_{\mathrm{ad}}(q)$	$-$
E_8	1	$E_8(q)$		$-$
A_{n-1}	2	$\mathrm{SU}_n(q)$	$\mathrm{PGU}_n(q)$	$Z_{d/e}.\mathrm{PSU}_n(q).Z_e$ $e\|d := (n, q+1)$
D_n	2	$\mathrm{Spin}_{2n}^-(q)$	$(\mathrm{PCO}_{2n}^\circ)^-(q)$	$\mathrm{SO}_{2n}^-(q)$
D_4	3	$^3D_4(q)$		$-$
E_6	2	$(^2E_6)_{\mathrm{sc}}(q)$	$(^2E_6)_{\mathrm{ad}}(q)$	$-$
B_2	2	$^2B_2(q^2) = \mathrm{Suz}(q^2),\ q^2 = 2^{2f+1}$		$-$
G_2	2	$^2G_2(q^2),\ q^2 = 3^{2f+1}$		$-$
F_4	2	$^2F_4(q^2),\ q^2 = 2^{2f+1}$		$-$

Example 22.9 We construct the two types of even-dimensional orthogonal groups.

(1) Recall the orthogonal group GO_{2n} with respect to the quadratic form

$$f(x_1, \ldots, x_{2n}) := x_1 x_{2n} + \cdots + x_n x_{n+1}$$

on $V = \overline{\mathbb{F}_q}^{2n}$. Clearly GO_{2n} is stable under the standard Frobenius map $F_q : GL_{2n} \to GL_{2n}$. We write $GO_{2n}^+(q) := GO_{2n}^{F_q}$ and call it the *general orthogonal group of plus type*. Similarly, $SO_{2n}^+(q) := SO_{2n}^{F_q}$ (where $SO_{2n} = GO_{2n}^\circ$). Note that the subspace of V spanned by the first n standard basis vectors is totally singular for $SO_{2n}^+(q)$.

(2) Now let

$$g := \begin{pmatrix} I_{n-1} & & & \\ & 0 & 1 & \\ & 1 & 0 & \\ & & & I_{n-1} \end{pmatrix} \in GO_{2n}.$$

As $\det(g) = -1$ we have $g \in GO_{2n} \setminus SO_{2n}$ when $\mathrm{char}(k) \neq 2$. In fact, this also holds when $\mathrm{char}(k) = 2$. An easy calculation shows that g interchanges the root subgroups $U_{n-1,n}$ and $U_{n-1,n+1}$ of SO_{2n} (see Example 11.7), hence the simple roots $\epsilon_{n-1} - \epsilon_n$ and $\epsilon_{n-1} + \epsilon_n$, and fixes all other simple roots.

Thus $F' := gF : GO_{2n} \to GO_{2n}$ is a Steinberg endomorphism which induces the non-trivial graph automorphism of order 2 on the Dynkin diagram of SO_{2n} of type D_n. The fixed point group $GO_{2n}^-(q) := GO_{2n}^{F'}$, is called the *general orthogonal group of minus type*, and $SO_{2n}^-(q) := SO_{2n}^{F'}$ the *special orthogonal group of minus type*, or *non-split orthogonal group*.

Equivalently, these non-split groups can be obtained as follows: When $\mathrm{char}(k) \neq 2$ let G be the isometry group of the $\mathbb{F}_q$-anisotropic quadratic form $f(x_1, x_2) := x_1^2 - w x_2^2$ on $V = \overline{\mathbb{F}_q}^2$, where $w \in \mathbb{F}_q^\times$ is a fixed non-square. Since all non-degenerate quadratic forms on V are equivalent (see [2, §21], for example), we have $G \cong GO_2$ and $G^\circ \cong SO_2$ by Definition 1.15. The standard Frobenius map $F_q : GL(V) \to GL(V)$ stabilizes G. But it is easily seen that $V^{F_q} = \mathbb{F}_q^2$ contains no non-zero isotropic vector for f, so $(G^\circ)^{F_q}$ is not conjugate to the special orthogonal group $SO_2^+(q)$ on V^{F_q}. In fact, we have $(G^\circ)^{F_q} \cong SO_2^-(q)$. More generally, let G be the isometry group of the quadratic form

$$f(x_1, \ldots, x_{2n}) := x_1 x_{2n} + \cdots + x_{n-1} x_{n+2} + x_n^2 - w x_{n+1}^2$$

on $V = \overline{\mathbb{F}_q}^{2n}$, isomorphic to GO_{2n}. Then $G^{F_q} \cong GO_{2n}^-(q)$ and $(G^\circ)^{F_q} \cong SO_{2n}^-(q)$. A similar construction, starting from any two-dimensional $\mathbb{F}_q$-anisotropic quadratic form, works when $\mathrm{char}(k) = 2$.

See the books of Dieudonné [21] or Grove [30], for example, for more information on the various types of finite classical groups.

The various possibilities for the finite groups of Lie type G^F can again be encoded in a purely combinatorial way. The root datum $(X, \Phi, Y, \Phi^\vee)$ of a semisimple group G is obtained by fixing a maximal torus T of G. If $F : G \to G$ is a Steinberg endomorphism, we may assume T to be F-stable. Then its Weyl group $W = N_G(T)/T$ is also F-stable, and this defines a semidirect product $W\langle F \rangle$. Any element wF in the coset $W.F$ of W in $W\langle F \rangle$ still stabilizes T, hence the root datum. By Corollary 21.8 it also defines a G-conjugate, hence isomorphic, finite group of fixed points. So, by Theorem 22.5, (G, F) is determined up to isomorphism by the root datum of G, the coset $W\phi$ and q, where $\phi \in \mathrm{Aut}(X_\mathbb{R})$ stabilizes $\Phi \subset X$ and also $\Phi^\vee \subset Y$, hence normalizes W, in the following sense: if (G, F), (G', F') both correspond to $(X, \Phi, Y, \Phi^\vee, W\phi)$ and the same q, there is an isomorphism $\sigma : G \to G'$ such that $F' \circ \sigma = \sigma \circ F$.

Definition 22.10 $\mathbb{G} := (X, \Phi, Y, \Phi^\vee, W\phi)$ as above is called a *complete root datum*.

The complete root datum $\mathbb{G}$, together with q, determines (G, F) and hence the finite group G^F up to isomorphism by Corollary 21.8. We write $G^F =: \mathbb{G}(q)$. Thus, $\mathbb{G}$ stands for a whole family of finite groups of Lie type, for example $\{\mathrm{SL}_n(q)\}$, $\{\mathrm{SU}_n(q)\}$ or $\{E_8(q)\}$, where q runs over all prime powers.

It is not true, though, that groups G^F for different complete root data are necessarily non-isomorphic. For example we have $\mathrm{SL}_3(2) \cong \mathrm{PGL}_3(2)$ (non-isomorphic but isogenous algebraic groups may lead to isomorphic finite groups), $\mathrm{Sp}_{2n}(2^f) \cong \mathrm{SO}_{2n+1}(2^f)$ (different root systems may lead to isomorphic finite groups), or $\mathrm{Sp}_4(3) \cong \mathrm{SU}_4(2)$ (here even the underlying characteristic differs). But this happens not very often and all such "exceptional" isomorphisms are known, see Remark 24.9.

Example 22.11 (Complete root data for $\mathrm{GL}_n(q)$ and $\mathrm{GU}_n(q)$)

(1) Let $G \le \mathrm{GL}_n$ be a connected reductive subgroup stable under the standard Frobenius map F_q on GL_n, and assume that F_q acts trivially on the Weyl group W of G. Then ϕ is trivial and the corresponding complete root datum is $\mathbb{G} = (\Gamma, W)$, where Γ is the root datum of G.

(2) Let $G = \mathrm{GL}_n$, $F_q : G \to G$ the standard Frobenius map, and $\sigma : G \to G$, $(a_{ij}) \mapsto ((a_{ij})^{-\mathrm{tr}})^w$, with w as in Example 21.14(2). Let $F := F_q\sigma$, so $G^F = \mathrm{GU}_n(q)$. Then $T := \mathrm{D}_n$ is an F-stable maximal torus, and σ acts on $X(T)$ as $-w$, hence F acts as $-wq$ on $X(T)$. Thus, in this case

$\mathbb{G} = (\Gamma, -W)$ is the complete root datum for the family $\{GU_n(q)\}$ (with Γ the root datum of GL_n and W its Weyl group).

This is sometimes expressed by saying that $GU_n(q)$ is obtained from $GL_n(q)$ by replacing q by $-q$, or "$GU_n(q)$ is *Ennola-dual* to $GL_n(q)$".

23

Weyl group, root system and root subgroups

We already saw how the structure of semisimple algebraic groups is controlled by their root system and its Weyl group. We now show how the root system of a connected reductive group with a Steinberg endomorphism gives rise to a root system and then also to root subgroups of the corresponding group of fixed points.

23.1 The root system

In order to investigate the action of Steinberg endomorphisms on the root system and its Weyl group we first need another consequence of the Lang–Steinberg Theorem. In general taking fixed points under F is not exact, that is, it does not preserve short exact sequences.

Example 23.1 Consider the exact sequence of algebraic groups

$$1 \longrightarrow Z(\mathrm{SL}_n) \longrightarrow \mathrm{SL}_n \longrightarrow \mathrm{PGL}_n \longrightarrow 1$$

with fixed points

$$1 \longrightarrow Z(\mathrm{SL}_n(q)) \longrightarrow \mathrm{SL}_n(q) \longrightarrow \mathrm{PGL}_n(q)$$

under the standard Frobenius map. Here, the last arrow is in general not surjective, the image $\mathrm{PSL}_n(q)$ has index $\gcd(n, q - 1)$ in $\mathrm{PGL}_n(q)$, see Section 24.1.

This phenomenon cannot occur if the kernel is connected, see Exercise 30.7:

Proposition 23.2 *Let G be a linear algebraic group with a Steinberg endomorphism $F : G \to G$, and H an F-stable closed connected normal subgroup of G. Then the natural map $G^F/H^F \to (G/H)^F$ is an isomorphism.*

Applied to the normalizer $N_G(T)$ of an F-stable maximal torus T of a connected reductive group G with Steinberg endomorphism $F : G \to G$, this shows that the F-fixed points on W satisfy

$$W^F = (N_G(T)/T)^F \cong N_{G^F}(T)/T^F.$$

Note that in general, though, $N_{G^F}(T) \neq N_{G^F}(T^F)$; for example we might have $T^F = 1$ while $T \neq 1$, which happens when T is maximally split and $q = 2$, so $T^F = (\mathbb{F}_2^\times)^{\dim T}$. In order to study the fixed point group W^F we need the following result about automorphisms of Weyl groups. For this, recall the standard parabolic subgroup W_I of W, for I a subset of the set of simple reflections S of W.

Lemma 23.3 *Let W be the Weyl group of a root system with set of simple reflections $S \subset W$ and ϕ an automorphism of W stabilizing S.*

(a) *For each ϕ-orbit $I \subseteq S$, $W_I^\phi = \langle s_I \rangle$ for a (unique) involution $s_I \in W_I$.*
(b) *The group of fixed points W^ϕ is generated by $\{s_I \mid I \subseteq S$ a ϕ-orbit$\}$.*

Proof As ϕ stabilizes S, it induces a graph automorphism on the Coxeter diagram of W. So according to the Classification Theorem 9.6, there are only the following two possibilities for I: either $I = \{s_1, \ldots, s_r\}$ are mutually commuting and ϕ permutes them cyclically, in which case $s_I := s_1 \cdots s_r$ is as claimed, or $I = \{s_1, s_2\}$ with $o(s_1 s_2) = m > 2$. In the latter case an easy computation in the dihedral group $W_I = \langle s_1, s_2 \rangle$ shows that $W_I^\phi = \langle s_I \rangle$, where $s_I := s_1 s_2 s_1 \cdots$ (m factors). For (b) (and a proof of (a) not referring to the classification of root systems) see Lemma C.1.   □

The proof shows that the element s_I can also be described as being the longest element of the parabolic subgroup W_I. The various possibilities for W^ϕ when W is irreducible and ϕ is non-trivial are displayed in Table 23.1. It turns out that W^ϕ is again a Coxeter group, and even a Weyl group except for W of type F_4 with ϕ acting non-trivially; in the latter case, W^ϕ is isomorphic to the dihedral group $I_2(8)$ of order 16, which is a real, but non-rational reflection group.

Table 23.1 *Automorphisms of irreducible Weyl groups*

Φ	A_{2n-1}	A_{2n}	D_n	D_4	E_6	B_2	G_2	F_4
$o(\phi)$	2	2	2	3	2	2	2	2
W^ϕ	C_n	B_n	B_{n-1}	G_2	F_4	A_1	A_1	$I_2(8)$

We next introduce a root system for a finite group of Lie type. Let G be a semisimple algebraic group with a Steinberg endomorphism $F : G \to G$. As we saw in Section 22.1, fixing an F-stable torus inside an F-stable Borel subgroup B of G with associated root system Φ induces an action of F on the set of corresponding root subgroups U_α, $\alpha \in \Phi^+$, and hence a permutation ρ of Φ^+. As in Proposition 22.2 we write $F|_{X_\mathbb{R}} = q\phi$ with $\phi \in \mathrm{Aut}(X_\mathbb{R})$ of finite order δ. Let $X_\mathbb{R}^\phi$ denote the fixed space of ϕ on $X_\mathbb{R}$, i.e., the eigenspace for the eigenvalue 1. Then the homomorphism

$$\pi : X_\mathbb{R} \longrightarrow X_\mathbb{R}^\phi, \qquad \chi \mapsto \frac{1}{\delta} \sum_{i=0}^{\delta-1} \phi^i(\chi),$$

clearly is the identity on $X_\mathbb{R}^\phi$, hence a projection. Let $S \subset W$ denote the set of simple reflections of W.

Lemma 23.4   *Let $\Omega := \{w(\Phi_I^+) \mid I \subseteq S \text{ an } F\text{-orbit}, \ w \in W^F\}$. Then:*

(a) *Ω defines a partition of Φ. If $\sim$ denotes the associated equivalence relation, then $\alpha \sim \beta$ if and only if $\pi(\alpha), \pi(\beta)$ are positive multiples of each other.*

(b) *Each $\omega \in \Omega$ is contained in either Φ^+ or in Φ^-.*

For the proof see Lemma C.2.

Definition 23.5   For each equivalence class ω for the equivalence relation $\sim$ defined in Lemma 23.4 let α_ω denote the vector of maximal length among the $\{\pi(\alpha) \mid \alpha \in \omega\}$. The set $\Phi_F := \{\alpha_\omega \mid \omega \in \Omega\}$ is called the *root system* of G^F. We let $\Delta_F := \{\alpha_\omega \mid \omega \subseteq \Delta\} \subseteq \Phi_F$.

By Theorem C.5 the set Φ_F satisfies axioms (R1)–(R3) of an abstract root system in $X_\mathbb{R}^\phi$, with base Δ_F and Weyl group W^ϕ (in its action on $X_\mathbb{R}^\phi$) with set of simple reflections $S_F := \{s_I \mid I \subseteq S \text{ an } F\text{-orbit}\}$. For Φ indecomposable, Φ_F also satisfies the integrality axiom (R4) unless Φ is of type F_4 and ρ is non-trivial. Apart from this exceptional case, the type of the root system Φ_F can then be read off from Table 23.1. These facts will not be needed in the sequel (see also Exercise C.2).

Example 23.6   Let's consider G of type D_4 with $F : G \to G$ inducing the triality automorphism (see Example 12.12). Here

$$\Phi^+ = \{\alpha_1, \alpha_2, \alpha_3, \alpha_4, \alpha_1 + \alpha_2, \alpha_2 + \alpha_3, \alpha_2 + \alpha_4, \alpha_1 + \alpha_2 + \alpha_3,$$
$$\alpha_1 + \alpha_2 + \alpha_4, \alpha_2 + \alpha_3 + \alpha_4, \alpha_1 + \alpha_2 + \alpha_3 + \alpha_4, \alpha_1 + 2\alpha_2 + \alpha_3 + \alpha_4\}$$

with the simple roots numbered as in the Dynkin diagram in Table 9.1. The second simple root is fixed, while the other three are permuted cyclically by triality. Let's set

$$\mu := \pi(\alpha_1) = \frac{1}{3}(\alpha_1 + \alpha_3 + \alpha_4), \quad \nu := \pi(\alpha_2) = \alpha_2.$$

Then $\pi(\alpha_1+\alpha_2) = \pi(\alpha_2+\alpha_3) = \pi(\alpha_2+\alpha_4) = \mu+\nu$, $\pi(\alpha_1+\alpha_2+\alpha_3) = 2\mu+\nu$, $\pi(\alpha_1 + \alpha_2 + \alpha_3 + \alpha_4) = 3\mu + \nu$ and $\pi(\alpha_1 + 2\alpha_2 + \alpha_3 + \alpha_4) = 3\mu + 2\nu$. Thus here Φ_F is a root system of type G_2, with base $\{\mu, \nu\}$ (see Example 9.5).

23.2 Root subgroups

Let G be connected reductive with Steinberg endomorphism $F : G \to G$. We next construct root subgroups in G^F.

Proposition 23.7 *For any equivalence class $\omega \in \Omega$ the group*

$$U_\omega := \langle U_\alpha \mid \alpha \in \omega \rangle$$

is F-stable connected unipotent and has a product decomposition

$$U_\omega = \prod_{\alpha \in \omega} U_\alpha$$

with uniqueness of expression, for any fixed order of the $\alpha \in \omega$.

Proof After conjugation by some $w \in W^F$ we may assume that $\omega = \Phi_I^+$ for some ϕ-orbit $I \subseteq S$. Then U_ω is just the unipotent radical U_I of a Borel subgroup of the standard Levi subgroup L_I corresponding to I, see Section 12.2. The product decomposition thus follows from Theorem 11.1. Since all U_α are connected, so is U_ω. It is clearly F-stable, being generated by an F-stable set of subgroups (see Proposition 22.2(a)). □

The fixed point subgroups U_ω^F, $\omega \in \Omega$, are called the *root subgroups* of G^F. We next determine the orders of these root subgroups. For this, recall the q_α attached to $\alpha \in \Phi^+$ from Proposition 22.2.

Proposition 23.8 *Let U be an F-stable subgroup of $R_u(B)$ normalized by T. Then $|U^F| = \prod_{\alpha \in \Sigma} q_\alpha$, where $\Sigma := \{\alpha \in \Phi^+ \mid U_\alpha \leq U\}$.*

Proof We use induction on $|\Sigma|$. Let $U' := [U, U]$, the derived subgroup of U and $\Sigma' := \{\alpha \in \Phi^+ \mid U_\alpha \leq U'\}$ the corresponding roots. By Proposition 11.5 both U and U' are generated by the U_α they contain, and moreover the abelian group U/U' is naturally isomorphic to the direct product of the U_α with $\alpha \in \Sigma_1 := \Sigma \setminus \Sigma'$. Clearly U' is normalized by T, F-stable and of strictly smaller dimension than U, so by induction $|(U')^F| = \prod_{\alpha \in \Sigma'} q_\alpha$. Furthermore, as U' is connected, $|U^F| = |(U')^F| \cdot |(U/U')^F|$ by Proposition 23.2, so we need to show that $|(U/U')^F| = \prod_{\alpha \in \Sigma_1} q_\alpha$.

The Steinberg endomorphism F acts on the direct product of the U_α according to Proposition 22.2(a). Considering a single ρ-orbit in Σ_1 at a time we are reduced to showing that the number of solutions $t_1, \ldots, t_n \in k$ to the system

$$c_1 t_1^{q_1} = t_2, \ c_2 t_2^{q_2} = t_3, \ \ldots, \ c_n t_n^{q_n} = t_1,$$

equals $\prod_i q_i$, where $c_i \in k^\times$ and q_i are positive powers of $\mathrm{char}(k)$. Solving the first $n-1$ equations for the t_i, $i \geq 2$, in terms of t_1 we are left with the single equation $c t_1^r = t_1$, with $c \in k^\times$ and $r = \prod_i q_i$ an integral power of $\mathrm{char}(k)$, which indeed has precisely r solutions in the algebraically closed field k. $\square$

Corollary 23.9 *Let G be simple, $\omega \in \Omega$. Then $|U_\omega^F| = q^{|\omega|}$.*

Proof By Lemma 22.3 either all $q_\alpha = q$ for all $\alpha \in \Phi^+$, or $\omega = \{\alpha, \beta\}$ with α long, β short and $q_\alpha q_\beta = q^2$. In both cases, the claim follows from Propositions 23.7 and 23.8. $\square$

The values of $q^{|\omega|}$ for the non-trivial automorphisms of indecomposable root systems are shown in Table 23.2. The underlying diagrams are those for the Weyl group W^F from Table 23.1, respectively the Coxeter group $I_2(8)$.

Table 23.2 $q^{|\omega|}$ *for twisted groups of Lie type*

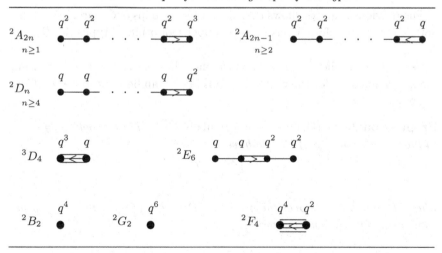

Example 23.10   Let's compute the structure of some root subgroups.

(1) Let G be semisimple, $F : G \to G$ a Steinberg endomorphism with trivial action on the Weyl group. Then $X_{\mathbb{R}}^{\phi} = X_{\mathbb{R}}$, the root system Φ_F equals Φ, and the root subgroups of G^F are just the $U_{\alpha}^F \cong \mathbb{F}_q^+$, for $\alpha \in \Phi$ (see Example 21.4(2)).

(2) Let $G = \mathrm{SL}_3$ and $F : G \to G$ the Steinberg endomorphism with $G^F = \mathrm{SU}_3(q)$ as in Example 21.14(2). Here $\Phi = \{\pm\alpha, \pm\beta, \pm(\alpha + \beta)\}$, and ρ interchanges α with β and fixes $\alpha + \beta$. Thus $\delta = 2$, $\pi(\alpha + \beta) = \alpha + \beta = 2\pi(\alpha) = 2\pi(\beta)$, and the equivalence classes under $\sim$ are just $\pm\{\alpha, \beta, \alpha + \beta\}$. For $\omega = \{\alpha, \beta, \alpha + \beta\}$ we have

$$U_{\omega} = U_{\alpha} U_{\beta} U_{\alpha+\beta} = \mathrm{U}_3,$$

the unipotent radical of the Borel subgroup T_3 of G. For our choice of F we get

$$U_{\omega}^F = \mathrm{U}_3^F = \left\{ \begin{pmatrix} 1 & a & b \\ 0 & 1 & -a^q \\ 0 & 0 & 1 \end{pmatrix} \;\middle|\; a, b \in \mathbb{F}_{q^2},\ b^q + b = -a^{q+1} \right\}$$

as (unique positive) root subgroup of $\mathrm{SU}_3(q)$ for the root $\alpha_{\omega} = \alpha + \beta$. It has order q^3, since for any value of $a \in \mathbb{F}_{q^2}$, there are q solutions to $x^q + x = -a^{q+1}$ in $\mathbb{F}_{q^2}$ (see also Corollary 23.9 and Table 23.2 for $^2A_2$).

(3) For G of type D_4 with $F : G \to G$ inducing the triality automorphism as in Example 23.6 we have $U_{\mu}^F \cong \mathbb{F}_{q^3}^+$ and $U_{\nu}^F \cong \mathbb{F}_q^+$ (see Table 23.2 for $^3D_4$).

The previous example shows that the root subgroups of G^F are, in general, not commutative. More precisely, this happens for orbits of types A_2, B_2 and G_2.

Using the Chevalley commutator formula (Theorem 11.8) for the algebraic group, an analogue for the root subgroups in G^F can be shown (see [73, Cor. of Lemma 62]):

Proposition 23.11 (Commutator formula in G^F)   *For all roots $\alpha_{\omega} \neq -\alpha_{\nu}$ in the root system Φ_F of G^F we have*

$$[U_{\omega}^F, U_{\nu}^F] \subseteq \prod_{\mu} U_{\mu}^F$$

where the product runs over the orbits $\mu \in \Omega$ such that $\alpha_{\mu} = m\alpha_{\omega} + n\alpha_{\nu}$ for some $m, n > 0$, in any fixed order.

24
A BN-pair for G^F

We now derive a Bruhat decomposition for the finite groups of Lie type by investigating the fixed points of a Steinberg endomorphism on the Bruhat decomposition for algebraic groups. This leads to a first order formula for groups of Lie type. A second, factored form can be derived from this using some standard results on the invariant theory of finite reflection groups. It turns out that the orders in a family of groups of Lie type belonging to the same complete root datum are given by a polynomial in q.

The Bruhat decomposition also gives rise to a BN-pair structure for finite groups of Lie type. Using general properties of BN-pairs this then allows us to conclude that most groups of Lie type derived from simple algebraic groups of simply connected type are quasi-simple, that is, simple modulo their center. Furthermore, we are able to determine their automorphism groups.

24.1 Bruhat decomposition and the order formula

Recall the double coset decomposition of a connected reductive group with respect to a Borel subgroup proved in Theorem 11.17. We obtain the following finite analogue:

Theorem 24.1 (Bruhat decomposition in G^F) *Let G be connected reductive with Steinberg endomorphism $F : G \to G$, $T \le B$ an F-stable maximal torus in an F-stable Borel subgroup of G, W the Weyl group of G with respect to T. Then*

$$G^F = \bigsqcup_{w \in W^F} (U_w^-)^F \dot{w} B^F$$

with uniqueness of expression. Here, for $w \in W^F$, $\dot{w}$ denotes a fixed preimage in $N_{G^F}(T)$.

Proof Theorem 11.17 gives the Bruhat decomposition

$$G = \bigsqcup_{w \in W} U_w^- \dot{w} B$$

with uniqueness. Now B and T are F-stable, so F acts on $W = N_G(T)/T$ and it permutes the U_w^-. Note that by Proposition 23.2 for each $w \in W^F$ we may choose an F-stable representative $\dot{w} \in N_{G^F}(T)$. Now if $g = u\dot{w}b \in G$ is F-stable, with $w \in W$, $u \in U_w^-$ and $b \in B$, then

$$u\dot{w}b = g = F(g) = F(u)F(\dot{w})F(b)$$

with $v = F(w) \in W$, $F(u) \in U_v^-$ and $F(b) \in B$. By uniqueness of the decomposition, this implies that $w \in W^F$ and so $F(\dot{w}) = \dot{w}$, and $F(b) = b$. Also, U_w^- is clearly F-stable for $w \in W^F$, so $F(u) = u$. It follows that $u \in (U_w^-)^F$, $w \in W^F$ and $b \in B^F$, so

$$G^F = \bigsqcup_{w \in W^F} (U_w^-)^F \dot{w} B^F. \qquad \square$$

As a first consequence we obtain a result on generation:

Corollary 24.2 *With the notation as in Theorem 24.1, the F-stable representatives $\dot{w}$ of $w \in W^F$ may be chosen in $\langle U^F, (U^-)^F \rangle$, where $U = R_u(B)$ and $U^- = U^{\dot{w}_0}$ is the unipotent radical of the opposite Borel subgroup. In particular $G^F = \langle U^F, (U^-)^F, T^F \rangle$.*

Proof Let $I \subseteq S$ be an F-orbit on S, $\nu \in \Omega$ the corresponding simple root with root subgroups $U_\nu \leq U$ and $U_{-\nu} = U_\nu^{\dot{w}_0} \leq U^-$. By Corollary 23.9 there exists $1 \neq u \in U_{-\nu}^F$. By the Bruhat decomposition in Theorem 24.1 applied to the reductive group $\langle U_\nu, U_{-\nu} \rangle^F$, this may be written as $u = u_1 \dot{w} u_2$ with $u_i \in U_\nu^F$ and $\dot{w}$ a preimage in $N_{G^F}(T)$ of $w \in W^F$. Since $U_\nu \cap U_{-\nu} = 1$ we have $w \neq 1$, so $w = w_I$ by Lemma 23.3, whence $\dot{w}_I \in \langle U^F, (U^-)^F \rangle$. Since W^F is generated by the w_I we may choose such representatives for all $w \in W^F$. The Bruhat decomposition for G^F then shows that $G^F = \langle U^F, (U^-)^F, T^F \rangle$. $\qquad \square$

We now determine the orders of the finite groups of Lie type. Recall the length function ℓ on W introduced in the proof of the Bruhat decomposition Theorem 11.17. According to Proposition A.21 we have the alternative characterization

$$\ell(w) = |\{\alpha \in \Phi^+ \mid w.\alpha \in \Phi^-\}| \qquad \text{for } w \in W.$$

Proposition 24.3 *Let G be simple with Steinberg endomorphism $F : G \to$ G. Let $T \le B$ be an F-stable maximal torus in an F-stable Borel subgroup, Φ the root system associated to B with set of positive roots Φ^+ and Weyl group W. Then*

$$|G^F| = |B^F| \sum_{w \in W^F} q^{\ell(w)} = q^{|\Phi^+|} |T^F| \sum_{w \in W^F} q^{\ell(w)}.$$

Proof The Bruhat decomposition

$$G^F = \bigsqcup_{w \in W^F} (U_w^-)^F \dot{w} B^F$$

from Theorem 24.1 implies that

$$|G^F| = \sum_{w \in W^F} |(U_w^-)^F| \cdot |B^F| = \sum_{w \in W^F} |(U_w^-)^F| \cdot |T^F| \cdot |U^F|$$

(using that $B^F = U^F \cdot T^F$). Now we have the factorization $U = R_u(B) = \prod_{\alpha \in \Phi^+} U_\alpha$ by Theorem 11.1 while $U_w^- = \prod_{\alpha \in \Phi^+, w.\alpha \in \Phi^-} U_\alpha$. Clearly, $\{\alpha \in \Phi^+ \mid w.\alpha \in \Phi^-\}$ is ϕ-invariant for $w \in W^F$, so by Corollary 23.9

$$|U^F| = q^{|\Phi^+|} \quad \text{and} \quad |(U_w^-)^F| = q^{|\{\alpha \in \Phi^+ \mid w.\alpha \in \Phi^-\}|} = q^{\ell(w)}$$

by the formula for $\ell(w)$ above. $\qquad\square$

Similarly, a formula for $|G^F|$ in terms of the q_α from Proposition 22.2 can be deduced for arbitrary connected reductive groups G, see Exercise 30.9.

The polynomial $\sum_{w \in W} x^{\ell(w)} \in \mathbb{Z}[x]$ is called the *Poincaré polynomial* of the Weyl group W. In order to obtain a factorization, we need another interpretation of this polynomial in terms of the invariant theory of W in its reflection representation on $V := X \otimes_\mathbb{Z} \mathbb{R}$. It is a fundamental property of *finite reflection groups*, i.e., finite linear groups generated by complex reflections, that their invariants form a polynomial ring:

Theorem 24.4 (Shephard–Todd, Chevalley) *Let $W \le \mathrm{GL}(V)$ be a finite reflection group on a real vector space V of dimension n. Then:*

(a) *The invariants $S(V)^W$ of W in the symmetric algebra $S(V)$ of V form a polynomial algebra.*

(b) *Let $f_1, \ldots, f_n$ denote algebraically independent generators of $S(V)^W$, homogeneous of degrees $d_i := \deg f_i$. Then the multiset $\{d_i \mid 1 \le i \le n\}$ is uniquely determined by W.*

(c) *We have $d_1 \cdots d_n = |W|$ and $d_1 + \cdots + d_n = N + n$, where N denotes the number of reflections in W.*

See [9, V, Thm. 5.3] for (a), [9, V, §5.1, Cor.] for (b), and [9, V, Cor. 5.3 and Prop. 5.3] for (c).

The $d_1, \ldots, d_n$ are called the *degrees* of the reflection group W. Note that for W the Weyl group of a root system Φ we have $N = |\Phi|/2 = |\Phi^+|$ since by axiom (R2), Proposition A.21 and Exercise A.6 there are precisely two roots for each reflection in W.

Returning to our algebraic group G, the Steinberg endomorphism F acts on V as a scalar times an automorphism ϕ of finite order δ, which normalizes W (see Proposition 22.2). Thus ϕ also acts naturally on the ring of invariants $S(V)^W$, and the algebraically independent homogeneous generators f_i may in fact be chosen to be eigenvectors of ϕ. We denote the corresponding eigenvalues by ϵ_i. Thus, the ϵ_i are roots of unity of order dividing δ. Clearly all $\epsilon_i = 1$ if F acts as a scalar on V. Using that the Poincaré polynomial is the Hilbert series of the coinvariant algebra of $S(V)$ one obtains (see [72, Thm. 2.1 and Cor. 2.9]):

Proposition 24.5 *Let $W \leq \mathrm{GL}(V)$ be a finite reflection group on a real vector space V of dimension n, $\phi \in N_{\mathrm{GL}(V)}(W)$ and f_i, $1 \leq i \leq n$, homogeneous generators of $S(V)^W$ of degrees d_i which are eigenvectors of ϕ with eigenvalues ϵ_i. Then the ϵ_i are also the eigenvalues of ϕ on V and we have*

$$\sum_{w \in W^\phi} x^{\ell(w)} = \prod_{i=1}^{n} \frac{x^{d_i} - \epsilon_i}{x - \epsilon_i}.$$

This leads to the following factorization of the order formula from Proposition 24.3 (see [72, Thm. 11.16]):

Corollary 24.6 *Let G be connected reductive, $F : G \to G$ a Steinberg endomorphism. Then*

$$|G^F| = q^{|\Phi^+|} \prod_{i=1}^{\mathrm{rk}(G)} (q^{d_i} - \epsilon_i),$$

where the d_i are the degrees of W and ϵ_i are as defined above.

Proof for the case that G is simple We evaluate the formula in Proposition 24.3. By Lemma 22.1(b) we have $|T^F| = |X/(F-1)X| = |\det_{X_{\mathbb{R}}}(F-1)|$ by the elementary divisor theorem. Using that $|\det(\phi)| = 1$ since ϕ is an automorphism of finite order, and that the characteristic polynomial of ϕ is real, so equal to the one of ϕ^{-1}, this equals

$$|T^F| = |\det_{X_{\mathbb{R}}}(q\phi - 1)| = |\det_{X_{\mathbb{R}}}(q - \phi^{-1})| = \prod_{i=1}^{n}(q - \epsilon_i).$$

Hence the claim follows with Proposition 24.5. □

It ensues that the order of G^F is already determined by the underlying complete root datum $\mathbb{G} = (X, \Phi, Y, \Phi^\vee, W\phi)$, together with q. We call

$$|\mathbb{G}| := x^{|\Phi^+|} \prod_{i=1}^{\text{rk}(\mathbb{G})} (x^{d_i} - \epsilon_i) \in \mathbb{Z}[x]$$

the *order polynomial* of the complete root datum $\mathbb{G}$. The preceding corollary can then be rephrased as saying that $|\mathbb{G}(q)| = |\mathbb{G}|(q)$, i.e., the order of G^F is the order polynomial evaluated at q. Note that $\deg|\mathbb{G}| = \dim G$ by Theorems 8.17(b) and 24.4(c).

A further immediate consequence of the above formula is the fact that $|\mathbb{G}(q)|$ is independent of the isogeny type of G, e.g., $|\text{SL}_n(q)| = |\text{PGL}_n(q)|$, and $|E_6(q)_{\text{sc}}| = |E_6(q)_{\text{ad}}|$. We also get a divisibility property of order polynomials, see Exercise 30.10:

Corollary 24.7 *Let G be connected reductive with Steinberg endomorphism $F : G \to G$, $H \le G$ a closed connected reductive F-stable subgroup, with complete root data $\mathbb{G}$, respectively $\mathbb{H}$. Then $|\mathbb{H}|$ divides $|\mathbb{G}|$ in $\mathbb{Z}[x]$.*

Example 24.8 Let's look at some examples of order polynomials.

(1) Let $G = \text{GL}_n$, F the standard Frobenius map. Here $|\Phi^+| = \binom{n}{2}$ by Example 9.8. The Weyl group acts on $V = X_\mathbb{R}$ as the symmetric group in its natural permutation representation, so the invariants $S(V)^W$ are generated by the elementary symmetric polynomials, of degrees $d_i = i$, $1 \le i \le n$. Since F acts as a scalar on X, $\epsilon_i = 1$ for all i. Therefore Corollary 24.6 gives the well-known formula

$$|\text{GL}_n(q)| = q^{\binom{n}{2}} \prod_{i=1}^{n} (q^i - 1).$$

(2) For $G = \text{SL}_n$ with the standard Frobenius map, the Weyl group acts via the deleted permutation module on $X_\mathbb{R}$, hence the invariants are generated by the elementary symmetric polynomials of degree at least 2. We obtain

$$|\text{SL}_n(q)| = q^{\binom{n}{2}} \prod_{i=2}^{n} (q^i - 1) = |\text{PGL}_n(q)|.$$

(3) For the special orthogonal group $G = \text{SO}_8$, with root system of type D_4 the degrees of W are $(d_1, d_2, d_3, d_4) = (2, 4, 6, 4)$ and $N = d_1 + d_2 +$

$d_3 + d_4 - n = 12$. If F acts trivially on the Weyl group, so $\epsilon_i = 1$ for $1 \le i \le 4$, then

$$|\mathrm{SO}_8^+(q)| = q^{12}(q^2 - 1)(q^4 - 1)^2(q^6 - 1).$$

For the triality graph automorphism on D_4, ϕ is of order 3, with eigenvalues $(\epsilon_1, \epsilon_2, \epsilon_3, \epsilon_4) = (1, \zeta_3, 1, \zeta_3^2)$ on suitable homogeneous generating invariants of degrees $(2, 4, 6, 4)$, where $\zeta_3 = \exp(2\pi i/3)$ is a primitive third root of unity. So

$$\begin{aligned}
|{}^3D_4(q)| &= q^{12}(q^2 - 1)(q^4 - \zeta_3)(q^6 - 1)(q^4 - \zeta_3^2) \\
&= q^{12}(q^2 - 1)(q^6 - 1)(q^8 + q^4 + 1).
\end{aligned}$$

The list of order polynomials for the various types is given in Table 24.1 (see for example [13, Prop. 10.2.5 and Thm. 14.3.2]).

Table 24.1 *Orders of finite groups of Lie type*

G^F	$|G^F|$	
$\mathrm{SL}_n(q)$	$q^{\binom{n}{2}}(q^2 - 1)(q^3 - 1)\cdots(q^n - 1)$	
$\mathrm{SU}_n(q)$	$q^{\binom{n}{2}}(q^2 - 1)(q^3 + 1)\cdots(q^n - (-1)^n)$	
$\mathrm{SO}_{2n+1}(q)$	$q^{n^2}(q^2 - 1)(q^4 - 1)\cdots(q^{2n} - 1)$	
$\mathrm{Sp}_{2n}(q)$	$q^{n^2}(q^2 - 1)(q^4 - 1)\cdots(q^{2n} - 1)$	
$\mathrm{SO}_{2n}^+(q)$	$q^{n^2-n}(q^2 - 1)(q^4 - 1)\cdots(q^{2n-2} - 1)(q^n - 1)$	
$\mathrm{SO}_{2n}^-(q)$	$q^{n^2-n}(q^2 - 1)(q^4 - 1)\cdots(q^{2n-2} - 1)(q^n + 1)$	
$G_2(q)$	$q^6(q^2 - 1)(q^6 - 1)$	
${}^3D_4(q)$	$q^{12}(q^2 - 1)(q^6 - 1)(q^8 + q^4 + 1)$	
$F_4(q)$	$q^{24}(q^2 - 1)(q^6 - 1)(q^8 - 1)(q^{12} - 1)$	
$E_6(q)$	$q^{36}(q^2 - 1)(q^5 - 1)(q^6 - 1)(q^8 - 1)(q^9 - 1)(q^{12} - 1)$	
${}^2E_6(q)$	$q^{36}(q^2 - 1)(q^5 + 1)(q^6 - 1)(q^8 - 1)(q^9 + 1)(q^{12} - 1)$	
$E_7(q)$	$q^{63}(q^2 - 1)(q^6 - 1)(q^8 - 1)(q^{10} - 1)(q^{12} - 1)(q^{14} - 1)(q^{18} - 1)$	
$E_8(q)$	$q^{120}(q^2 - 1)(q^8 - 1)(q^{12} - 1)(q^{14} - 1)(q^{18} - 1)(q^{20} - 1)(q^{24} - 1)$ $\cdot(q^{30} - 1)$	
${}^2B_2(q^2)$	$q^4(q^2 - 1)(q^4 + 1)$	$(q^2 = 2^{2f+1})$
${}^2G_2(q^2)$	$q^6(q^2 - 1)(q^6 + 1)$	$(q^2 = 3^{2f+1})$
${}^2F_4(q^2)$	$q^{24}(q^2 - 1)(q^6 + 1)(q^8 - 1)(q^{12} + 1)$	$(q^2 = 2^{2f+1})$

Remark 24.9 It is easy to see from the formulas in Table 24.1 that most of the time the defining prime occurs to a much higher power in $|G^F|$ than any other prime. From this, and using Zsigmondy's Theorem (see Theorem 28.3), it is straightforward to check for coincidences of orders among the (generally

simple, see Theorem 24.17) groups $G^F/Z(G^F)$ for G simple of simply connected type. Apart from those coming from an isomorphism of the underlying root system (like $A_3 \cong D_3$ or $B_2 \cong C_2$, for example), or from the isogeny between type B_n and C_n in characteristic 2 (viz. $\mathrm{Sp}_{2n}(2^f) \cong \mathrm{SO}_{2n+1}(2^f)$, see Remark 18.14(a)) there are only the following three cases of isomorphisms

$$\mathrm{PSL}_2(4) \cong \mathrm{PSL}_2(5), \ \mathrm{PSL}_2(7) \cong \mathrm{PSL}_3(2), \ \mathrm{PSU}_4(2) \cong \mathrm{PSp}_4(3),$$

and the following infinite series of order coincidences between non-isomorphic groups:

$$|\mathrm{PSp}_{2n}(q)| = |\mathrm{SO}_{2n+1}(q)'| \qquad \text{for } q \text{ odd, } n \geq 3$$

(see for example [44, §2.9]).

24.2 BN-pair, simplicity and automorphisms

The Bruhat decomposition allows us to exhibit a BN-pair in G^F (see Definition 11.15):

Theorem 24.10 *Let G be connected reductive with Steinberg endomorphism $F : G \to G$, $T \leq B$ an F-stable maximal torus in an F-stable Borel subgroup of G, $N := N_G(T)$. Then B^F, N^F is a BN-pair in G^F with Weyl group W^F.*

Proof The Bruhat decomposition in Theorem 24.1 shows that G^F is generated by B^F and N^F, whence (BN1). Next, $B^F \cap N^F = (B \cap N)^F = T^F$ is normal in $N^F = N_{G^F}(T)$, showing (BN2). By Proposition 23.2 we have

$$N^F/(B^F \cap N^F) = N^F/T^F \cong (N/T)^F = W^F,$$

which is generated by the set of involutions $\{s_I \mid I \subseteq S \text{ a } \phi\text{-orbit}\}$ by Lemma 23.3. For (BN4) let $w \in W^F$, and s_I corresponding to a ϕ-orbit $I \subseteq S$, so $s_I = s_1 \cdots s_r$ for suitable $s_i \in I$. Then an easy induction, using (BN4) for the group G, shows that

$$B\dot{w}B \cdot B\dot{s}_I B \subseteq \bigcup_{v \in W_I} B\dot{w}vB.$$

A double coset $B\dot{w}vB$ on the right-hand side is F-stable if and only if wv is, that is, if $v \in W^F$. But by Lemma 23.3 the only F-stable elements in W_I are 1 and s_I, whence taking F-fixed points we find (BN4).

Finally, let $s = s_I$ with preimage $\dot{s} \in N^F$ and set $U := U_s^-$. Then

$$\dot{s}U\dot{s} = \langle U_{-\alpha} \mid \alpha \in \Phi^+, s(\alpha) \in \Phi^- \rangle \leq \prod_{\alpha \in \Phi^-} U_\alpha \leq B^{\dot{w}_0},$$

so $B \cap \dot{s} U \dot{s} \leq B \cap B^{\dot{w}_0} = T$ by Corollary 11.18. On the other hand, U is F-stable, so U^F contains non-trivial unipotent elements by Proposition 23.8. This shows that $\dot{s} B^F \dot{s} \geq \dot{s} U^F \dot{s}$ cannot be contained in B^F, thus (BN5) holds. $\qquad\square$

This means that all general results on groups with a BN-pair which were derived in Sections 11.2 and 12.1 also hold for the group G^F. In particular, there is a natural notion of parabolic subgroups and Levi subgroups of G^F, coming from its BN-pair structure. We will investigate the relation between these and the F-fixed points of parabolic subgroups of G later in Section 26.1. For the moment, let's just observe the following:

Corollary 24.11 *In the situation of Theorem 24.10, let $U = R_u(B)$. Then U^F is a Sylow p-subgroup of G^F, with normalizer $N_{G^F}(U^F) = B^F$.*

Proof By Proposition 12.2, $N_{G^F}(U^F)$ is a parabolic subgroup of G^F. By (BN5), it does not contain any simple reflection, so it equals B^F. By Proposition 23.2 we have

$$|B^F : U^F| = |B^F / U^F| = |(B/U)^F| = |T^F|.$$

Since T consists of semisimple elements, this cardinality is prime to the field characteristic p. So the p-subgroup U^F of G^F has index prime to p in its normalizer. By elementary group theory this implies that U^F is a Sylow p-subgroup (see, for example [2, (9.10)]). $\qquad\square$

One may be tempted to expect that for G a simple algebraic group, G^F is again close to being a (finite) simple group. In fact, most of the argument in the proof of Proposition 12.5 only depends on the BN-pair axioms, and we obtain:

Lemma 24.12 *Let G be a simple algebraic group with Steinberg endomorphism $F : G \to G$. Any normal subgroup H of G^F either satisfies $B^F H = G^F$ or $H \leq Z(G^F)$.*

Proof Let $T \leq B$ denote an F-stable maximal torus inside an F-stable Borel subgroup of G. Since G^F has a BN-pair by Theorem 24.10, the arguments given in the proof of Proposition 12.5 apply to the present situation to show that any proper normal subgroup H of G^F either lies in B^F or satisfies $B^F H = G^F$.

If $H \leq B^F$, then it is contained in all G^F-conjugates of B^F. The set of simple roots Δ with respect to B is F-stable, hence so is the unique element w_0 sending it to $-\Delta$ (the longest element of W). But $B \cap B^{\dot{w}_0} = T$ by Corollary 11.18. So any proper normal subgroup $H \leq B^F$ of G^F lies in the abelian group T^F. Let $U := R_u(B)$. Then $[U^F, H] \leq U^F \cap H \leq U \cap T = 1$, whence $H \leq Z(B^F)$, and similarly $H \leq Z((B^{\dot{w}_0})^F)$. Since $G^F = \langle B^F, (B^F)^{\dot{w}_0} \rangle$ by Corollary 24.2 this shows that $H \leq Z(G^F)$. $\square$

Corollary 24.13 *Let G be a simple algebraic group with Steinberg endomorphism $F : G \to G$. Then $Z(G^F) = Z(G)^F$.*

In particular, for G of simply connected type, $Z(G^F)$ is as given in Table 24.2. Here, an entry d stands for a cyclic group of that order, while $(2, q-1)^2$ denotes a group $Z_d \times Z_d$, with $d = (2, q-1)$.

Table 24.2 *Centers of groups of simply connected type*

G^F	$Z(G^F)$	G^F	$Z(G^F)$
$\mathrm{SL}_n(q)$, $n \geq 2$	$(n, q-1)$	$^2B_2(2^{2f+1})$	1
$\mathrm{SU}_n(q)$, $n \geq 3$	$(n, q+1)$	$^2G_2(3^{2f+1})$	1
$\mathrm{Spin}_{2n+1}(q)$, $n \geq 3$	$(2, q-1)$	$G_2(q)$	1
$\mathrm{Sp}_{2n}(q)$, $n \geq 2$	$(2, q-1)$	$^3D_4(q)$	1
$\mathrm{Spin}_{2n}^+(q)$, $n \geq 4$ even	$(2, q-1)^2$	$^2F_4(2^{2f+1})$	1
$\mathrm{Spin}_{2n}^+(q)$, $n \geq 5$ odd	$(4, q-1)$	$F_4(q)$	1
$\mathrm{Spin}_{2n}^-(q)$, $n \geq 4$ even	$(2, q-1)$	$E_6(q)$	$(3, q-1)$
$\mathrm{Spin}_{2n}^-(q)$, $n \geq 5$ odd	$(4, q+1)$	$^2E_6(q)$	$(3, q+1)$
		$E_7(q)$	$(2, q-1)$
		$E_8(q)$	1

Proof We have $Z(G)^F \leq Z(G^F) \leq T^F$, the latter inclusion by the previous proof. Now assume that $t \in T^F \setminus Z(G)^F$. Then by Theorem 8.17 there exists $\alpha \in \Phi$ with $\alpha(t) \neq 1$. But then also $F^i(\alpha)(t) = \alpha(F^i(t)) = \alpha(t) \neq 1$ for all i, so t acts fixed-point freely on the root subgroup U_ω, with ω the equivalence class of α as in Lemma 23.4. Thus, $t \notin Z(G^F)$ by Corollary 23.9.

By Proposition 9.15 the center $Z(G)$ of a simple group G of simply connected type is isomorphic to the p'-part of the fundamental group Λ of the root system of G, as given in Table 9.2. The center $Z(G^F)$ can then be computed by the formula in the first part. $\square$

Corollary 24.14 *Let G be a simple algebraic group with Steinberg endomorphism $F : G \to G$. If G^F is perfect, then $G^F/Z(G^F)$ is a finite simple group.*

Proof By Lemma 24.12 any normal subgroup H of G^F lies in $Z(G^F)$ or satisfies $B^F H = G^F$. In the latter case

$$G^F/H = B^F H/H \cong B^F/(B^F \cap H),$$

with the right-hand side being solvable since B is, while G^F is perfect, so $H = G^F$. The result follows. ☐

This criterion leads one to ask under which conditions G^F is perfect, and how this can be verified. For this, let's set $G_1 := \langle G_u^F \rangle$, the normal (in fact, characteristic) subgroup of G^F generated by its unipotent elements.

Theorem 24.15 (Steinberg) *Let G be a simply connected semisimple linear algebraic group with Steinberg endomorphism $F : G \to G$. Then $G^F = G_1$, that is, G^F is generated by its unipotent elements.*

Proof Let $B \le G$ be an F-stable Borel subgroup of G with F-stable maximal torus T and unipotent radical U, and $U^- = U^{\dot{w}_0}$ the unipotent radical of the opposite Borel subgroup. By Corollary 24.2 we have that $G^F = \langle U^F, (U^-)^F, T^F \rangle$. Since T^F normalizes both U^F and $(U^-)^F$, we conclude that $G_1 = \langle U^F, (U^-)^F \rangle$.

Thus it remains to show that $T^F \le \langle U^F, (U^-)^F \rangle$. Now for $\alpha \in \Delta$ let $G_\alpha = \langle U_\alpha, U_{-\alpha} \rangle$ and $T_\alpha = T \cap G_\alpha$, a one-dimensional torus (see Section 8.3). Since G is simply connected, the coroots $\{\alpha^\vee : k^\times \to T_\alpha \le T \mid \alpha \in \Delta\}$ generate $Y(T)$, so $T = \prod_{\alpha \in \Delta} T_\alpha$. Collecting according to F-orbits we obtain the F-invariant decomposition $T = \prod_{I \subseteq S} T_I$, where I runs over F-orbits in S and $T_I := \prod_{\alpha \in \Delta_I} T_\alpha$, so $T^F = \prod_{I \subseteq S} T_I^F$.

We are thus reduced to the case that S is a single F-orbit. Moreover, by Exercise 30.2 we may assume that G is simple. But then Φ has root system of type A_1, A_2, B_2 or G_2. In the second case, $G^F = \mathrm{SU}_3(q)$ and the claim can be checked by direct computation, as in Exercise 30.11. In the other three cases, $W^F = \{1, w = w_S\}$. Let $t \in T^F$ and $t_1 \in T$ with $t_1^2 = t$. Then $\dot{w} t_1 \dot{w}^{-1} = t_1^{-1}$, so $t_1^{-1} \dot{w} t_1 = \dot{w} t$. Recall from Corollary 24.2 that we may choose $\dot{w} \in G_1$. If $\mathrm{char}(k) \ne 2$ then $\dot{w}$ and $\dot{w} t$ are semisimple elements of G^F which are G-conjugate, so also G^F-conjugate by Theorem 26.7(c) below. In particular, as we chose $\dot{w} \in G_1$, $\dot{w} t \in G_1$ and so $t \in G_1$. If $\mathrm{char}(k) = 2$ this also holds, since then $|T^F|$ is odd, so t_1 may already be chosen in T^F. ☐

Example 24.16 Let $G = \mathrm{SL}_2$ with Steinberg endomorphism F, so $G^F = \mathrm{SL}_2(q)$ for some prime power q by the classification of Steinberg endomorphisms. We claim that $G_1 = \langle G_u^F \rangle$ is perfect for $q > 3$ and equal to G^F. Indeed, by Theorem 24.15 $\mathrm{SL}_2(q)$ is generated by its root subgroups $U_{\pm\alpha}^F = \{u_{\pm\alpha}(c) \mid c \in \mathbb{F}_q\}$, where

$$u_\alpha(c) := \begin{pmatrix} 1 & c \\ 0 & 1 \end{pmatrix}, \qquad u_{-\alpha}(c) := \begin{pmatrix} 1 & 0 \\ c & 1 \end{pmatrix},$$

so $G^F = G_1$. Since $q > 3$ there exists $a \in \mathbb{F}_q^\times \setminus \{\pm 1\}$. Set $t := \mathrm{diag}(a, a^{-1}) \in$

G^F. Then $tu_{\pm\alpha}(c)t^{-1} = u_{\pm\alpha}(a^{\pm 2}c)$, so by the choice of a the map

$$U_{\pm\alpha}^F \longrightarrow U_{\pm\alpha}^F, \qquad u_{\pm\alpha}(c) \mapsto [t, u_{\pm\alpha}(c)] = u_{\pm\alpha}((a^{\pm 2} - 1)c),$$

is surjective. Thus $U_{\pm\alpha}^F \leq [G^F, G^F]$, and so G^F is perfect. Note that $SL_2(q)$ is solvable for $q \leq 3$.

Similarly using Example 23.10(2) one can check that $SU_3(q)$ is perfect for $q > 2$ (see Exercise 30.11).

This example generalizes to yield perfectness and hence simplicity modulo center in the simply connected case.

Theorem 24.17 (Tits)   *Let G be a simply connected simple linear algebraic group with Steinberg endomorphism $F : G \to G$. Then, unless G^F is one of*

$$SL_2(2), \ SL_2(3), \ SU_3(2), \ Sp_4(2), \ G_2(2), \ {}^2B_2(2), \ {}^2G_2(3), \ {}^2F_4(2),$$

$G_1 = G^F$ is perfect, and thus $G^F/Z(G^F)$ is simple.

Proof when F is not very twisted and $q > 3$ If F is not very twisted, then an orbit ω of F on the set of simple roots of G either consists of mutually orthogonal roots, so is of type $A_1 \times \cdots \times A_1$, or it is of type A_2. In the first case, the various $\langle U_{\pm\alpha}\rangle$, for $\alpha \in \omega$, commute pairwise and are permuted transitively by F, so

$$\langle U_{\pm\alpha} \mid \alpha \in \omega \rangle^F \cong \langle U_{\pm\beta}\rangle^{F^m}$$

for $m = |\omega|$ and any $\beta \in \omega$, by Exercise 30.2. But $G_\beta := \langle U_{\pm\beta}\rangle$ is a central quotient of SL_2 by Section 8.2 and Theorem 8.17(f), so an application of Example 24.16 shows that $(G_\beta)_1$ is perfect, and similarly when ω is of type A_2 (see Exercise 30.11). Now by Proposition A.11 any root subgroup of $G^F = G_1$ is conjugate to a root subgroup corresponding to a simple root. So $[G_1, G_1]$ contains all root subgroups of G_1, hence a Sylow p-subgroup by Corollary 24.11, and thus equals G_1. Now apply the previous theorem and Corollary 24.14. $\qquad\square$

A slight variation of the above argument applies when $q \leq 3$. The groups of type ${}^2B_2$ and ${}^2G_2$ have to be treated differently. Then, the groups of type ${}^2F_4$ can again be handled as before, see [82].

Remark 24.18   The groups $SL_2(2), SL_2(3), SU_3(2), {}^2B_2(2)$ occurring in the exceptions of Theorem 24.17 are solvable, the groups $Sp_4(2) \cong \mathfrak{S}_6$, $G_2(2) \cong \text{Aut}(PSU_3(3))$, ${}^2G_2(3) \cong \text{Aut}(PSL_2(8))$ are almost simple, and ${}^2F_4(2)$ contains a normal simple subgroup of index 2 which does not occur elsewhere in the classification of finite simple groups, the so-called *Tits group* (see [82, 4.3]).

Remark 24.19 The fact that G^F in Theorem 24.17 is perfect and $S = G^F/Z(G^F)$ is simple implies by definition that G^F is a so-called covering group of S. In fact, whenever G^F is perfect then, except for the finitely many cases in Table 24.3 it is the full covering group of the simple group S, that is, $Z(G^F)$ as given in Table 24.2 is the Schur multiplier $M(S)$ of S (see e.g. [29, Table 6.1.3]). Here $O_7(3) = \mathrm{Spin}_7(3)/Z(\mathrm{Spin}_7(3))$, which is also isomorphic to the derived subgroup of $SO_7(3)$ (see Proposition 24.21 below).

Table 24.3 *Non-generic Schur multipliers*

S	$M(S)$	S	$M(S)$	S	$M(S)$
$PSL_2(4)$	Z_2	$PSU_4(3)$	$Z_{12} \times Z_3$	$^2B_2(8)$	$Z_2 \times Z_2$
$PSL_2(9)$	Z_6	$PSU_6(2)$	$Z_6 \times Z_2$	$G_2(3)$	Z_3
$PSL_3(2)$	Z_2	$Sp_6(2)$	Z_2	$G_2(4)$	Z_2
$PSL_3(4)$	$Z_{12} \times Z_4$	$O_7(3)$	Z_6	$F_4(2)$	Z_2
$PSL_4(2)$	Z_2	$SO_8^+(2)$	$Z_2 \times Z_2$	$^2E_6(2)$	$Z_6 \times Z_2$
$PSU_4(2)$	Z_2				

Some of the exceptional multipliers are "explained" by exceptional automorphisms, see Remark 24.9. It may be noted that the generic Schur multiplier always has order prime to the characteristic p, while the exceptional part has order a power of p.

In order to investigate the general, non-simply connected case, we need the following elementary result (see Proposition 23.2 for a related statement):

Lemma 24.20 *Let H be a group, $Z \le H$ a central subgroup, and $F : H \to H$ an automorphism normalizing Z. Then the short exact sequence*

$$1 \longrightarrow Z \longrightarrow H \longrightarrow H/Z \longrightarrow 1$$

induces a long exact sequence

$$1 \longrightarrow Z^F \longrightarrow H^F \longrightarrow (H/Z)^F \longrightarrow (L(H) \cap Z)/L(Z) \longrightarrow 1,$$

where $L : H \to H$ is defined by $L(h) := F(h)h^{-1}$.

For the proof, see Exercise 30.12.

Proposition 24.21 *Let G be a simple algebraic group, $\pi : G_{\mathrm{sc}} \to G$ the natural isogeny from a group of simply connected type with central kernel from Proposition 9.15. Let F be a Steinberg endomorphism on G_{sc} normalizing $Z := \ker(\pi)$. Then:*

(a) $G^F/\pi(G_{\mathrm{sc}}^F) \cong Z/L(Z)$ *and*
(b) $\pi(G_{\mathrm{sc}}^F) \cong G_{\mathrm{sc}}^F/Z^F$.

In particular, if G_{sc}^F is perfect then $\pi(G_{\mathrm{sc}}^F) = [G^F, G^F] \cong G_{\mathrm{sc}}^F/Z^F$, and $[G_{\mathrm{ad}}^F, G_{\mathrm{ad}}^F]$ is simple.

Proof We apply Lemma 24.20 with $H = G_{\mathrm{sc}}$. As G_{sc} is connected the map $L : G_{\mathrm{sc}} \to G_{\mathrm{sc}}$, $g \mapsto F(g)g^{-1}$, is surjective by Theorem 21.7 and we obtain the exact sequence

$$1 \longrightarrow Z^F \longrightarrow G_{\mathrm{sc}}^F \overset{\pi}{\longrightarrow} G^F \longrightarrow Z/L(Z) \longrightarrow 1,$$

which yields (a) and (b). Clearly, if G_{sc}^F is perfect, then so is its image $\pi(G_{\mathrm{sc}}^F)$, which must hence be contained in $[G^F, G^F]$. On the other hand, $G^F/\pi(G_{\mathrm{sc}}^F)$ is abelian by (a), so we have equality $[G^F, G^F] = \pi(G_{\mathrm{sc}}^F)$.

Finally, if $G = G_{\mathrm{ad}}$ is of adjoint type, then $\ker(\pi) = Z(G_{\mathrm{sc}})$ (see Section 9.2), so $[G^F, G^F] \cong G_{\mathrm{sc}}^F/Z^F = G_{\mathrm{sc}}^F/Z(G_{\mathrm{sc}}^F)$ (see Corollary 24.13) is simple by Theorem 24.17. $\qquad\square$

Example 24.22 Let $G = \mathrm{PGL}_n$ with standard Frobenius map F so that $G^F = \mathrm{PGL}_n(q)$. This is a solvable group for $(n, q) \in \{(2, 2), (2, 3)\}$. Otherwise, using that $G_{\mathrm{sc}} = \mathrm{SL}_n$ we see from Proposition 24.21 that the derived subgroup equals

$$[\mathrm{PGL}_n(q), \mathrm{PGL}_n(q)] \cong \mathrm{SL}_n(q)/Z(\mathrm{SL}_n(q)) = \mathrm{PSL}_n(q) = G_1.$$

Since $|\mathrm{SL}_n(q)| = |\mathrm{PGL}_n(q)|$, this has index $|Z(\mathrm{SL}_n)^F| = |Z(\mathrm{SL}_n(q))| = \gcd(n, q-1)$ in G^F and is simple (compare with the situation over k where $\mathrm{PGL}_n \cong \mathrm{SL}_n/Z(\mathrm{SL}_n)$ as abstract groups).

The automorphisms of the simple group $[G_{\mathrm{ad}}^F, G_{\mathrm{ad}}^F]$ induced by G_{ad}^F are called *diagonal* (since for $G_{\mathrm{ad}} = \mathrm{PGL}_n$ they are induced by conjugation with diagonal matrices).

Further automorphisms of finite groups of Lie type arise as follows: Let G be semisimple. The endomorphism $F_p : G \to G$ from Theorem 16.5 is an $\mathbb{F}_p$-split Steinberg endomorphism of G (see Example 22.6) and thus all $\mathbb{F}_q$-split Steinberg endomorphisms, where $q = p^f$, are obtained as powers $F = F_p^f$. Then clearly F_p stabilizes G^F; the induced automorphism $F_p : G^F \to G^F$ and its powers are called *field automorphisms* of G^F. Furthermore, as stated in Example 22.6 the graph automorphisms $\sigma : G \to G$ from Theorem 11.12 induced by symmetries of the Dynkin diagram of G can be chosen so as to commute with F_p, so restrict to automorphisms $\sigma : G^F \to G^F$; these are called the *graph automorphisms* of G^F. Clearly, any automorphism of G^F descends to the characteristic subgroup $[G^F, G^F]$.

In fact, these already generate the full automorphism group. Let's first consider the case of groups of rank 1:

Proposition 24.23 *Any automorphism of $S = \mathrm{PSL}_2(p^f)$ is the product of an inner, a diagonal and a field automorphism.*

This is an easy exercise, see for example [37, Aufg. 11.15].

Theorem 24.24 (Steinberg) *Let G_{sc} be simple of simply connected type, $F : G_{\mathrm{sc}} \to G_{\mathrm{sc}}$ a split Frobenius map, $G = G_{\mathrm{ad}}$ the adjoint type image under the natural isogeny, and assume that G_{sc}^F is perfect (and hence $S := [G^F, G^F]$ is simple). Then any element of $\mathrm{Aut}(S)$ is the product of an inner, a diagonal, a field and a graph automorphism.*

Proof for the case that all roots of G have the same length Let $T \le B$ be a maximal torus inside a Borel subgroup of G with root system Φ and set of simple roots Δ. As pointed out above, we may assume that $F = F_p^f$ for the endomorphism F_p from Theorem 16.5, for some $f \ge 1$, and T, B are F-stable and F acts trivially on Φ. Let $U = R_u(B)$ be the unipotent radical of B. Then $P := U^F$ is a Sylow p-subgroup of G^F with normalizer B^F by Corollary 24.11. By Proposition 24.21, G^F/S consists of semisimple elements, so P is a Sylow p-subgroup of S. Thus, by Sylow's Theorem any $\sigma \in \mathrm{Aut}(S)$ can be multiplied by an inner automorphism so as to stabilize P.

We claim that we may modify σ further so as to also stabilize $P^- := (U^-)^F = (U^{\dot{w}_0})^F$, a second Sylow p-subgroup of S. Indeed, $\sigma(P^-)$ must be S-conjugate to P, by an element $g = u\dot{w}b \in S$ with $w \in W^F$, $u \in (U_w^-)^F$ and $b \in B^F$ (see Theorem 24.1); as B normalizes U we have $\sigma(P^-) = u\dot{w}P\dot{w}^{-1}u^{-1}$. Now

$$u\dot{w}P\dot{w}^{-1}u^{-1} \cap P = \sigma(P^-) \cap \sigma(P) = \sigma(P^- \cap P) \le \sigma(U^- \cap U) = 1$$

by Corollary 11.18, whence $\dot{w}P\dot{w}^{-1} \cap P = 1$ by conjugation by $u \in (U_w^-)^F \le U^F = P$. If $\alpha \in \Phi^+$ is not made negative by w, then $\dot{w}U_\alpha^F\dot{w}^{-1} \le P$, so w must send all positive roots to negative ones and thus $w = w_0$. So $\sigma(P^-) = u\dot{w}_0P\dot{w}_0^{-1}u^{-1} = uP^-u^{-1}$. Changing σ by the inner automorphism induced by u (which stabilizes P) then gives $\sigma(P^-) = P^-$.

Since σ fixes B^F, it permutes the minimal parabolic subgroups above B^F, which are of the form $B^F \cup B^F\dot{s}_\alpha B^F$ for some $\alpha \in \Delta$ by Proposition 12.2 (and the subsequent remark). Now $U^- \cap B\dot{s}_\alpha B = U_{-\alpha}$, so σ induces a permutation on the set $\{U_{-\alpha}^F \mid \alpha \in \Delta\}$, and similarly a permutation ρ on $\{U_\alpha^F \mid \alpha \in \Delta\}$. Note that these two permutations must agree since $U_\alpha, U_{-\beta}$ commute for distinct $\alpha, \beta \in \Delta$, by the commutator formulas in Theorem 11.8, while $[U_\alpha^F, U_{-\alpha}^F] \ne 1$.

Furthermore, ρ induces a graph automorphism of the Dynkin diagram since α, β are linked in the Dynkin diagram if and only if $[U_\alpha^F, U_\beta^F] \neq 1$ (recall that we assume that all roots have the same length). By Theorem 11.12 and the remarks preceding this theorem there exists a graph automorphism of G^F corresponding to the inverse permutation ρ^{-1}; its composition with σ fixes the U_α^F, $\alpha \in \pm\Delta$.

So we now have $\sigma(u_\alpha(1)) = u_\alpha(a_\alpha)$ for all $\alpha \in \Delta$ and suitable $a_\alpha \in \mathbb{F}_q$. We claim that we may change σ by a diagonal automorphism so that $a_\alpha = 1$ for $\alpha \in \Delta$. Indeed, since Δ is linearly independent, there exists $t \in T$ with $\alpha(t) = a_\alpha$ for all $\alpha \in \Delta$. Now $F(\alpha(t)) = F(a_\alpha) = a_\alpha = \alpha(t)$ for all $\alpha \in \Delta$, so $F(\chi)(t) = \chi(t)$ for all $\chi \in \mathbb{Z}\Delta = X$ (since G is of adjoint type), whence $t \in T^F$. Replacing σ by its composition with the diagonal automorphism induced by $t \in G^F$ we thus obtain that $\sigma(u_\alpha(1)) = u_\alpha(1)$ for $\alpha \in \Delta$.

The case of rank 1 is settled by Proposition 24.23. Otherwise, let $\alpha_1, \alpha_2 \in \Delta$ be linked in the Dynkin diagram, so that $\alpha_3 = \alpha_1 + \alpha_2 \in \Phi$. Then

$$\sigma(u_{\alpha_i}(c)) = u_{\alpha_i}(f_i(c))$$

for suitable functions $f_i : \mathbb{F}_q \to \mathbb{F}_q$, $1 \le i \le 3$, with $f_1(1) = f_2(1) = 1$. Since u_{α_i} is additive, so is f_i. Moreover, by the commutator relations in Theorem 11.8 and Exercise 20.6 we have

$$[u_{\alpha_1}(c_1), u_{\alpha_2}(c_2)] = u_{\alpha_3}(\pm c_1 c_2),$$

so application of σ gives that

$$f_3(\pm c_1 c_2) = \pm f_1(c_1) f_2(c_2)$$

for all $c_1, c_2 \in \mathbb{F}_q$, so $f_3(c_1 c_2) = f_1(c_1) f_2(c_2)$. Choosing $c_1 = 1$ we get that $f_3(c_2) = f_2(c_2)$, and similarly $f_3(c_1) = f_1(c_1)$, hence $f_1 = f_2 = f_3$ is also multiplicative. So in fact f_i is a field automorphism of $\mathbb{F}_q$. Composing σ with the inverse field automorphism of S, we reach that $\sigma(u_\alpha(c)) = u_\alpha(c)$ for all $\alpha \in \Delta$, $c \in \mathbb{F}_q$.

Thus for $\alpha \in \Delta$, σ stabilizes $\langle U_\alpha^F, U_{-\alpha}^F \rangle$ and is the identity on U_α^F. By Proposition 24.23 this forces σ to also centralize $U_{-\alpha}^F$, hence some representative $\dot{s}_\alpha$ of the simple reflection $s_\alpha \in W$. As W^F is generated by the s_α, and U^F, $(U^-)^F$ are products of suitable W^F-conjugates of the U_α^F, σ centralizes S by Theorem 24.15. $\square$

See [73, Thm. 30] or [13, Thm. 12.5.1] for the case of different root lengths. If F is twisted, the statement must be modified in so far as no graph automorphisms do arise, see [73, Thm. 36] for the statement and its proof in that case.

25

Tori and Sylow subgroups

We now turn to the subgroup structure of the finite groups of Lie type, first discussing maximal tori and then some aspects of Sylow subgroups. Let G be connected reductive with Steinberg endomorphism F, B an F-stable Borel subgroup of G. Recall that $p = \text{char}(k)$. By Corollary 24.11 (or by the order formula in Corollary 24.6) $U^F = R_u(B)^F$ is a Sylow p-subgroup of G^F. Some results on the structure of U^F were obtained in Section 23.2; see also Theorem 26.5 below. Here we will be concerned with Sylow subgroups of G^F for the other prime divisors of its order. Again, there is a close connection to the structure of the algebraic group G.

25.1 F-stable tori

Let $\Phi_d(x) \in \mathbb{Z}[x]$ be the *dth cyclotomic polynomial* over $\mathbb{Q}$, i.e., the monic irreducible polynomial whose zeros are the primitive dth roots of unity, so that $x^m - 1 = \prod_{d|m} \Phi_d(x)$. Then we may factorize the order formula from Corollary 24.6 in $\mathbb{Z}[x]$ to obtain

$$|\mathbb{G}| = x^{|\Phi^+|} \prod_{i=1}^{\text{rk}(G)} (x^{d_i} - \epsilon_i) = x^{|\Phi^+|} \prod_{d \geq 1} \Phi_d(x)^{a(d)}$$

for suitable integers $a(d) \geq 0$. Thus the irreducible factors of the generic order $|\mathbb{G}|$ are the $\Phi_d(x)$ with $a(d) > 0$, so the $\Phi_d(q)$ can in some sense be considered as the "generic primes" dividing $|\mathbb{G}|(q) = |G^F|$. In this section we will substantiate this statement. In order to do this, we first need more precise information on the classification of F-stable maximal tori and the orders of their F-fixed points.

Proposition 25.1 *Let G be connected reductive with Steinberg endomorphism $F : G \to G$, $T \le G$ an F-stable maximal torus with Weyl group W. There is a natural bijection*

$$\left\{ \begin{matrix} G^F\text{-classes of } F\text{-stable} \\ \text{maximal tori of } G \end{matrix} \right\} \overset{1\text{-}1}{\longleftrightarrow} \left\{ \phi\text{-conjugacy classes in } W \right\}.$$

Proof Apply Theorem 21.11 with G acting on the set of maximal tori of G. The stabilizer $N_G(T)$ is closed, and $N_G(T)^\circ = T$ by Theorem 3.10 and Corollary 8.13, so the G^F-classes of F-stable conjugates of T are in natural bijection with the F-classes (hence the ϕ-classes, where $F|_{X(T)} = q\phi$) in $N_G(T)/T = W$. $\quad\square$

More precisely, according to the proof of Theorem 21.11, if the conjugate gTg^{-1} of T is F-stable, then it corresponds to the element $w := g^{-1}F(g)T \in N_G(T)/T = W$. We'll write $T_w := gTg^{-1}$ for such an F-stable maximal torus of G corresponding to the ϕ-class of $w \in W$.

We next compute the orders of F-fixed points of maximal tori.

Proposition 25.2 *Let G be connected reductive with Steinberg endomorphism $F : G \to G$, $T \le G$ an F-table maximal torus. Then*

$$|T^F| = |\det{}_{X_\mathbb{R}}(F-1)| = \det{}_{X_\mathbb{R}}(q - (\phi)^{-1}),$$

where $X_\mathbb{R} := X(T) \otimes_\mathbb{Z} \mathbb{R}$, and $F|_X = q\phi$.

Proof By Lemma 22.1 we have $T^F \cong X(T)/(F-1)X(T)$. The first equality is thus a direct consequence of the elementary divisor theorem. With $F|_X = q\phi$ we get $\det(F-1) = \det(\phi)\det(q - (\phi)^{-1})$. Since ϕ has finite order, all its eigenvalues are roots of unity. So $|\det(\phi)| = 1$, and moreover all eigenvalues of $q - (\phi)^{-1}$ have real part at least $q - 1 > 0$, whence the real number $\det(q - (\phi)^{-1})$ is positive. $\quad\square$

The parametrization in Proposition 25.1 will now allow us to relate the orders of all maximal tori to the characteristic polynomials of elements on the character group of one fixed reference torus, but this translation (conjugation) introduces an element $w \in W$.

Proposition 25.3 *In the situation of Proposition 25.1 let X denote the character group of the F-stable maximal torus T. Then we have:*

(a) $N_{G^F}(T_w)/T_w^F \cong C_W(w\phi)$.
(b) $T_w^F \cong X/(wF - 1)X$.
(c) $|T_w^F| = |\det_{X \otimes \mathbb{R}}(wF - 1)| = \det_{X \otimes \mathbb{R}}(q - (w\phi)^{-1})$.

Proof Note that $T = T_1$ with the identity element $1 \in W$ in our above parametrization. For $T_w = gT_1g^{-1}$ with $w = g^{-1}F(g)T_1 \in W$ we have $N_G(T_w) = gN_G(T_1)g^{-1}$, and $N_{G^F}(T_w)/T_w^F \cong (N_G(T_w)/T_w)^F$ by Proposition 23.2. An element $gng^{-1}T_w \in N_G(T_w)/T_w$, with $n \in N_G(T_1)$, is F-stable if and only if

$$ng^{-1}F(g)T_1 = g^{-1}F(g)F(n)T_1.$$

Writing $\bar{n}$ for the image of n in $N_G(T_1)/T_1 = W$ this is the case if and only if $\bar{n}w = w\phi(\bar{n})$, so if and only if $\bar{n}$ lies in the centralizer $C_W(w\phi)$, giving (a).

For (b) note that F sends $gtg^{-1} \in gT_1g^{-1} = T_w$ to $gwF(t)w^{-1}g^{-1}$, so it acts on T_w like wF acts on T_1. The claim is then just the assertion of Lemma 22.1(b). Part (c) is Proposition 25.2.   □

In particular, since F acts on $X(T_w)$ as wF acts on $X(T)$, we see that if (G, F) has complete root datum $(X, \Phi, Y, \Phi^\vee, W\phi)$ with respect to T, then the torus T_w has complete root datum $(X, \emptyset, Y, \emptyset, w\phi)$.

Recall that $N_{G^F}(T_w) \neq N_{G^F}(T_w^F)$ in general, see the remarks after Proposition 23.2.

Example 25.4   We describe the classes of maximal tori in $\mathrm{GL}_n(q)$.

(1) For $G = \mathrm{GL}_n$ with $F_q : G \to G$ the standard Frobenius map with fixed point group $G^{F_q} = \mathrm{GL}_n(q)$ we let $T = \mathrm{D}_n$ denote the maximally split torus of diagonal matrices. For $g \in \mathrm{GL}_n$ with

$$g^{-1}F_q(g) = w := \begin{pmatrix} 0 & \cdots & 0 & 1 \\ 1 & & & 0 \\ 0 & \ddots & & \vdots \\ 0 & 0 & 1 & 0 \end{pmatrix} \in \mathrm{GL}_n$$

let $T_w = gTg^{-1}$, an F_q-stable maximal torus of G. By Proposition 25.3, F_q acts on T_w like $F = wF_q$ acts on T. Now for $t = \mathrm{diag}(t_1, \ldots, t_n) \in T = \mathrm{D}_n$ we have $F(t) = \mathrm{diag}(t_n^q, t_1^q, \ldots, t_{n-1}^q)$. So

$$T^F = \left\{ \begin{pmatrix} t_1 & & 0 \\ & \ddots & \\ 0 & & t_n \end{pmatrix} \,\middle|\, t_2 = t_1^q,\ t_3 = t_1^{q^2}, \ldots, t_n = t_1^{q^{n-1}},\ t_1 = t_1^{q^n} \right\} \cong \mathbb{F}_{q^n}^\times,$$

as predicted by the order formula $|T^F| = |\det(wq - 1)| = q^n - 1$ from Proposition 25.2. Now w is the preimage of an n-cycle in the Weyl group $W = N_G(T)/T \cong \mathfrak{S}_n$ of G, so its centralizer in W is cyclic of order n. According to Proposition 25.3(a) we then have $N_{\mathrm{GL}_n(q)}(T) = T^F \cdot Z_n$. The generators of this cyclic torus of $\mathrm{GL}_n(q)$ are sometimes called *Singer cycles*.

Let's construct this subgroup even more explicitly. As $\mathbb{F}_{q^n}/\mathbb{F}_q$ is a field extension of degree n, the additive group $\mathbb{F}_{q^n}^+ \cong (\mathbb{F}_q)^n$ is an n-dimensional vector space over $\mathbb{F}_q$. Multiplication by $a \in \mathbb{F}_{q^n}^\times$ is an invertible linear map on this vector space, thus any choice of basis defines an embedding $\mathbb{F}_{q^n}^\times \hookrightarrow \mathrm{GL}_n(q)$. Also, $\mathrm{Gal}(\mathbb{F}_{q^n}/\mathbb{F}_q) \cong Z_n$ acts by linear maps on this space. Putting these together, we have obtained an embedding of the semidirect product

$$\mathbb{F}_{q^n}^\times \rtimes \mathrm{Gal}(\mathbb{F}_{q^n}/\mathbb{F}_q) \hookrightarrow \mathrm{GL}_n(q)$$

whose image, by comparing with the structure obtained above, is all of $N_{G^F}(T)$.

(2) More generally, let $n_1 + \cdots + n_r = n$ be a partition of n and $\mathrm{GL}_{n_1} \times \cdots \times \mathrm{GL}_{n_r} \le \mathrm{GL}_n$ a Levi complement of a corresponding parabolic subgroup (see Example 12.9). In each factor GL_{n_i} choose a maximal F-stable torus T_i with $|T_i^F| = q^{n_i} - 1$ as in (1) above, and let $T := T_1 \times \cdots \times T_r \le \mathrm{GL}_n$. As $\dim T_i = n_i$, T is a maximal, F-stable torus of G with

$$|T^F| = \prod_{i=1}^r |T_i^F| = \prod_{i=1}^r (q^{n_i} - 1),$$

corresponding under the parametrization in Proposition 25.1 to the class of permutations $v \in \mathfrak{S}_n$ of cycle shape $(n_1, \ldots, n_r)$.

As a consequence of our description of F-stable maximal tori, we can now derive a formula for their number:

Theorem 25.5 (Steinberg) *Let G be connected reductive with Steinberg endomorphism $F : G \to G$. Then the number of F-stable maximal tori of G is q^{2N} where $N := |\Phi^+|$.*

Proof By Proposition 25.1 the G^F-classes of F-stable maximal tori of G are in bijection with the ϕ-conjugacy classes in the Weyl group W, where $F|_{X_\mathbb{R}} = q\phi$. The G^F-class of a maximal torus T_w parametrized by the F-class of $w \in W$ contains $|G^F : N_{G^F}(T_w)|$ distinct conjugates. Thus the number in question equals

$$\sum_{w \in W} \frac{1}{|[w\phi]|} |G^F : N_{G^F}(T_w)|,$$

where $[w\phi]$ denotes the ϕ-class of w. Here, $|N_{G^F}(T_w)| = |C_W(w\phi)| \cdot |T_w^F|$ by Proposition 25.3(a). Now by Proposition 25.3(c) and using that $|[w\phi]| =$

$|W|/|C_W(w\phi)|$ (see Exercise 30.3), the above quantity equals

$$\frac{1}{|W|} \sum_{w \in W} \frac{|G^F|}{|T_w^F|} = \frac{|G^F|}{|W|} \sum_{w \in W} \frac{1}{|\det(wF - 1)|}$$

$$= \frac{|G^F|}{|W|} \sum_{w \in W} \frac{1}{q^l \det(1 - (qw\phi)^{-1})},$$

where $l = \dim X_{\mathbb{R}} = \dim(T_w) = \mathrm{rk}(G)$. By Molien's formula (see Exercise 30.15) this can be rewritten as

$$\frac{|G^F|}{q^l} \prod_{i=1}^{l} \frac{1}{1 - \epsilon_i^{-1} q^{-d_i}} = \frac{q^N}{q^l} \prod_{i=1}^{l} \frac{q^{d_i} - \epsilon_i}{1 - \epsilon_i^{-1} q^{-d_i}} = q^{N + d_1 + \ldots + d_l - l} = q^{2N},$$

where we've also used Theorem 24.4(c) and Corollary 24.6. (Note that the multiset $\{(d_i, \epsilon_i)\}$ is invariant under complex conjugation, since ϕ is a real endomorphism.)   □

Torus orders are products of cyclotomic polynomials, which naturally leads to the following notion:

Definition 25.6   Let G be a connected reductive group with a Steinberg endomorphism $F : G \to G$. An F-stable torus $S \leq G$ is called a d-*torus* if its generic order equals $|\mathbb{S}| = \Phi_d(x)^a$ for some $a \geq 0$, where $\mathbb{S}$ denotes the complete root datum corresponding to (S, F). So in particular $|S^F| = |\mathbb{S}|(q) = \Phi_d(q)^a$. We call S a *Sylow d-torus* of G if $|\mathbb{S}| = \Phi_d(x)^{a(d)}$ is the precise power of $\Phi_d(x)$ dividing the generic order $|\mathbb{G}|$.

In view of Lemma 22.1(b) a torus S is a d-torus if and only if the characteristic polynomial of ϕ on its character group (as well as on its cocharacter group) is a power of $\Phi_d(x)$.

The structure of d-tori is very simple:

Proposition 25.7   *Let S be a d-torus of G of rank r. Then S is a direct product of $r/\varphi(d)$ F-stable tori of generic order Φ_d, where $\varphi(d) = \deg(\Phi_d)$.*

Proof   By our previous remark $\Phi_d(\phi) = 0$ on the character group X of S. We may thus define an action of the ring $R := \mathbb{Z}[\zeta]$ on X by letting ζ act as ϕ, where ζ denotes a primitive dth root of unity. This endows X with the structure of a projective module of finite rank over R. Since R is the ring of integers in $\mathbb{Q}(\zeta)$, hence a Dedekind domain, X is a direct sum of projective submodules of rank 1 over R, hence of ϕ-stable submodules of rank $\varphi(d)$ over $\mathbb{Z}$. (See, for example [25, Chap. 2] for the theory of Dedekind domains.)

Since ζ acts as ϕ we have by Lemma 22.1

$$S^F \cong X/(F-1)X \cong X/(q\zeta-1)X \cong X/(q-\zeta^{-1})X$$

as R-modules.

So from now on we may assume that X is a projective R-module of rank 1, hence can be identified with an ideal of R. Then multiplying by its inverse we obtain $X/(q-\zeta^{-1})X \cong R/(q-\zeta^{-1})R$, whose order equals the norm of the principal ideal $(q-\zeta^{-1})R$, so that $|S^F| = \Phi_d(q)$ in this case. If $n \in \mathbb{N}$ with $nX \subseteq (q-\zeta^{-1})X$, then $q-\zeta^{-1}$ divides n in R. As $\mathbb{Z}[\zeta] = \mathbb{Z}[x]/(\Phi_d)$ this implies $n \equiv q-x \pmod{\Phi_d}$, so $n \equiv 0 \pmod{\Phi_d(q)}$. Hence S^F is cyclic of order $\Phi_d(q)$. $\qquad\square$

Example 25.8 For $G = \mathrm{GL}_n$ with $F = F_q$ the standard Frobenius, $T = D_n$ is a maximally split torus. Since $|T^F| = (q-1)^n$ and $|\mathrm{GL}_n(q)| = q^{\binom{n}{2}} \prod_{i=1}^n (q^i - 1)$ by Example 24.8(1), we see that T is a Sylow 1-torus.

More generally, maximally split tori always contain a Sylow 1-torus, and conversely an F-stable maximal torus containing a Sylow 1-torus is maximally split. Tori always possess Sylow d-tori (see Exercise 30.16):

Lemma 25.9 *Let (T, F) be a torus with corresponding complete root datum* $\mathbb{T} = (X, \emptyset, Y, \emptyset, \phi)$. *Then* $\mathbb{S} = (X', \emptyset, Y', \emptyset, \phi)$, *with* $Y' = \ker_Y(\Phi_d(\phi))$ *and* $X' = X/\mathrm{Ann}(Y')$, *is the complete root datum of the Sylow d-torus of T.*

We now show why the name Sylow tori is justified. For this we need a further fundamental result from the theory of finite reflection groups (see [65, 3.4 and 6.2]):

Theorem 25.10 (Springer) *Let $W \le \mathrm{GL}(V)$ be a finite complex reflection group, $\phi \in N_{\mathrm{GL}(V)}(W)$ and $\zeta \in \mathbb{C}$ a root of unity. For $w \in W$ let*

$$V(w\phi, \zeta) := \{v \in V \mid w\phi v = \zeta v\}$$

denote the ζ-eigenspace of $w\phi$. Let $a(\zeta) := \max_{w \in W} \dim V(w\phi, \zeta)$. Then we have:

(a) *For all $w_1, w_2 \in W$ with $\dim V(w_1\phi, \zeta) = \dim V(w_2\phi, \zeta) = a(\zeta)$ there exists $w \in W$ with $w.V(w_1\phi, \zeta) = V(w_2\phi, \zeta)$.*

(b) *For any $w \in W$ there exists $w' \in W$ with $V(w\phi, \zeta) \le V(w'\phi, \zeta)$ and such that $\dim V(w'\phi, \zeta) = a(\zeta)$ is maximal.*

This implies the following result (see [10, Thm. 3.4], and also Exercise 30.17 for the case of GL_n):

Theorem 25.11 (Generic Sylow Theorems)   *Let G be a connected reductive group, $F : G \to G$ a not very twisted Steinberg endomorphism and $d \geq 1$.*

(a) *There exist Sylow d-tori in G and they are all G^F-conjugate.*
(b) *Any d-torus of G is contained in some Sylow d-torus.*

Proof   Let W denote the Weyl group of G with respect to an F-stable maximal torus T with character group X and cocharacter group Y, such that $F|_X = q\phi$. By the remark after Proposition 25.3 the complete root data of F-stable maximal tori of G are of the form $(X, \emptyset, Y, \emptyset, w\phi)$ for $w \in W$. By Example 9.12 the complete root datum of an arbitrary F-stable torus S is then of the form $\mathbb{S} = (X/\mathrm{Ann}(Y_S), \emptyset, Y_S, \emptyset, w\phi)$ for some $w\phi$-stable sublattice Y_S of Y, and by Proposition 25.3(c) its generic order is given by the characteristic polynomial of $w\phi$ on $Y_S \otimes_{\mathbb{Z}} \mathbb{R}$. Now first note that by Molien's formula in Exercise 30.15(d) there exist F-stable maximal tori in G whose generic order is divisible by the same power of Φ_d as the generic order of G, and so by Lemma 25.9 there exist d-tori of that order, which proves the first half of (a).

Write $V := Y \otimes_{\mathbb{Z}} \mathbb{C}$ and $V_S := Y_S \otimes_{\mathbb{Z}} \mathbb{C} \leq V$ and let ζ_d be a primitive dth root of unity. Then S is a d-torus if and only if the characteristic polynomial of $w\phi$ on $Y(S)$ is a power of Φ_d, hence if and only if
$$V_S = \bigoplus\nolimits_{i:\gcd(i,d)=1} V_S(w\phi, \zeta_d^i).$$

By Proposition 25.1, conjugacy of tori is translated into ϕ-conjugacy in the Weyl group. Thus we may rephrase assertions (a) and (b) as assertions about elements of W as follows: all maximal eigenspaces $V(w\phi, \zeta_d)$ are conjugate under W, and the ζ_d-eigenspace of any element $w\phi$ is contained in a ζ_d-eigenspace of maximal possible dimension. But these are just the statements of Theorem 25.10. $\qquad\square$

Example 25.12   For $G = \mathrm{GL}_n$ or SL_n with the standard Frobenius map, a maximal torus parametrized by $w \in W = \mathfrak{S}_n$ (as in Proposition 25.1) contains a Sylow d-torus if and only if the cycle decomposition of the permutation w contains a disjoint d-cycles, where $n = ad + b$ with $b < d$, since according to Example 25.4(2) these are precisely the tori whose generic order is divisible by the maximal possible power of Φ_d.

For the very twisted groups of types $^2B_2$, $^2G_2$ or $^2F_4$ excluded in the statement above, the right "generic primes" to consider are divisors of the order polynomial over a suitable quadratic extension of $\mathbb{Z}$. With this modification the assertions of Theorem 25.11 remain true in these cases as well, see Broué–Malle [10, 3F] for details, and for an analogue of the third Sylow Theorem.

25.2 Sylow subgroups

We now discuss the structure of Sylow subgroups for true primes, their centralizers and normalizers. We give statements and proofs only for not very twisted Steinberg endomorphisms, but analogous statements hold after suitable modifications in the other cases as well.

Consider a prime r such that r divides $|G^F|$ and $r \neq \operatorname{char}(k) = p$. (For the defining prime $r = p$ see Corollary 24.11.) As $|G^F| = q^N \prod_d \Phi_d(q)^{a(d)}$ this implies that $r | \Phi_d(q)$ for some d with $a(d) > 0$. Let us set

$$e := e_r(q) := \text{the order of } q \text{ modulo } r,$$

so that $r | q^e - 1$, but r does not divide $q^d - 1$ for any positive integer $d < e$. We require the following standard number theoretical lemma (see [56, Lemma 5.2], for example).

Lemma 25.13 *If $r | \Phi_d(q)$, r prime, then $d = e_r(q) \, r^f$ for some $f \geq 0$.*

Theorem 25.14 *Let G be connected reductive with a not very twisted Steinberg endomorphism $F : G \to G$, $r \neq \operatorname{char}(k)$ a prime dividing $|G^F|$ with $r \nmid |W\langle\phi\rangle|$. Then:*

(a) *There exists a unique d such that $a(d) > 0$ and $r | \Phi_d(q)$. Then any Sylow d-torus of G contains a Sylow r-subgroup of G^F.*

(b) *The Sylow r-subgroups of G^F are homocyclic with $a(d)$ cyclic factors of order $|\Phi_d(q)|_r$.*

In particular, in this situation all Sylow r-subgroups of G^F are abelian.

Recall that a group is *homocyclic* if it is the direct product of isomorphic cyclic groups. So in the situation of Theorem 25.14 the Sylow r-subgroups are isomorphic to $Z_m \times \cdots \times Z_m$ ($a(d)$ times), where $m = |\Phi_d(q)|_r$ denotes the r-part of $\Phi_d(q)$.

Proof By Corollary 24.6 the generic order $|\mathbb{G}|$ of (G, F) is a product of cyclotomic factors $\Phi_d^{a(d)}$, where d divides some δd_i, with δ the order of ϕ on $X_{\mathbb{R}}$ and d_i the degrees of W. By Lemma 25.13, if $r | \Phi_{e_1}(q)$, $r | \Phi_{e_2}(q)$, with $e_1 < e_2$, then $r | e_2$. Thus, if $a(e_1) > 0$, $a(e_2) > 0$, that is, if both Φ_{e_1}, Φ_{e_2} divide $|\mathbb{G}|$ then $r | \delta d_i$ for some i, whence it divides $|W\langle\phi\rangle| = \delta d_1 \cdots d_n$ (see Theorem 24.4(c)), which was excluded.

Now let S be a Sylow d-torus of G, of generic order $\Phi_d^{a(d)}$. Then the claim in (b) follows from Proposition 25.7. $\qquad\square$

For general primes r the Sylow r-subgroups of G^F may be non-abelian, so we cannot expect to embed them into tori. But it turns out that Sylow r-subgroups do embed in the normalizer of a maximal torus.

Definition 25.15 An algebraic group S is *supersolvable* if there exists a series $1 = S_0 \lhd S_1 \lhd \cdots \lhd S_n = S$ of closed normal subgroups $S_i \lhd S$ such that S_i/S_{i-1} is either cyclic or a torus for all $i = 1, \ldots, n$.

For example, finite p-groups, having non-trivial center, are supersolvable, hence so are finite nilpotent groups.

Theorem 25.16 *Let G be a connected reductive group, $F : G \to G$ a Steinberg endomorphism, $S \le G$ an F-stable supersolvable subgroup such that each $S_i \lhd S$ is F-stable and consists of semisimple elements. Then there exists an F-stable maximal torus $T \le G$ with $S \le N_G(T)$.*

Proof We argue by induction on the length of the chain $1 = S_0 \lhd \cdots \lhd S_n = S$. If $S = 1$, then the existence of an F-stable maximal torus T is guaranteed by Corollary 21.12. Secondly, if $S \ne 1$ but $C_G(S_1) = G$, then $S_1 \le Z(G)$, so S_1 lies in every maximal torus of G by Corollary 8.13(b). In this case we may argue with a shorter chain in G/S_1 by induction.

Otherwise, $G_1 := C_G(S_1)^\circ$ is a proper connected F-stable subgroup of G, reductive by Corollary 8.13(a) (if S_1 is a torus) respectively by Theorem 14.2 (if S_1 is cyclic), and the maximal tori of G_1 are maximal tori of G by the same references. Now, by the previous case there exists an F-stable maximal torus of G_1 which is normalized by S. This is a maximal torus of G. $\square$

Corollary 25.17 *Let r be a prime, $r \ne p$. Then any Sylow r-subgroup of G^F lies in $N_{G^F}(T)$ for some F-stable maximal torus $T \le G$.*

Proof As the Sylow r-subgroups are nilpotent and consist of semisimple elements, and finite nilpotent groups are supersolvable, this follows immediately from Theorem 25.16 by taking F-fixed points. $\square$

Example 25.18 Let $G = \mathrm{GL}_n$, F the standard Frobenius map with respect to $\mathbb{F}_q$. The normalizer in G^F of the maximally split torus D_n of G equals $N_{G^F}(\mathrm{D}_n) = \mathrm{D}_n^F.W = \mathrm{D}_n^F.\mathfrak{S}_n$ (see Example 8.2(1)). It can be checked from the order formula that for $q \equiv 1 \pmod 4$ this contains a Sylow 2-subgroup of $\mathrm{GL}_n(q)$. On the other hand, if $q \equiv 3 \pmod 4$ then the order of $N_{G^F}(\mathrm{D}_n)$ is not divisible by the full 2-part of the order of $\mathrm{GL}_n(q)$, so in particular it cannot contain a Sylow 2-subgroup. In this case, though, the normalizer of a Sylow 2-torus contains a Sylow 2-subgroup.

We now show that centralizers of Sylow subgroups are strongly related to centralizers of Sylow tori.

Theorem 25.19 *Let G be simple with a not very twisted Steinberg endomorphism $F : G \to G$, let r be a prime divisor of $|G^F|$, $r \neq p$. Then every semisimple element $g \in G^F$ which centralizes a Sylow r-subgroup of G^F lies in an F-stable torus containing a Sylow d-torus of G, where $d = e_r(q)$. In particular, g centralizes a Sylow d-torus.*

Proof for $r > 3$ Let $g \in G$ be semisimple, centralizing a Sylow r-subgroup of G, and set $C := C_G(g)$. We proceed in several steps, where for the first two G may be an arbitrary connected reductive group.

(1) Assume for a moment that the index $|C : C^\circ|$ is prime to r. Then $(C^\circ)^F$ contains a Sylow r-subgroup of G^F. The generic orders of G and C° are both monic polynomials, with the second dividing the first, see Corollary 24.7. Hence, all cyclotomic polynomials Φ_i with $r | \Phi_i(q)$ dividing the generic order of G also divide the generic order of C°, to the same power. In particular, a Sylow d-torus of C° is a Sylow d-torus of G. Thus, g centralizes a Sylow d-torus of G. By Proposition 14.1 we even have $g \in C^\circ$, so $g \in Z(C^\circ)$, whence g is contained in a maximal torus containing a Sylow d-torus.

(2) Now assume that the order of g is prime to r. By Proposition 14.20 the exponent of C/C° divides the order of g, so is prime to r. Hence the claim follows from the previous consideration.

(3) Next assume that g has r-power order. Let $\pi : G_{\mathrm{sc}} \to G$ denote the natural isogeny from a group of simply-connected type (see Proposition 9.15). Again by Proposition 14.20 the factor group C/C° is isomorphic to a subgroup of $\ker(\pi) \leq Z(G_{\mathrm{sc}})$, hence we are done by step (1) if the order of the fundamental group is prime to r. This argument is valid for arbitrary semisimple groups G.

Consulting the list of possible orders of fundamental groups for simple groups in Table 9.2 we see, as $r > 3$, that the only cases not covered by the above argument are G of type A_{n-1} with r dividing n.

For G of type A_{n-1} we use the explicit knowledge of the Sylow structure of G^F. First assume that G^F is untwisted. Then, by Corollary 25.17 a Sylow r-subgroup R of G^F is contained in the normalizer of a Sylow d-torus T of order $\Phi_d(q)^a$ of G^F, where $0 \leq n - ad < d$, with normalizer quotient $W_d = N_{G^F}(T)/C_{G^F}(T)$ a wreath product $Z_d \wr \mathfrak{S}_a$ (see Exercise 30.17). The action of W_d on T^F is given as follows: By Proposition 25.7, $T = T_1 \times \cdots \times T_a$ is a direct product of F-stable d-tori $T_i \cong T_1$ of generic order Φ_d. The Z_d-factors of W_d now act as field automorphisms of $\mathbb{F}_{q^d}/\mathbb{F}_q$ on the cyclic subgroups T_i^F of T^F of order $\Phi_d(q)$, while the symmetric group $\mathfrak{S}_a$ permutes the coordinates, see Example 25.4. Now the r-elements in $\mathbb{F}_{q^d}^\times$ are not contained in a proper subfield, by the definition of $e_r(q)$, hence not centralized by non-trivial field automorphisms. In particular, the normalizer quotient W_d acts faithfully on the Sylow r-subgroup of T, so every r-central element of R is already contained in the maximal torus T, as claimed. A similar reasoning applies if G^F is a unitary group.

(4) Finally, in the general case, let $g = g_1 g_2$ be the decomposition of g into its r- and r'-part. Let $C_1 := C_G(g_1)$, the centralizer in G of the r-part of g_1. By step (3) its connected component C_1° contains a Sylow d-torus of G. Since $C_G(g) \geq C_{C_1^\circ}(g_2)$ and C_1° is reductive by Theorem 14.2, the result now follows from (2) applied to the r'-element g_2 of C_1°. $\Box$

In the case $r \leq 3$, some further configurations have to be considered in step (3) of the above proof, see [56, Thm. 5.9]. A suitably modified statement remains true for very twisted F, see [56, Thm. 8.2].

We end this section by citing the following extension of Corollary 25.17 to normalizers of Sylow subgroups ([56, Thm. 5.14]; see also [29, Thm. 4.10.3]):

Theorem 25.20 *Let G be a semisimple group, $F : G \to G$ a not very twisted Steinberg endomorphism, $r > 3$ a prime different from p, $R \in \mathrm{Syl}_r(G^F)$, $d = e_r(q)$. Then there exists a Sylow d-torus S of G such that $N_{G^F}(R) \leq N_{G^F}(S)$.*

About the proof By Corollary 25.17, $R \leq N_G(T)$ for some F-stable maximal torus T of G. As $r \geq 5$ one can show (using that T^F is not a so-called failure of factorization module for $N_G(T)/T$, see Cabanes [12]) that R contains a unique maximal elementary abelian toral r-subgroup A which, moreover, lies in a unique Sylow d-torus S of G. So any $g \in N_{G^F}(R)$ stabilizes A, hence stabilizes S, so lies in $N_{G^F}(S)$. $\Box$

Example 25.21 This example is to show that the conclusion of Theorem 25.20 can fail for $r \leq 3$. Let $G = \mathrm{SL}_2$, F the standard Frobenius with $G^F = \mathrm{SL}_2(q)$. Assume $q \equiv 3, 5 \pmod 8$, and let $r = 2$. Then a Sylow 2-subgroup R of G^F is a quaternion group Q_8 and $N_{G^F}(R) \cong \mathrm{SL}_2(3)$. But the normalizers $N_{G^F}(T) \cong Z_{q-1}.2$ or $Z_{q+1}.2$ of the two distinct F-stable maximal tori (up to G^F-conjugation) clearly cannot contain such a subgroup.

Similar examples exist for all $\mathrm{Sp}_{2n}(q)$, $q \equiv 3, 5$ (8) and $r = 2$, and for $r = 3$ in $\mathrm{SL}_3(q)$, $\mathrm{SU}_3(q)$ and $G_2(q)$. See [56, Thm. 5.14, 5.19] for the precise statements, and [56, Thm. 8.4] for the case of Suzuki and Ree groups.

See [29, Thm. 4.10.6] for a completely different approach, due to Aschbacher, to the Sylow 2-subgroups of groups of Lie type over fields of odd characteristic.

26

Subgroups of maximal rank

We transfer results in Chapter 12 on parabolic subgroups and Levi complements in a reductive algebraic group G to the finite group G^F of fixed points under a Steinberg endomorphism and compare with the BN-pair structure in G^F. We also derive several results on conjugacy classes and centralizers in G^F from the corresponding statements for G in Chapters 13 and 14.

26.1 Parabolic subgroups and Levi subgroups

Let G be a connected reductive group, $T \leq B$ a maximal torus inside a Borel subgroup and $N := N_G(T)$. By Theorem 11.16, B, N form a BN-pair in G, with Weyl group W generated by a distinguished set of involutions S. Now assume in addition that we are given a Steinberg endomorphism $F : G \to G$ and choose B, T both F-stable. Then we observed in Theorem 24.10 that B^F, N^F form a BN-pair in G^F, with Weyl group W^F generated by a set S_F of involutions in natural bijection with the F-orbits on S (see Lemma 23.3).

We compare the BN-pairs in G and G^F. Let $I \subseteq S$ and P_I the corresponding parabolic subgroup of G with standard Levi complement L_I. Then by the double coset decomposition in Proposition 12.2(a) P_I is F-stable if and only if I is F-invariant, i.e., if I is a union of F-orbits in S, so corresponds to a subset $I_F \subseteq S_F$ of the set of F-orbits on S. We write P_{I_F} for the parabolic subgroup of G^F defined by I_F.

Proposition 26.1 *Let G be connected reductive with a Steinberg endomorphism $F : G \to G$ and let $I \subseteq S$ be F-invariant.*

(a) *There is a unique G^F-conjugacy class of F-stable G-conjugates of P_I, and we have $P_I^F = P_{I_F}$.*

(b) *There is a unique P_I^F-conjugacy class of F-stable Levi complements to $U_I = R_u(P_I)$ in P_I, and we have $P_I^F = U_I^F \rtimes L_I^F$.*

Proof The first claim follows exactly as in the special case $I = \emptyset$, that is, $P_I = B$, in Corollary 21.12. Indeed, P_I is self-normalizing by Proposition 12.2(a), so there is a unique G^F-conjugacy class of F-stable G-conjugates of P_I by Theorem 21.11(b). Taking F-fixed points in the double coset decomposition for P_I in Proposition 12.2 we obtain that $P_I^F = P_{I_F}$.

For part (b) we use that the standard Levi complement L_I of P_I is generated by an F-stable subset of root groups and so is F-stable, and that all Levi complements in P_I are P_I-conjugate (see Proposition 12.6). It is clear by this same result that $U_I^F \cap L_I^F \leq U_I \cap L_I = \{1\}$. Now let $p \in P_I^F$. Then $p = ul$ for unique $u \in U_I$ and $l \in L_I$. Since both U_I, L_I are F-invariant, this implies that u and l are both F-stable, whence $P_I^F = U_I^F \rtimes L_I^F$. $\square$

That is, the standard parabolic subgroups of the BN-pair G^F arise as the F-fixed points of F-stable standard parabolic subgroups of G. Note however that in the finite case it is no longer true in general that all complements to U_I^F in P_I^F are conjugate to L_I^F. Over small finite fields, the central torus of L_I used in the proof of Proposition 12.6 may have a trivial fixed point group, so that the argument given there fails in this case; see the following example.

Example 26.2 Let $G = \mathrm{SL}_4$, $\mathrm{char}(k) = 2$, and F the 2-power map on G. Set $P = \left\{ \begin{pmatrix} A & * \\ 0 & * \end{pmatrix} \,\middle|\, A \in \mathrm{GL}_3 \right\} \cap G$. One complement of the normal unipotent group U^F in P^F is the subgroup

$$L_1 := \left\{ \begin{pmatrix} A & 0 \\ 0 & \det(A)^{-1} \end{pmatrix} \,\middle|\, A \in \mathrm{GL}_3^F \right\}.$$

A second non-conjugate complement is obtained as the image of a four-dimensional indecomposable representation of the group $\mathrm{SL}_3(2) \cong \mathrm{PSL}_2(7)$, over the field $\mathbb{F}_2$, in which $\mathrm{SL}_3(2)$ stabilizes and acts irreducibly on a 3-space. (This representation is obtained by decomposing the natural permutation representation of $\mathrm{PSL}_2(7)$ acting on the cosets of a Borel subgroup.)

Let's turn to the parametrization of F-stable Levi subgroups L of G. They are not necessarily contained in F-stable parabolic subgroups of G. Indeed, already in the case $I = \emptyset$, where L is a maximal torus, we saw that not all F-stable G-conjugates of L lie in an F-stable Borel subgroup (see Proposition 25.1).

Example 26.3   Let $G = \mathrm{GL}_n$ with the Steinberg endomorphism $F : G \to G$ from Example 21.14(2) such that $\mathrm{GL}_n^F = \mathrm{GU}_n(q)$. Recall from Example 12.4 the set $S = \{s_1, \ldots, s_{n-1}\}$ of simple reflections generating the Weyl group of G.

(1) The parabolic subgroup $P_I := \left\{ \left(\begin{array}{c|c|c} * & * & * \\ \hline 0 & A & * \\ \hline 0 & 0 & * \end{array} \right) \;\middle|\; A \in \mathrm{GL}_{n-2} \right\}$ of G

corresponding to $I = \{s_2, \ldots, s_{n-2}\}$ is F-stable, with fixed point group

$$P_I^F = \left\{ \left(\begin{array}{c|c|c} * & * & * \\ \hline 0 & A & * \\ \hline 0 & 0 & * \end{array} \right) \;\middle|\; A \in \mathrm{GU}_{n-2}(q) \right\} \cap \mathrm{GU}_n(q)$$

and F-stable Levi complement L_I with F-fixed points $L_I^F \cong \mathbb{F}_{q^2}^\times \times \mathrm{GU}_{n-2}(q)$.

(2) On the other hand, for $n \geq 3$ the maximal parabolic subgroup

$$P_J := \left\{ \left(\begin{array}{c|c} * & * \\ \hline 0 & A \end{array} \right) \;\middle|\; A \in \mathrm{GL}_{n-1} \right\}$$

of G corresponding to $J = \{s_2, \ldots, s_{n-1}\}$ has no F-stable G-conjugate, since J is not F-invariant. Nevertheless, its Levi complement

$$L_J := \left\{ \left(\begin{array}{c|c} * & 0 \\ \hline 0 & A \end{array} \right) \;\middle|\; A \in \mathrm{GL}_{n-1} \right\} \cong \mathbf{G}_{\mathrm{m}} \times \mathrm{GL}_{n-1}$$

possesses F-stable G-conjugates. This is most easily seen by replacing F by the Steinberg endomorphism F' as in Example 21.2 (which replaces G^F by a G-conjugate, see Corollary 21.8). Then L_J is F'-stable.

As $W_J \cong \mathfrak{S}_{n-1}$ is self-normalizing in $W \cong \mathfrak{S}_n$, so is L_J in G by Corollary 12.11. So there is just one G^F-conjugacy class of F-stable conjugates of L_J; the fixed point group is isomorphic to $\mathbb{F}_{q^2}^\times / \mathbb{F}_q^\times \times \mathrm{GU}_{n-1}(q)$. It is not contained in any proper parabolic subgroup of G^F, since, for example, its order is larger than that of any proper Levi complement of G^F.

See Exercise 30.20 for geometric interpretations of the two subgroups P_I^F and $L_J^{F'}$ appearing above.

The situation here is described by the obvious generalization of Propositions 25.1 and 25.3:

Proposition 26.4   *Let G be connected reductive with Steinberg endomorphism $F : G \to G$, $L \leq G$ an F-stable Levi complement of some parabolic subgroup of G, with Weyl group W_L.*

(a) *The G^F-classes of F-stable G-conjugates of L are in bijection with ϕ-classes of $N_W(W_L)/W_L$, with ϕ as in Proposition 22.2(b).*

(b) *Write L_w for a Levi subgroup corresponding to the ϕ-class of w under the map in (a). Then $N_{G^F}(L_w)/L_w^F \cong N_W(W_L w\phi)/W_L$.*

This is Exercise 30.21.

Let's close our discussion of parabolic subgroups with the following extension of the theorem of Borel and Tits to finite groups of Lie type:

Theorem 26.5 (Borel–Tits) *Let G be connected reductive with Steinberg endomorphism $F : G \to G$. Let $U \leq G^F$ be a unipotent subgroup of G^F. Then there exists an F-stable parabolic subgroup P of G with $U \leq R_u(P)^F$ and $N_{G^F}(U) \leq P^F$.*

Proof We first claim that U lies in some F-stable Borel subgroup of G. Indeed, U is unipotent, hence a p-group, where p is the underlying characteristic of G. Thus, by Sylow's Theorem, it lies in a Sylow p-subgroup of G^F. Corollary 24.11 shows that for any F-stable Borel subgroup B of G, B^F contains a Sylow p-subgroup of G^F. Thus, U lies in a G^F-conjugate of B^F, whence of B.

Thus, we are in the situation of Theorem 17.10. As U is F-stable, all the groups $U_i, N_i, P(U)$ constructed in the proof there are F-stable as well, hence $N_G(U)$ is contained in the F-stable parabolic subgroup $P := P(U)$, with $U \leq R_u(P)$. □

26.2 Semisimple conjugacy classes

The Lang–Steinberg Theorem easily allows one to deduce properties of conjugacy classes and centralizers in the finite groups G^F.

Let's first observe the following analogue of Corollary 6.11(a):

Proposition 26.6 *Let G be connected reductive with Steinberg endomorphism $F : G \to G$, and $s \in G^F$ semisimple. Then s lies in an F-stable maximal torus of G.*

Proof By Theorem 14.2, the centralizer $C_G(s)^\circ$ is connected reductive, F-stable since s is. Thus, by Corollary 21.12 it contains an F-stable maximal torus $T \leq C_G(s)^\circ$. Now Proposition 14.1 shows that $s \in T$. □

Note that any conjugacy class of an algebraic group G with Steinberg endomorphism $F : G \to G$ is F^m-stable for some sufficiently large m. Indeed, embed $G \hookrightarrow \mathrm{GL}_n$ such that F^i is the restriction to G of a standard Frobenius map F_q on GL_n. Then any $s \in G$ is represented by a matrix with entries in some finite subfield of k, hence fixed under some power F_q^j of the standard Frobenius map. Thus $s \in G^{F^m}$ with $m := ij$; in particular, the conjugacy class of s is F^m-stable.

Theorem 26.7 *Let G be a connected reductive group, $F : G \to G$ a Steinberg endomorphism. Then we have:*

(a) *Every F-stable conjugacy class of semisimple elements of G contains an F-stable element.*

(b) *Let $[s]_G$ be an F-stable semisimple conjugacy class of G. Then the G^F-classes in $[s]_G^F$ correspond bijectively to the F-classes in $C_G(s)/C_G(s)^\circ$.*

(c) *Assume that the derived group G' of G is simply connected. Then there is a natural bijective correspondence*

$$\begin{Bmatrix} F\text{-stable semisimple} \\ classes\ in\ G \end{Bmatrix} \overset{1-1}{\longleftrightarrow} \begin{Bmatrix} G^F\text{-classes of semisimple} \\ elements\ in\ G^F \end{Bmatrix}.$$

Proof Parts (a) and (b) follow directly from Theorem 21.11 applied to the conjugation action of G on itself, since the centralizer $C_G(s)$ of any semisimple element $s \in G$ is closed by Proposition 5.2. In the situation of (c), all centralizers $C_G(s)$ of semisimple elements are connected by Theorem 14.16, so the claim follows from parts (a) and (b). $\qquad\square$

Example 26.8 How many semisimple conjugacy classes are there in the general linear group $\mathrm{GL}_n(q)$? A conjugacy class is determined by its Jordan canonical form, hence semisimple classes are parametrized by the characteristic polynomial of their elements. Any monic polynomial over $\mathbb{F}_q$ of degree n and with non-vanishing constant coefficient occurs as characteristic polynomial, and there are precisely $(q-1)q^{n-1}$ of these, so this is the number of semisimple conjugacy classes of $\mathrm{GL}_n(q)$.

Steinberg found a natural generalization of this observation; its proof requires the following result on the natural representation of a finite reflection group, which is also due to Steinberg [72, Thm. 14.4] (see also [9, V, §5, Ex. 2.3(b)]):

Lemma 26.9 *Let $W \le \mathrm{GL}(V)$ be a finite reflection group on a finite-dimensional real vector space V, with trivial fixed space $V^W = 0$ and $\sigma \in N_{\mathrm{GL}(V)}(W)$. Then*

$$\frac{1}{|W|} \sum_{w \in W} \det{}_V(x\,\mathrm{id}_V - \sigma w) = x^{\dim V}.$$

Sketch of proof The sum in question is the scalar product on the coset $W.\sigma$ of the class function $\det_V(x\,\mathrm{id}_V - \sigma w)$ with the trivial character (see [22, I.6] for the elementary properties of such scalar products). Now note that the characteristic polynomial of any $z \in \mathrm{GL}(V)$ is given by

$$\det{}_V(x\,\mathrm{id}_V - z) = \sum_{i=0}^{n} (-1)^i \chi_i(z)\, x^{n-i},$$

where $n = \dim V$ and χ_i is the character of $\mathrm{GL}(V)$ on the ith exterior power $\wedge^i(V)$. Using the assumptions on W one shows that W has trivial fixed space on all $\wedge^i(V)$, $i > 0$, so only the term for $i = 0$ in the sum contributes to that scalar product, yielding the value x^n. $\square$

Theorem 26.10 (Steinberg) *Let G be a connected reductive algebraic group, $G' := [G, G]$, $l := \mathrm{rk}\, G'$, $Z := Z(G)^\circ$, and $F : G \to G$ a Steinberg endomorphism. Then:*

(a) *The number of F-stable semisimple conjugacy classes of G equals $q^l\,|Z^F|$.*
(b) *If the derived group G' is simply connected, then this is also the number of G^F-classes of semisimple elements in G^F.*

Proof By Proposition 14.6 the F-stable semisimple conjugacy classes of G are in bijection with the set $(T/W)^F$ of F-stable W-orbits on an F-stable maximal torus T of G. Thus, we are counting $t \in T$ satisfying $F(t) = t^w$ for some $w \in W$, up to W-conjugation. Hence

$$|(T/W)^F| = \frac{1}{|W|} \sum_{t \in T} \sum_{\substack{w \in W \\ F(t)=t^w}} 1 = \frac{1}{|W|} \sum_{w \in W} \sum_{\substack{t \in T \\ F(t)=t^w}} 1 = \frac{1}{|W|} \sum_{w \in W} |T^{w^{-1}F}|.$$

Now $|T^{w^{-1}F}| = |Z^F| \cdot |S^{w^{-1}F}|$ with $Z = Z(G)^\circ$ and $S = T \cap G'$ by Exercise 30.22, and by Proposition 25.2 we have $|S^{w^{-1}F}| = \det_V(q - \phi^{-1}w)$, where $V = X(S) \otimes_{\mathbb{Z}} \mathbb{R}$, giving

$$|(T/W)^F| = \frac{|Z^F|}{|W|} \sum_{w \in W} \det{}_V(q - \phi^{-1}w).$$

Now (a) follows from Lemma 26.9, (b) follows from this by Theorem 26.7(c).
$\square$

Example 26.11 (Centralizers of some isolated elements in $E_6(q)$) We continue Example 14.21, with G of simply connected type E_6, $\mathrm{char}(k) \neq 3$ and H the subsystem subgroup of type A_2^3 corresponding to the subset $\{0, 1, 2, 3, 5, 6\}$ of the extended Dynkin diagram (see Table 13.1). By Theorem 22.5 there exists a Steinberg endomorphism $F : G \to G$ which acts trivially on W, so F is $\mathbb{F}_q$-split for some power q of char(k) and $G^F = E_6(q)$. Then the set J is F-stable, hence so is H.

Let $s \in Z(H) \backslash Z(G)$, with centralizer $C := C_G(s) = H$. As C is connected, according to Theorem 26.7 there is a unique G^F-class in $[s]_G^F$, and we now assume that $s \in G^F$. In order to determine the centralizer of s in G^F, we consider the possible F-fixed points of $C = H$. By Theorem 21.11 there exist three G^F-classes of F-stable conjugates of H, parametrized by the conjugacy classes in $W(H) = N_G(H)/H \cong \mathfrak{S}_3$. As $W(H)$ acts on H by permuting the three A_2-factors, the structure of H_w^F, for $w \in W(H)$ can be determined using Exercise 30.2; see Table 26.1 where we've also given the order of the center of H_w^F, which can be determined from the action of F on $Z(H_w)$.

Table 26.1 *Some centralizers in $E_6(q)$*

w	H_w^F	$N_G(H_w)^F/H_w^F$	$|Z(H_w^F)|$	$|Z(\bar{H}_w^F)|$
$(\ )$	$A_2(q)^3$	$\mathfrak{S}_3$	$\gcd(3, q-1)^2$	$\gcd(3, q-1)$
$(1\ 2)$	$A_2(q^2) \cdot {}^2A_2(q)$	Z_2	$\gcd(3, q^2-1)$	$\gcd(3, q+1)$
$(1\ 2\ 3)$	$A_2(q^3)$	Z_3	$\gcd(3, q-1)$	$\gcd(3, q-1)$

Note that $Z(G^F) \leq Z(H_w^F)$ for all w. As $|Z(G^F)| = \gcd(3, q-1)$ this shows that in the simply connected case, $E_6(q)$ contains a class of semisimple elements of order 3 with centralizer of type $A_2(q)^3$ when $q \equiv 1 \pmod{3}$, respectively of type $A_2(q^2) \cdot {}^2A_2(q)$ when $q \equiv 2 \pmod{3}$.

Now consider $\bar{G} = G/Z(G)$, of adjoint type, with the induced split Frobenius map. Here, for $1 \neq s \in Z(\bar{H})$ the centralizer $C := C_{\bar{G}}(s)$ is disconnected, with $C/C^\circ \cong Z_3$, by Example 14.21. Now, for $q \equiv 1 \pmod{3}$, F acts trivially on C/C°, so by Theorem 26.7(b) there exist three $\bar{G}^F$-classes of semisimple elements in $[s]_{\bar{G}}^F$. For $q \equiv 2 \pmod{3}$, F acts non-trivially on C/C°, so there is just one F-class in C/C°, hence also just one $\bar{G}^F$-class in $[s]_{\bar{G}}^F$.

Looking again at the various F-stable conjugates of $\bar{H}$ and their centers, as given in the last column of Table 26.1, we see that for $q \equiv 1 \pmod{3}$, there is one class with centralizer $A_2(q)^3.Z_3$ and two with centralizer $A_2(q^3).Z_3$, while for $q \equiv 2 \pmod{3}$, there is one class with centralizer $A_2(q^2) \cdot {}^2A_2(q)$.

Taking for F the non-split Steinberg endomorphism on G instead, we obtain the Ennola dual parametrization of classes, i.e., "with q suitably replaced by $-q$" throughout.

Maximal subgroups of finite classical groups

The aim of this chapter is to study the subgroup structure of the finite classical groups. The classical groups are defined as certain subgroups of the isometry groups of the zero form or of non-degenerate bilinear, sesquilinear or quadratic forms f on finite-dimensional vector spaces over finite fields. According to the classification of such forms and their isometry groups (which can be found for example in [2, §21]), there exist the following non-degenerate possibilities: f is symplectic on an even-dimensional vector space, f is quadratic of maximal Witt index on an odd-dimensional vector space, f is quadratic on a $2n$-dimensional vector space, of Witt index n or $n-1$, or f is unitary on a vector space over a field with a subfield of index 2. Here, the *Witt index* of a quadratic form is the maximal dimension of a totally singular subspace.

The various possibilities for the isometry groups are collected in Table 27.1.

Table 27.1 *Finite classical groups*

V	form	isometry group	$\mathrm{Cl}(V)$
$\mathbb{F}_q^n$	zero	$\mathrm{GL}_n(q)$	$\mathrm{SL}_n(q)$
$\mathbb{F}_q^{2n}$	symplectic	$\mathrm{Sp}_{2n}(q)$	$\mathrm{Sp}_{2n}(q)$
$\mathbb{F}_q^{2n+1}$	quadratic	$\mathrm{GO}_{2n+1}(q)$	$\mathrm{SO}_{2n+1}(q)$
$\mathbb{F}_q^{2n}$	quadratic of Witt index n	$\mathrm{GO}_{2n}^+(q)$	$\mathrm{SO}_{2n}^+(q)$
$\mathbb{F}_q^{2n}$	quadratic of Witt index $n-1$	$\mathrm{GO}_{2n}^-(q)$	$\mathrm{SO}_{2n}^-(q)$
$\mathbb{F}_{q^2}^n$	unitary over $\mathbb{F}_{q^2}$	$\mathrm{GU}_n(q)$	$\mathrm{SU}_n(q)$

Definition 27.1 We shall refer to groups in the last column of the above table as *classical groups*, and we shall denote them by $\mathrm{Cl}(V)$.

27.1 The theorem of Liebeck and Seitz

The key to the understanding of maximal subgroups of the groups $\mathrm{Cl}(V)$ will be to relate them to algebraic groups. In all but the last case of Table 27.1, extending scalars to $k = \overline{\mathbb{F}_q}$ we obtain the corresponding isometry group G over the algebraically closed field k, acting on $\bar{V} = k^d$ such that the original group consists of the fixed points G^F on $V = \bar{V}^F$ under the standard Frobenius on $\bar{V}$ with respect to $\mathbb{F}_q$, with compatible $\mathbb{F}_q$-structures on G and on $\bar{V}$, as indicated:

G^F	$\mathrm{SL}_n(q)$	$\mathrm{Sp}_{2n}(q)$	$\mathrm{SO}_{2n+1}(q)$	$\mathrm{SO}_{2n}^+(q)$	$\mathrm{SO}_{2n}^-(q)$
G	SL_n	Sp_{2n}	SO_{2n+1}	SO_{2n}	SO_{2n}

This is clear from our definition of the classical algebraic groups in Section 1.2, respectively from the construction of $\mathrm{SO}_{2n}^-(q)$ in Example 22.9. (In the case of orthogonal groups in characteristic 2 this can serve as the definition of the finite special orthogonal groups.)

A similar construction is not possible for the classical groups $\mathrm{SU}_n(q)$, despite the fact that they also arise as fixed points of SL_n under a Steinberg endomorphism, as the concept of a sesquilinear form is tied to the field in question. So it is not possible to obtain compatible $\mathbb{F}_q$-structures on SL_n and the underlying natural n-dimensional k-vector space k^n. The $\mathrm{SU}_n(q)$-case is therefore more technical and we will not go into the proofs for this case here.

We want to apply techniques from algebraic groups to obtain results on classical groups. For this, we need the following standard result (see for example [66, §11.1]).

Proposition 27.2 *Let $V = k^n$ and $F : V \to V$, $(v_1, \ldots, v_n) \mapsto (v_1^q, \ldots, v_n^q)$, the standard Frobenius map. Assume that $V_1 \leq V$ is an F-stable subspace. Then V_1^F contains a k-basis of V_1; in particular $\dim_{\mathbb{F}_q} V_1^F = \dim_k V_1$.*

Proposition 27.3 *Let $V = k^n$, $H \leq \mathrm{GL}(V)$ be stable under a standard Frobenius map F of $\mathrm{GL}(V)$. Then:*

(a) *F permutes the H-invariant subspaces of V.*

(b) *If H is (absolutely) irreducible on V and preserves a non-degenerate bilinear or quadratic form on V, then the corresponding classical group is F-stable.*

Proof Assertion (a) is clear since $V_1 \leq V$ being H-invariant implies that $F(V_1) \leq V$ is H-invariant.

In (b) let f denote an H-invariant non-degenerate bilinear form on V and define

$$\tilde{f} : V \times V \to k, \qquad \tilde{f}(v,w) := F^{-1}(f(F(v), F(w))).$$

Then $\tilde{f}$ is bilinear, non-degenerate and H-invariant. But H, acting irreducibly, can fix at most one non-degenerate form on V, up to scalars, so $\tilde{f} = af$ for some $a \in k^\times$. Thus f and $\tilde{f}$ have the same isometry group, which is therefore F-stable. One argues similarly for quadratic forms. $\qquad\square$

We now state a refined version of the Reduction Theorem 18.6 for maximal subgroups of classical algebraic groups from Part II which accommodates for the presence of a Steinberg endomorphism, see [50, Thm. 1']. This will allow us later to descend to the finite groups G^F. The proof is very similar in that one basically has to go through the proof of Theorem 18.6 once again, but now paying attention to F-stability.

Recall the subgroup classes $\mathcal{C}_1, \ldots, \mathcal{C}_5$ from Section 18.1.

Theorem 27.4 (Liebeck–Seitz) *Let $G \le \mathrm{SL}(V)$ be a simple classical group on V, invariant under a standard Frobenius map F. If $H \le G$ is an F-stable closed subgroup, then one of the following occurs.*

(1) *H is contained in some F-stable member of $\mathcal{C}_1 \cup \ldots \cup \mathcal{C}_5$, or,*

(2) *$H/Z(H)$ is almost simple and H^∞ acts irreducibly on V. Moreover, if H is infinite, then H° acts tensor indecomposably on V.*

Sketch of proof Let $H \le G$ be as in the statement, and first assume that there is a proper H- and F-invariant subspace $0 \ne V_1 < V$ of V, which we choose of minimal possible dimension. If $G = \mathrm{Sp}(V)$ or $\mathrm{SO}(V)$, then $V_1^\perp$ is also H- and F-invariant and by minimality of V_1, $V_1 \cap V_1^\perp \in \{0, V_1\}$. In the first case, V_1 is non-degenerate, in the second it is totally isotropic. The argument given in the proof of Proposition 18.4 shows that in the case $(G, \mathrm{char}(k)) = (\mathrm{SO}(V), 2)$ either V_1 is totally singular or else one-dimensional non-singular. So, in any case, H is contained in an F-stable member of $\mathcal{C}_1$.

Thus, we may assume that there are no non-trivial proper H- and F-invariant subspaces in V. By Proposition 27.3(a), the socle of $V|_H$ is F-invariant, so equal to all of V. Hence H acts completely reducibly on V. Let $V_1, \ldots, V_t$ denote the homogeneous components of $V|_H$. Then F must permute the V_i, transitively since otherwise we obtain a proper F-stable H-submodule.

Assume that $t \ge 2$. Now, if $G = \mathrm{SL}(V)$ then $H \le \mathrm{Stab}_G(V_1 \oplus \cdots \oplus V_t)$, which lies in class $\mathcal{C}_2$ and is F-stable. For symplectic and orthogonal type, one can argue further as in the proof of Theorem 18.6 that in fact $V = V_1 \perp$

$\dots \perp V_t$ is an orthogonal decomposition, so again H will be contained in an F-stable member of $\mathcal{C}_2$.

Thus we may assume that $V|_H$ is homogeneous, and $V = V_1 \oplus \dots \oplus V_t$ with pairwise isomorphic irreducible H-modules V_i, of dimension $\dim(V_i) > 1$ since otherwise $H \leq Z(G)$, lying in an F-stable member of $\mathcal{C}_1$. Let now first $G = \mathrm{SL}(V)$ and set $C := C_G(H)$. Then C is also F-invariant, and by Proposition 18.1 there exists a decomposition $V \cong V_1 \otimes \mathrm{Hom}_H(V_1, V)$ such that $C = (1 \otimes \mathrm{GL}(U)) \cap G$ with $\dim(U) = t$ and $C_G(C) = (\mathrm{GL}(V_1) \otimes 1) \cap G$, which is again F-stable. So, if $t > 1$, then $V = U \otimes V_1$ is an F-invariant tensor decomposition of V stabilized by H, and H lies in the F-invariant member $N_G(C \circ C_G(C)) = N_G(\mathrm{SL}(V_1) \otimes \mathrm{SL}(U))$ of $\mathcal{C}_4$.

For $\mathrm{Sp}(V)$ and $\mathrm{SO}(V)$, the argument given in the proof of Theorem 18.6 can be adapted to show that H must be contained in an F-invariant member of $\mathcal{C}_3 \cup \mathcal{C}_4$.

So we may assume that $V|_H$ is irreducible. Let's first discuss the case when H is infinite. We saw in the proof of Theorem 18.6 that then the irreducibility assumption implies that H° is semisimple, say $H^\circ = H_1 \cdots H_t$. By Corollary 18.2 we then have a tensor product decomposition $V = \bigotimes_{i=1}^t V_i$ with $H^\circ \leq \bigotimes_{i=1}^t \mathrm{GL}(V_i)$ and $C_{\mathrm{GL}(V)}(H_j) = \bigotimes_{i \neq j} \mathrm{GL}(V_i)$. Since H is F-invariant, so is H°, and F permutes the H_i. Thus, if $G = \mathrm{SL}(V)$ then the *double centralizers*

$$CC(H_i) := C_G(C_G(H_i)) = \mathrm{GL}(V_i) \cap \mathrm{SL}(V) \qquad (1 \leq i \leq t)$$

are permuted by F and hence F normalizes $\bigotimes_{i=1}^t \mathrm{SL}(V_i)$, an F-invariant member of $\mathcal{C}_4$ when $t > 1$. On the other hand, for $G = \mathrm{Sp}(V)$ or $G = \mathrm{SO}(V)$, V is self-dual, hence so are the V_i by Corollary 18.2(b). So $H_i \leq \mathrm{Cl}(V_i)$ and $\prod_i CC(H_i) \leq \prod_i \mathrm{Cl}(V_i)$, an F-invariant group in $\mathcal{C}_4$ when $t > 1$. Otherwise, $t = 1$ and H° is simple and irreducible on V as in conclusion (2).

It remains to consider the case when H is finite. Let $H_1 \circ \dots \circ H_t$ be the product of the (quasi-simple) components of H, a characteristic subgroup. By the same argument as in the previous case, either $t \leq 1$ or H lies in a member of $\mathcal{C}_4$. If $t = 1$ then $C_G(H_1) = Z(G)$ by irreducibility, so $H_1 Z(G)/Z(G)$ is simple and thus $HZ(G)/Z(G)$ is almost simple as in (2).

Finally, if $t = 0$, all minimal normal subgroups of $\bar{H} := H/Z(H)$ are abelian and H contains a normal subgroup R as in class $\mathcal{C}_5$. Then H lies in a member of $\mathcal{C}_5$, F-stable since maximal elements of $\mathcal{C}_5$ are the full normalizers of the F-stable groups R. $\qquad \square$

27.2 The theorem of Aschbacher

We now study case (1) of the preceding theorem in some more detail in order to define certain natural collections of subgroups of the finite classical groups G^F, where $G = \mathrm{Cl}(V)$ with $\dim(V) = n$. If an F-stable subgroup H of G is contained in some F-stable group $M \in \mathcal{C}_i$, then clearly $H^F \leq M^F$. What possibilities can arise for M^F?

If $M \in \mathcal{C}_1$ then M stabilizes an F-stable non-trivial subspace $0 \neq V_1 < V$. By Proposition 27.2 then $0 \neq V_1^F < V^F$. This leads to the following class of subgroups:

Class $\mathcal{C}_1^F$: stabilizers in G^F of subspaces $0 \neq V_1^F < V^F$, non-degenerate or totally singular, or non-singular of dimension 1 when $(G, \mathrm{char}(k)) = (\mathrm{SO}(V), 2)$.

Now assume that $M \in \mathcal{C}_2$ is F-stable. Then M stabilizes an orthogonal decomposition $V = \bigoplus_{i=1}^t V_i$ into $t > 1$ mutually isometric subspaces V_i. As F stabilizes $M = G_{\{V_1,\dots,V_t\}} = \prod_{i=1}^t \mathrm{Cl}(V_i).\mathfrak{S}_t$, it permutes the V_i. So $\langle M^F, F \rangle$ induces a permutation group Δ_t on $\{1,\dots,t\}$. We may distinguish several cases:

1. Δ_t acts intransitively. Then M^F stabilizes a proper F-stable subspace of V corresponding to an orbit of Δ_t, so M^F lies in a group in $\mathcal{C}_1^F$.
2. Δ_t is transitive, but imprimitive. Then we may group together the V_i according to the blocks of imprimitivity of this action to get another member of $\mathcal{C}_2$, with smaller value of t, but still ≥ 2.
3. Δ_t acts primitively. Clearly, Δ_t commutes with F, so the orbits of (any power of) F form a block system. Hence, either F acts trivially, i.e., all V_i are F-stable, or t is a prime and F acts as a t-cycle, in which case Δ_t is the cyclic group of order t generated by the image of F.

In the third case, if F acts trivially, M^F lies in

Class $\mathcal{C}_2^F$: stabilizers in G^F of orthogonal decompositions $V^F = V_1^F \perp \dots \perp V_t^F$ into $t \geq 2$ mutually isometric F-stable subspaces V_i, isomorphic to $\mathrm{Cl}_m(q) \wr \mathfrak{S}_t$ where $m = \dim(V_i)$, so $n = mt$,

while otherwise we are in the situation of Exercise 30.2, so M^F lies in

Class $\mathcal{C}_2'^F$: field extension subgroups $\mathrm{Cl}_m(q^t).t$ with $n = mt$ where t is prime.

In the case that all V_i are one-dimensional, this leads to the normalizer of the Singer cycle as in Example 25.4(1).

Now assume that $M \in \mathcal{C}_3$ is F-stable, so $G = \mathrm{Sp}(V)$ or $\mathrm{SO}(V)$, M fixes a totally singular decomposition $V = W \oplus W'$, and $M \cong \mathrm{GL}_m$ or $\mathrm{GL}_m.2$. Then $M^F \in \{\mathrm{GL}_m(q).a, \mathrm{GU}_m(q).a\}$ with $a \leq 2$. This gives

Class $\mathcal{C}_3^F$: $\mathrm{GL}_m(q).a$ or $\mathrm{GU}_m(q).a$ with $a \leq 2$, in symplectic or orthogonal groups G^F where $n = 2m$.

Now assume that $M \in \mathcal{C}_4$ is F-stable, the stabilizer of a tensor product decomposition, and either $V = V_1 \otimes V_2$ with V_1, V_2 not isometric, or $V = \bigotimes_{i=1}^{t} V_i$ with $t \geq 2$ mutually isometric V_i. In the first case, M^F falls into

Class $\mathcal{C}_4^F$: stabilizers in G^F of tensor product decompositions $V^F = V_1^F \otimes V_2^F$ with V_1, V_2 not isometric, of the form $N_{G^F}(\mathrm{Cl}_{n_1}(q) \circ \mathrm{Cl}_{n_2}(q))$ where $n = n_1 n_2$ (see Proposition 18.1).

In the second case, $M = N_G(\prod_{i=1}^{t} \mathrm{Cl}(V_i))$ with F permuting the factors. Again, we look at the permutation group Δ_t induced by the action of $\langle M^F, F \rangle$ on the set of factors and argue as for class $\mathcal{C}_2$ above: if Δ_t is intransitive, we obtain a member of $\mathcal{C}_4^F$. If Δ_t is imprimitive, grouping together the V_i according to the blocks, we obtain another member of $\mathcal{C}_4$ with smaller $t \geq 2$. Finally, if Δ_t is primitive, then either F is trivial or t is prime and Δ_t is cyclic of order t. This leads to

Class $\mathcal{C}_4'^F$: stabilizers of tensor product decompositions $V^F = \bigotimes_{i=1}^{t} V_i^F$ into $t \geq 2$ isometric subspaces V_i^F, with $M = N_{G^F}(\prod_{i=1}^{t} \mathrm{Cl}_m(q))$, where $n = m^t$, or

Class $\mathcal{C}_4''^F$: twisted tensor product subgroups $\mathrm{Cl}_m(q^t).t$, where $n = m^t$ with t prime.

Here, a tensor product structure is preserved over $\overline{\mathbb{F}_q}$, but not over $\mathbb{F}_q$, like for example (modulo scalars) $\mathrm{PSL}_m(q^t) < \mathrm{PSL}_{m^t}(q)$, $\mathrm{PSL}_m(q^2) < \mathrm{PSU}_{m^2}(q)$ or $\mathrm{PSp}_m(q^t) < \mathrm{PSp}_{m^t}(q)$.

Next, if $M \in \mathcal{C}_5$ is F-stable, then it has a characteristic (and hence also F-stable) normal r-subgroup R which is extraspecial or a central product of such with Z_4. Here the F-fixed points lie in class

Class $\mathcal{C}_5^F$: $N_{G^F}(R)$ with $N_G(R)$ in $\mathcal{C}_5$, as in Table 27.2 (see [50, p. 430] or [44, Tab. 4.6.A]).

This completes our discussion of F-stable members of $\mathcal{C}_1, \ldots, \mathcal{C}_5$. If $H \leq G$ is F-stable but not contained in any F-stable member of $\mathcal{C}_1 \cup \ldots \cup \mathcal{C}_5$, then by Theorem 27.4 it is an almost simple subgroup (modulo $Z(G)$), with H° acting irreducibly on V. Let's single out two further subclasses in this situation:

Table 27.2 *Normalizers of extraspecial subgroups R*

G	$N_G(R)/Z(G)$
SL_{r^m}	$r^{2m}.\mathrm{Sp}_{2m}(r)$
Sp_{2m}	$2^{2m}.\mathrm{GO}_{2m}^-(2)$
SO_{2m}	$2^{2m}.\mathrm{GO}_{2m}^+(2)$

Class $\mathcal{C}_6^F$: subgroups of G^F which are classical on V^F (respectively on V^{F^2}, for G^F a unitary group).

Examples of this type are the normalizers of $\mathrm{Sp}_n(q)$ or $\mathrm{SO}_n^{(\pm)}(q)$ inside $\mathrm{SL}_n(q)$ and $\mathrm{SU}_n(q)$, the normalizer of $\mathrm{SU}_n(q^{1/2})$ inside $\mathrm{SL}_n(q)$, and the normalizer of $\mathrm{SO}_n(q)$ inside $\mathrm{Sp}_n(q)$ if $p = 2$.

Finally, for subfields $\mathbb{F}_{q'} < \mathbb{F}_q$ of prime index, we get:

Class $\mathcal{C}_7^F$: subfield subgroups $N_{G^F}(\mathrm{Cl}_n(q'))$, with $|\mathbb{F}_q : \mathbb{F}_{q'}|$ prime, where $G^F = \mathrm{Cl}_n(q)$.

We can now state the principal reduction result on maximal subgroups of finite classical groups, first proved by Aschbacher [1]; here we present the approach by Liebeck and Seitz [50] using descent from the corresponding algebraic groups.

Theorem 27.5 (Aschbacher, Liebeck–Seitz) *Let $G \le \mathrm{SL}(V)$ be a classical group on the finite-dimensional k-vector space V, stable under a Steinberg endomorphism $F : \mathrm{SL}(V) \to \mathrm{SL}(V)$, so that G^F is a finite classical group on V^F (respectively V^{F^2}), and let $H \le G^F$. Then one of the following holds:*

(1) *H lies in a member of $\mathcal{C}_1^F, \ldots, \mathcal{C}_7^F$, or*
(2) *H lies in class $\mathcal{S}$, that is, $HZ(G)/Z(G)$ is almost simple, H^∞ acts absolutely irreducibly and tensor indecomposably on V^F (respectively V^{F^2}), the representation cannot be realized over a proper subfield, and if $G^F = \mathrm{SL}_n(q)$ then H fixes no non-degenerate bilinear or unitary form on V.*

Proof for G^F not unitary Apply Theorem 27.4 and our previous discussion. Since H is F-stable, either $H \le M^F$ for an F-stable member $M \in \mathcal{C}_1^F \cup \ldots \cup \mathcal{C}_5^F$, or $HZ(G)/Z(G)$ is almost simple and H^∞ acts absolutely irreducibly (since otherwise it fixes a direct sum decomposition over $\overline{\mathbb{F}_q}$ and H lies in a member of $\mathcal{C}_2'^{F}$). If H lies in no member of $\mathcal{C}_7^F$, then it cannot be defined over a proper subfield, and if it lies in no member of $\mathcal{C}_6$, it fixes no non-degenerate form on V^F other than the one defining G. $\square$

Example 27.6   Assume that $G^F = \mathrm{SL}_n(q)$ for n a prime. Then only the classes $\mathcal{C}_1^F$, $\mathcal{C}_5^F$, $\mathcal{C}_6^F$, $\mathcal{C}_7^F$ and $\mathcal{S}$ can occur, since for the other classes n has to be composite (see Proposition 28.1 for arbitrary n).

In fact Liebeck and Seitz [50] prove a more general version of the reduction theorem for normalizers of classical groups G^F inside $\Gamma L(V^F)$, the extension of $\mathrm{GL}(V^F)$ by the group of field automorphisms, where additional classes $\mathcal{C}_1'^F$ and $\mathcal{C}_7'^F$ can arise. We shall not attempt to describe this here. This general version then allows one to obtain a similar reduction theorem for maximal subgroups of all isogeny types of classical groups. Indeed, for our given classical group $G = \mathrm{Cl}(V)$ let G_{sc} be the corresponding group of simply connected type. By Proposition 9.15 there exists a natural isogeny $\pi : G_{\mathrm{sc}} \to G$ which restricts to a homomorphism $G_{\mathrm{sc}}^F \to G^F$ with central kernel and whose image contains $[G^F, G^F]$ by Proposition 24.21. Now note that π induces a bijection between maximal subgroups of G_{sc}^F and those of its image $\pi(G_{\mathrm{sc}}^F)$ in G^F. In this way, the results on maximal subgroups of $[G^F, G^F]$ can be transferred to those of G_{sc}^F.

On the other hand, if G_{ad} denotes the corresponding group of adjoint type, then the natural isogeny $G \to G_{\mathrm{ad}}$ induces a homomorphism $G^F \to G_{\mathrm{ad}}^F$ whose image contains $[G_{\mathrm{ad}}^F, G_{\mathrm{ad}}^F]$. Now, G_{ad}^F is an extension of $[G_{\mathrm{ad}}^F, G_{\mathrm{ad}}^F]$ by certain (diagonal) automorphisms, so the above-mentioned general version of the reduction theorem applies to the maximal subgroups of G_{ad}^F as well.

Remark 27.7 (Relationship between the classes $\mathcal{C}_i^F$ and Aschbacher's and Kleidman–Liebeck's classes $\mathcal{C}_i$)   The classes $\mathcal{C}_1^F, \ldots, \mathcal{C}_7^F$ of natural subgroups defined above do not agree completely with the classes $\mathcal{C}_i$ introduced by Aschbacher in his original paper [1]. Moreover, the latter were already modified in the book of Kleidman and Liebeck [44] in which the structure, conjugacy and maximality of members in these classes were determined. The following table indicates the relationship between the classes $\mathcal{C}_i^F$ here and in [50], and the classes $\mathcal{C}_i$ in [44]:

[50]	$\mathcal{C}_1^F, \mathcal{C}_1'^F$	$\mathcal{C}_2^F, \mathcal{C}_2'^F, \mathcal{C}_3^F$	$\mathcal{C}_4^F$	$\mathcal{C}_4'^F$	$\mathcal{C}_5^F$	$\mathcal{C}_6^F$	$\mathcal{C}_7^F, \mathcal{C}_7'^F$	$\mathcal{C}_4''^F, \mathcal{S}$
[44]	$\mathcal{C}_1$	$\mathcal{C}_2, \mathcal{C}_3$	$\mathcal{C}_4$	$\mathcal{C}_7$	$\mathcal{C}_6$	$\mathcal{C}_8$	$\mathcal{C}_5$	$\mathcal{S}$

28
About the classes $\mathcal{C}_1^F, \ldots, \mathcal{C}_7^F$ and $\mathcal{S}$

The theorem of Aschbacher in the previous chapter still leaves several important questions open:

1. What is the structure of members in the classes $\mathcal{C}_1^F, \ldots, \mathcal{C}_7^F$?
2. Which of the groups in classes $\mathcal{C}_1^F, \ldots, \mathcal{C}_7^F$ are actually maximal in $\mathrm{Cl}(V)$?
3. What kind of groups can occur in class $\mathcal{S}$?
4. Which of the groups in class $\mathcal{S}$ are actually maximal?

In this section we comment on the status of these questions at the time of writing and on some of the methods used to treat them.

28.1 Structure and maximality of groups in $\mathcal{C}_i^F$

The answer to the first question on the structure of members in $\mathcal{C}_i^F$ is relatively straightforward, given their explicit geometric description. This has been laid out in the book of Kleidman and Liebeck [44].

The second question on maximality of members in the classes $\mathcal{C}_i^F$ is already much more difficult. Again, the complete answer has been obtained by Kleidman and Liebeck [44] at least in the case where the dimension of the classical module V of $\mathrm{Cl}(V)$ is at least 13.

It turns out that essentially all members in the classes $\mathcal{C}_i^F$ are maximal, apart from a few well-understood situations. As an illustration we present the result for $G^F = \mathrm{SL}_n(q)$, taken from [44, Table 3.5.A] (compare with the corresponding Table 18.2 for the algebraic group SL_n); here, as in Table 18.2, P_m denotes the standard parabolic subgroup of SL_n corresponding to the set $S \setminus \{s_m\}$ of simple reflections of its Weyl group:

Proposition 28.1 *The maximal members in the classes $\mathcal{C}_1^F, \ldots, \mathcal{C}_7^F$ for*

$G^F = \mathrm{SL}_n(q)$, $n \geq 3$, *are as given in Table 28.1. Here, for the classes* $\mathcal{C}_2^F, \mathcal{C}_2'^F, \mathcal{C}_4^F, \mathcal{C}_4'^F$, *the maximal subgroup is obtained as the normalizer in* $\mathrm{SL}_n(q)$ *of the intersection of the corresponding entry in the table with* $\mathrm{SL}_n(q)$.

Table 28.1 *The classes $\mathcal{C}_i^F$ in $\mathrm{SL}_n(q)$*

class	structure	conditions	non-max.
$\mathcal{C}_1^F$	P_m^F	$1 \leq m \leq n-1$	–
$\mathcal{C}_2^F$	$\mathrm{GL}_m(q) \wr \mathfrak{S}_t$	$n = mt, t \geq 2$	$m = 1, q \leq 4$
			or $m = q = 2$
$\mathcal{C}_2'^F$	$\mathrm{GL}_m(q^t).t$	$n = mt, t$ prime	–
$\mathcal{C}_4^F$	$\mathrm{GL}_{n_1}(q) \otimes \mathrm{GL}_{n_2}(q)$	$n = n_1 n_2, 2 \leq n_1 < n_2$	$n_1 = q = 2$
$\mathcal{C}_4'^F$	$\mathrm{GL}_m(q) \wr \mathfrak{S}_t$	$n = m^t, m \geq 3, t \geq 2$	–
$\mathcal{C}_5^F$	$r^{2m+1}.\mathrm{Sp}_{2m}(r)$	see below[1]	–
$\mathcal{C}_6^F$	$\mathrm{Sp}_n(q)$	n even	–
	$\mathrm{SO}_n(q)$	nq odd	–
	$\mathrm{SO}_n^\pm(q)$	n even, q odd	–
	$\mathrm{SU}_n(q_0)$	$q = q_0^2$	–
$\mathcal{C}_7^F$	$\mathrm{GL}_n(q_0)$	$q = q_0^t, t$ prime	–

[1] In $\mathcal{C}_5^F$, $r \neq p$ is prime, $q = p^f$ with f odd and minimal subject to $p^f \equiv 1$ (mod $2\gcd(2,r)$).

About the proof It is clear from the description in Table 18.2 of the maximal members in the classes $\mathcal{C}_1, \mathcal{C}_2, \mathcal{C}_4, \mathcal{C}_6$ that there exist F-stable representatives, with F-fixed points in $\mathcal{C}_1^F, \mathcal{C}_2^F, \mathcal{C}_4^F, \mathcal{C}_4'^F, \mathcal{C}_6^F$ as given in Table 28.1. Moreover, multiplying F by an inner element of G of order t which interchanges the factors GL_m in a group of type $\mathcal{C}_2$ cyclically leads to the fixed point group of type $\mathcal{C}_2'^F$ (using Exercise 30.2 and Corollary 21.8). The existence of extraspecial subgroups with the stated normalizer is well-known. Finally, the subgroups in class $\mathcal{C}_7^F$ clearly occur.

It then remains to work out for which values of the parameters any of the above groups is contained in a member of some other class. For example, since $\mathrm{SL}_n(q)$ has a BN-pair, the parabolic subgroups in class $\mathcal{C}_1^F$ are always maximal by Proposition 12.2.   □

In all classes and for all classical groups there are only three families of examples in which maximal members of a class $\mathcal{C}_i^F$ lie in some group in the class $\mathcal{S}$. They all originate from the fact that in characteristic 2 the symplectic groups are isomorphic to odd-dimensional orthogonal groups in one higher dimension. The three families of examples are listed in Table 28.2 (see [44, Table 3.5.I]).

Table 28.2 *Groups* $H \in \mathcal{C}_i^F$ *lying in members* $K \in \mathcal{S}$

G^F	class	H	K	conditions
$\mathrm{SO}_{2t}^+(q)$	$\mathcal{C}_4'{}^F$	$\mathrm{Sp}_2(q) \wr \mathfrak{S}_t$	$\mathrm{Sp}_{2t}(q)$	$q \geq 4$ even, $t \geq 4$
$\mathrm{SO}_{2t}^+(q)$	$\mathcal{C}_4'{}^F$	$\mathrm{Sp}_2(q) \wr \mathfrak{S}_t$	$\mathrm{SO}_{2t+2}^+(q)$	$q \geq 4$ even, $t \geq 5$ odd
$\mathrm{SO}_{4t}^+(q)$	$\mathcal{C}_4'{}^F$	$\mathrm{Sp}_4(q) \wr \mathfrak{S}_t$	$\mathrm{Sp}_{4t}(q)$	q even

Example 28.2 Let $G^F = \mathrm{SO}_{2t}^+(q)$, with $t \geq 4$ and $q \geq 4$ even. Then the class $\mathcal{C}_4'{}^F$ contains the tensor product subgroup $H := \mathrm{Sp}_2(q) \wr \mathfrak{S}_t$. On the other hand H is also contained in the symplectic group $K := \mathrm{Sp}_{2t}(q)$ as an imprimitive subgroup from class $\mathcal{C}_2^F$. But the 2^t-dimensional spin representation of $\mathrm{SO}_{2t+1}(q) \cong \mathrm{Sp}_{2t}(q)$ embeds K as an absolutely irreducible and tensor indecomposable subgroup, hence as a member from $\mathcal{S}$, into $G^F = \mathrm{SO}_{2t}^+(q)$. This gives the first family of examples in Table 28.2. The other two are obtained by a similar construction.

28.2 On the class $\mathcal{S}$

At the time of writing this book, the two questions on groups contained in the class $\mathcal{S}$ and their maximality (mentioned at the beginning of this section) are far from being resolved, and probably a complete answer will never be obtained. We content ourselves here with giving some first indications on how these problems can be attacked, and where the difficulties lie. At this point it seems unavoidable to use the classification of the finite simple groups.

Determining the members of $\mathcal{S}$ is tantamount to finding the degrees of all absolutely irreducible, tensor indecomposable representations of finite quasi-simple groups, and their invariant forms. This is far from being solved at present. Nevertheless, for any classical group of fixed dimension, a list of candidates for groups in $\mathcal{S}$ can be compiled.

The goal must be to eliminate as many candidates as possible by simple arguments. The first very helpful observation relies on the order formulas and is applicable in particular to embeddings of groups of Lie type in the same characteristic. We encountered the order polynomial of a finite group of Lie type and its factorization into cyclotomic polynomials, which in some sense behaved like generic prime divisors of the order. There is one further aspect to this factorization, given by the theorem of Zsigmondy (see [38, IX, Thm. 8.3]):

Theorem 28.3 *Let* $q > 1$ *be a prime power. Then there exists a prime divisor* ℓ *of* $q^n - 1$ *which does not divide* $q^i - 1$ *for* $1 \leq i \leq n - 1$, *unless one of:*

(1) $n = 2$ *and* $q = 2^f - 1$ *is a Mersenne prime, or*
(2) $n = 6$ *and* $q = 2$.

A prime divisor $\ell = \ell(n, q)$ of $q^n - 1$ as in the theorem is called a *primitive prime divisor of* $q^n - 1$. Note that $\ell(n,q)|\Phi_n(q)$, but does not divide $\Phi_i(q)$ for $i < n$.

Example 28.4 As a simple application we claim that the smallest degree d of a faithful representation of the finite symplectic group $\mathrm{Sp}_{2n}(q)$ over $\mathbb{F}_q$ is $2n$. Indeed, any such representation defines an embedding $\mathrm{Sp}_{2n}(q) \leq \mathrm{GL}_d(q)$. Since $|\mathrm{Sp}_{2n}(q)|$ is divisible by a primitive prime divisor $\ell(2n,q)$, and $|\mathrm{GL}_d(q)|_{q'} = (q-1)\cdots(q^d-1)$ (see Table 24.1), we conclude from Theorem 28.3 that $d \geq 2n$, except possibly for $\mathrm{Sp}_6(2)$ (where the claim can still be shown to hold).

As a further application, we obtain the following result on values of cyclotomic polynomials:

Corollary 28.5 *Let* $f = \prod_{i=1}^{r} \Phi_{d_i}(x), g = \prod_{i=1}^{s} \Phi_{e_i}(x) \in \mathbb{Z}[x]$ *with not necessarily distinct* $d_i, e_i \in \mathbb{N}$. *If* $f(q) = g(q)$ *for some prime power* $q > 1$, *then* $f = g$ *unless one of:*

(1) $q = 3$, $f/g = (x-1)^{2a}/(x+1)^a$ *with* $a \in \mathbb{Z}$, *or*
(2) $q = 2$, $f/g = (x-1)^a(x+1)^b(x^2 - x + 1)^{-b}$ *with* $a, b \in \mathbb{Z}$.

Proof Dividing both f, g by their greatest common divisor, we may assume that they are coprime in $\mathbb{Z}[x]$. Let n be maximal subject to $\Phi_n(x)|(fg)$. If there exists a primitive prime divisor $\ell(n, q)$, then $\Phi_n(x)$ has to divide both f and g, which is not possible. So, by the theorem of Zsigmondy either $q = 2$ and $n = 6$, or $n \leq 2$. In the latter case, $(q-1)^a = (q+1)^b$ with $a + b \geq 1$ implies that $q + 1 = 4$ and $a = 2b$. If $n = 6$ and $q = 2$, then clearly $f/g = (x-1)^a(x+1)^b(x^2 - x + 1)^{-b}$ for some $a, b \in \mathbb{Z}$.   □

Zsigmondy primes can often be used to rule out the existence of certain embeddings of quasi-simple groups of Lie type in the same characteristic, like for example $B_8(q) \not\leq E_8(q)$ for all q (using $\ell(16, q)$). To deal with embeddings in different characteristic, it is useful to know the minimal dimensions of faithful representations of groups of Lie type in cross characteristic. These turn out to be exponential in the rank of the group, and thus much larger than dimensions of representations in the defining characteristic. We give one example of such a result:

Proposition 28.6 (Landazuri–Seitz) *Let K be a field of characteristic $r \neq p$.*

(a) *Assume $n \geq 3$, and let $\rho \colon \mathrm{SL}_n(q) \to \mathrm{GL}_d(K)$ be a representation of $\mathrm{SL}_n(q)$ over K with central kernel, where $q = p^f$. Then $d \geq q^{n-1} - 1$.*

(b) *Assume $n \geq 2$, and let $\rho \colon \mathrm{Sp}_{2n}(q) \to \mathrm{GL}_d(K)$ be a representation of $\mathrm{Sp}_{2n}(q)$ over K with central kernel, where $q = p^f$ is odd. Then $d \geq q^{n-1}(q-1)/2$.*

Proof For (a), let $P \leq \mathrm{SL}_n(q)$ be the maximal parabolic subgroup corresponding to the subset $I = \{s_2, \ldots, s_{n-1}\}$ of the set of simple reflections, with Levi decomposition

$$P = U.L = \left\{ \begin{pmatrix} 1 & 0 \\ v & I_{n-1} \end{pmatrix} \;\middle|\; v \in \mathbb{F}_q^{n-1} \right\} \rtimes \left\{ \begin{pmatrix} a & 0 \\ 0 & A \end{pmatrix} \;\middle|\; a \det A = 1 \right\}.$$

Thus, the Levi complement L is isomorphic to $\mathrm{GL}_{n-1}(q)$, while the normal subgroup U is elementary abelian of order q^{n-1}. Moreover, the Levi factor L acts on U by conjugation via

$$l.u = lul^{-1} = \begin{pmatrix} a & 0 \\ 0 & A \end{pmatrix} \begin{pmatrix} 1 & 0 \\ v & 1 \end{pmatrix} \begin{pmatrix} a^{-1} & 0 \\ 0 & A^{-1} \end{pmatrix} = \begin{pmatrix} 1 & 0 \\ a^{-1}Av & 1 \end{pmatrix}$$

(for $l \in L$, $u \in U$), so U is the natural module for $\mathrm{SL}_{n-1}(q) \leq [L, L]$. (See Section 17.1 for the corresponding algebraic group situation.)

In particular, since $n \geq 3$ we see that L acts transitively on the non-zero vectors of $U \cong \mathbb{F}_q^{n-1}$, whence it also acts transitively on the set of non-trivial homomorphisms $\mathrm{Hom}(U, \bar{K}^\times) \setminus \{1_U\}$ from U into $\bar{K}^\times$, where $\bar{K}$ denotes an algebraic closure of K. By assumption, U is not contained in $\ker \rho$, so $\rho|_U$ contains one, hence by the action of L, all $\mu \in \mathrm{Hom}(U, \bar{K}^\times) \setminus \{1_U\}$. As $|U|$ is prime to the characteristic of K, $|\mathrm{Hom}(U, \bar{K}^\times)| = |U| = q^{n-1}$, which shows that $d = \dim \rho \geq q^{n-1} - 1$.

The proof of (b) is similar and is left as an exercise (see Exercise 30.23). $\square$

These so-called Landazuri–Seitz bounds show that subgroups of Lie type in the wrong characteristic inside finite groups of Lie type are very tiny. Indeed, if for example $\mathrm{SL}_n(q) \leq \mathrm{GL}_d(2)$ with q odd, then $|\mathrm{SL}_n(q)| \sim q^{n^2-1}$

by the order formula (see Example 24.8(2)), while the order of $|\mathrm{GL}_d(2)|$ is of the order of magnitude

$$2^{d^2} = 2^{(q^{n-1}-1)^2} = 2^{q^{2n-2}}$$

by the bound for d in Proposition 28.6. So indeed these are extremely small subgroups.

We close the chapter with some comments on our fourth question, the maximality of the groups in class $\mathcal{S}$. Here, one has to investigate the following situation: we have embeddings $H < K < G^F = \mathrm{Cl}(V)^F$ (respectively in $\mathrm{SU}(q^2)$), both H, K absolutely irreducible on V^F (respectively V^{F^2}), with $HZ(G)/Z(G)$ and $KZ(G)/Z(G)$ almost simple. By the classification of finite simple groups, H, K are either sporadic, alternating, or of Lie type. Moreover, in the latter case, the characteristics of H, K may coincide or not, and may or may not equal the characteristic of G. This leads to quite a number of different possible configurations for $(H, K, \mathrm{Cl}(V))$ to consider.

For example, if K is one of the 26 sporadic groups, the classification of triples $H < K < G^F$ is essentially a question on maximal subgroups of sporadic simple groups. In the situation where H is sporadic, one has to determine all absolutely irreducible representations of sporadic groups and their covering groups. Both are finite, but nevertheless very difficult problems which have not yet been solved for all groups.

If K is an alternating group $\mathfrak{A}_n$, one has to decide which absolutely irreducible representations of covering groups of $\mathfrak{A}_n$ restrict irreducibly to (almost simple) subgroups. The cases where H and K are of Lie type in different characteristics can essentially be dealt with using the Landazuri–Seitz bounds.

This short overview should have convinced the reader that the methods to study the groups in class $\mathcal{S}$ are of a representation theoretic nature; in particular they lie outside the scope of this text.

29

Finite exceptional groups of Lie type

In this chapter we are concerned with the determination of the maximal subgroups of the finite exceptional groups of Lie type. According to Table 22.1 these are the members of the ten infinite series

$$^2B_2(2^{2f+1}),\ ^2G_2(3^{2f+1}),\ G_2(q),\ ^3D_4(q),\ ^2F_4(2^{2f+1}),$$

$$F_4(q),\ E_6(q),\ ^2E_6(q),\ E_7(q),\ E_8(q),$$

where f may be an arbitrary positive integer, and q an arbitrary prime power. At first this might seem easier than the case of classical groups, since there are just a finite number of underlying complete root data. And indeed, complete lists of maximal subgroups have been obtained by Suzuki [78], Cooperstein [18], Kleidman [42, 43], respectively Malle [55], for the first five families in the above list. Thus, one may hope that eventually complete lists will also be obtained for the remaining five families of large rank. Except for some very "small" subgroups this has in fact been reached.

On the other hand the task is more difficult here since there is no convenient natural representation in which to study these groups. For example, the smallest faithful representation of $G = E_8$ in its defining characteristic is the adjoint one on its Lie algebra, of dimension $\dim G = 248$. Representations in any other characteristic are even much larger than this.

29.1 Maximal subgroups

Again, as for classical groups, it seems reasonable to attempt to use the results on maximal subgroups of simple exceptional algebraic groups (see Theorem 19.2), and study F-stability. Throughout we assume that G is an exceptional simple algebraic group of adjoint type, so $Z(G) = 1$. (For the other isogeny types the discussion before Remark 27.7 applies; by Table 9.2 this is only relevant for groups of types E_6 and E_7.) The first result here is (see [48, Thm. 1]):

Theorem 29.1 (Liebeck–Seitz) *Let G be a simple exceptional algebraic group of adjoint type, $F : G \to G$ a Steinberg endomorphism, and $M \le G$ an F-stable closed subgroup, maximal among positive-dimensional F-stable closed subgroups. If M° is simple, assume that $p = 0$ or $p > 7$. Then one of the following holds:*

(1) M *is a parabolic subgroup,*

(2) M *is the normalizer of some connected reductive subgroup of maximal rank,*

(3) M *is the normalizer of some semisimple subgroup not of maximal rank as in Theorem 19.1(3),*

(4) $G = E_7$, $p \ne 2$, $M = (Z_2^2 \times D_4).\mathfrak{S}_3$,

(5) $G = E_8$, $p \ne 2, 3, 5$, $M = A_1 \times \mathfrak{S}_5$, *or*

(6) $G = E_8$, $p \ne 2$, $M = (A_1 G_2 G_2).\mathfrak{S}_2$.

In the next step, one needs to descend to the finite groups G^F.

Let $H \le G^F$ be maximal. If H normalizes some non-trivial F-stable connected proper subgroup D of G, then HD lies in some closed F-stable maximal subgroup $M \le G$ of positive dimension, and maximality implies $H = M^F$.

For the types occurring in the preceding theorem, the possible classes of fixed point groups M^F can be worked out, as seen for tori, Levi subgroups, and so on, in the preceding chapters, by using Theorem 21.11.

So we may assume that H normalizes no non-trivial F-stable connected proper subgroup of G. Let's first consider the case that H normalizes a non-trivial F-stable r-subgroup $R \le G$ for some prime r, that is, H is an r-local subgroup. If $r = \operatorname{char}(k)$, then H lies in a proper F-stable parabolic subgroup of G, by Theorem 26.5, which is of positive dimension, so we have $r \ne \operatorname{char}(k)$. Since $N_G(R) \le N_G(R_0)$ for any characteristic normal subgroup R_0 of R, we may (and will) assume by maximality of H that R is elementary abelian and moreover that it has no proper H-invariant non-trivial subgroups.

Definition 29.2 Let G be a simple algebraic group. An elementary abelian r-subgroup R of G, with $r \ne \operatorname{char}(k)$, is called a *Jordan subgroup* of G if it satisfies the following conditions:

1. $C_G(R)$ (and hence $N_G(R)$) is finite,

2. R is a minimal normal subgroup of $N_G(R)$,

3. $N_G(R)$ is maximal subject to conditions 1. and 2., and

4. there is no non-trivial connected $N_G(R)$-invariant proper subgroup of G.

Thus, by what we said above the investigation of local maximal subgroups naturally leads us to the study of Jordan subgroups. For classical groups these are precisely the images in G_{ad} of extraspecial groups with normalizer in class $\mathcal{C}_5$. For simple groups of exceptional type, some further interesting examples occur (see [8, Thm. 1] and [17, Thm. 1]). In the next statement, the notation r^a, with r a prime and $a \geq 1$, denotes an elementary abelian r-group of this order, while r^{a+b} stands for an (unspecified) extension of an elementary abelian r-group of order r^a by one of order r^b.

Theorem 29.3 (Borovik, Cohen–Liebeck–Saxl–Seitz) *Let G be a simple exceptional algebraic group of adjoint type with Steinberg endomorphism $F :$ $G \to G$. Then the Jordan subgroups R in G^F and their normalizers $H = N_{G^F}(R)$ are given as follows:*

(1) $G = G_2$, $R = 2^3$ and $H = 2^3.\mathrm{SL}_3(2)$,

(2) $G = F_4$, $R = 3^3$, $H = 3^3.\mathrm{SL}_3(3)$,

(3) $G = E_6$, $R = 3^3$, $H = 3^{3+3}.\mathrm{SL}_3(3)$, *or*

(4) $G = E_8$, $R = 2^5$, $H = 2^{5+10}.\mathrm{SL}_5(2)$, *or* $R = 5^3$, $H = 5^3.\mathrm{SL}_3(5)$.

In each case, these are unique up to G^F-conjugation.

The normalizers $N_{G^F}(R)$ of these Jordan subgroups are called *exotic local subgroups* of exceptional groups of Lie type.

Sketch of proof First note that a Jordan subgroup $R \leq G^F$ cannot be contained in a torus T of G. Otherwise, $C_G(R)$ would contain T and hence be infinite, contrary to our assumption that R is a Jordan subgroup. Now let $\pi : G_{sc} \to G$ be the natural isogeny from a simply connected group of the same type as G as in Proposition 9.15. If $|\ker(\pi)|$ is prime to r, then $\pi^{-1}(R) = \tilde{R} \times \ker(\pi)$ for some subgroup $\tilde{R} \leq G_{sc}$ isomorphic to R (so $R = \pi(\tilde{R})$), which itself cannot be contained in any torus of G_{sc}. It then follows from Corollary 14.17 that r must be a torsion prime for G and moreover $|R| \geq r^3$. For simple groups of exceptional type the condition that $\ker(\pi)$ is prime to r is satisfied unless $(G, r) \in \{(E_6, 3), (E_7, 2)\}$ (see Table 9.1). In the latter two cases, we still have that r is torsion. Thus, by Table 14.1 we have $r = 2$ for G_2, $r \leq 3$ for F_4, E_6, E_7, and $r \leq 5$ for E_8.

The various possibilities are now analyzed case-by-case by using information on the centralizers of semisimple elements. Assume for example that R is a Jordan subgroup in $G = G_2$ and thus $r = 2$. Let $s \in R \setminus \{1\}$. Then one can show that $C_G(s) \cong SL_2 \circ SL_2$, the subgroup constructed in Example 14.19(2). By Example 14.19(3) both factors contain a quaternion subgroup of order 8. Their central product in $C_G(s)$ is then extraspecial of order 32 and contains a self-centralizing elementary abelian subgroup R of order 8, with the stated normalizer. Similarly, for $G = F_4$ and $r = 3$, respectively $G = E_8$ and $r = 5$ there is $s \in R \setminus \{1\}$ with centralizer $SL_3 \circ SL_3$, respectively $SL_5 \circ SL_5$, corresponding to maximal rank subsystems of type A_2A_2, respectively A_4A_4, inside which R can be constructed. □

Note that there do exist maximal local subgroups which are not covered by Theorem 29.3: they are normalizers of some positive-dimensional semisimple subgroups, as for example in Theorem 29.1(4)!

To complete our investigation of maximal subgroups now assume that $H \leq G^F$ is not local, and does not normalize a proper connected subgroup of positive dimension, that is, it is not covered by Theorem 29.1. Then its generalized Fitting subgroup is a direct product $F^*(H) = H_1 \times \cdots \times H_t$ of non-abelian simple groups H_i. We aim to show that $t = 1$ (and hence H is almost simple), so instead assume $t > 1$. For each i, let $CC(H_i) := C_G(C_G(H_i))$ be the double centralizer of H_i.

Lemma 29.4 *In the above situation the following holds:*

(a) *The $CC(H_i)$, $1 \leq i \leq t$, mutually commute.*
(b) *If $t > 1$, the commuting product $H^* := \prod_{i=1}^{t} CC(H_i)$ is finite.*

Proof Clearly for each $j \neq i$ we have $H_j \subseteq C_G(H_i)$, so $C_G(H_j) \supseteq CC(H_i)$ and $CC(H_j) \subseteq C_G(CC(H_i))$, whence (a).

Thus $H^* := \prod_{i=1}^{t} CC(H_i)$ is an F-stable commuting product, normalized by H (because H normalizes each H_i, so each $C(H_i)$, so each $CC(H_i)$). Moreover, $H^* < G$ as $t > 1$, thus H^* is finite by assumption. □

Now for any $g \in H_i$ we have $C_G(g) \supseteq C_G(H_i)$, so $Z(C_G(g))^\circ \subseteq CC(H_i)$. As H^* is finite by the lemma, we must have dim $Z(C_G(g))^\circ = 0$. Taking g semisimple and of prime order, $C_G(g)^\circ$ is a connected reductive subgroup containing a maximal torus and with finite center, so it is semisimple and of maximal rank. By the classification of maximal rank subgroups in Chapter 13 we see that in exceptional groups this can only happen for elements of order 2, 3 or 5 (the latter only in E_8). So if $G \neq E_8$, only the primes 2, 3 can divide $|F^*(H)|$, whence $F^*(H)$ is solvable by Burnside's $p^a q^b$-theorem, which contradicts our assumption that H is non-local.

Thus $t > 1$ necessarily forces $G = E_8$, $p > 5$, and the H_i are 2, 3, 5-groups. By the classification of finite simple groups, this implies that H_i is one of

$\mathfrak{A}_5, \mathfrak{A}_6$, or $\mathrm{PSU}_4(2)$. Finally one can show that the only possibility is $G = E_8$ and $F^*(H) = \mathfrak{A}_5 \times \mathfrak{A}_6$, see Borovik [8]. Otherwise, $t = 1$ and so $H \leq N_G(H_1)$ is almost simple.

We have thus sketched the proof of the following reduction theorem (see [48, Thm. 2]):

Theorem 29.5 (Borovik, Liebeck–Seitz) *Let G be an exceptional simple algebraic group, $F : G \to G$ a Steinberg endomorphism, and $H < G^F$ maximal. Then one of the following holds:*

(1) $H = M^F$ *for M as in Theorem 29.1,*
(2) H *is an exotic local subgroup as in Theorem 29.3,*
(3) $G = E_8$, $p > 5$, $H = (\mathfrak{A}_5 \times \mathfrak{S}_6).2$, *or*
(4) H *is almost simple.*

This is a first important reduction step. In order to complete the determination of all maximal subgroups of the finite exceptional groups of Lie type, one needs to investigate groups which might occur in conclusion (4) of the theorem. Here, two essentially different situations arise: if H is of Lie type in characteristic p, one would like to use the results on the subgroup structure of the corresponding simple algebraic groups and the representation theory of the finite groups as in the second part. To what extent such embeddings arise from embeddings of algebraic groups is discussed in the next section. One class of examples here is given by centralizers of graph-, field- and graph-field automorphisms of G.

On the other hand, if H is not of Lie type in characteristic p, then one can use the bounds of Landazuri–Seitz type (see Proposition 28.6) in order to get down to a finite (but still lengthy) list of possibilities for H. This can further be reduced by comparing the maximal r-rank of G to that of H for various primes r.

The complete list of candidates in conclusion (4) has at present not yet been obtained. Generally speaking, the smaller H is, the more difficult it is to find all embeddings into G^F up to conjugation. See [52] for a survey of recent results.

29.2 Lifting result

We derived information on the subgroup structure of finite groups of Lie type by considering fixed points of closed subgroups of an algebraic group under a Steinberg endomorphism. Now consider the converse question: given a subgroup of a finite group of Lie type, does it arise as the group of fixed points under a Steinberg endomorphism of some closed positive-dimensional subgroup of the associated algebraic group? In particular, when considering the configurations of Theorem 29.5(4), when H is again a finite group of Lie type in characteristic p, one is led to ask whether H is the fixed point subgroup of some maximal positive-dimensional closed connected subgroup?

Let's make this question a bit more precise. Let G, H be simple algebraic groups over $k = \overline{\mathbb{F}_p}$. Let $F : G \to G$, $F_1 : H \to H$ be Steinberg endomorphisms, so G^F, H^{F_1} are finite groups of Lie type. Let $\varphi \colon H^{F_1} \to G^F$ be a group homomorphism. Then under what conditions can φ be lifted to an appropriate morphism of algebraic groups, that is, when does there exist $\overline{\varphi} \colon H \to G$, a morphism of algebraic groups, such that $\overline{\varphi}|_{H^{F_1}} = \varphi$? The first major result in this direction is due to Steinberg [70, Thm. 1.3]:

Theorem 29.6 (Steinberg) *Let V be a finite-dimensional vector space over $\mathbb{F}_q$, $q = p^a$. Let $\varphi \colon H^{F_1} \to \mathrm{GL}(V)$ be an absolutely irreducible representation of H^{F_1}, i.e., H^{F_1} acts irreducibly on $V \otimes_{\mathbb{F}_q} k$. Then there exists a rational representation $\overline{\varphi} \colon H \to \mathrm{GL}(V \otimes_{\mathbb{F}_q} k)$ such that $\overline{\varphi}|_{H^{F_1}} = \varphi$. In particular, absolutely irreducible representations of finite groups of Lie type over fields of characteristic p are also parametrized by highest weights.*

Remark 29.7 In fact Steinberg also shows how to obtain a complete set of absolutely irreducible representations for H^{F_1}. In case F_1 is a standard Frobenius morphism, say a q-power map, then

$$\{L(\lambda) \mid 0 \le \langle \lambda, \alpha^\vee \rangle < q \text{ for all } \alpha \in \Delta\}$$

forms a complete set of non-isomorphic irreducible kH^{F_1}-modules.

The following example shows that the condition of absolute irreducibility is necessary.

Example 29.8 Let $G^F = \mathrm{SL}_2(3)$ (isomorphic to a 2-fold central extension of $\mathfrak{A}_4$) with $N \trianglelefteq G^F$ the normal Sylow 2-subgroup, so G^F/N is cyclic of order 3, generated by $\overline{c}$. Consider the representation

$$G^F/N \to \mathrm{GL}_3(\overline{\mathbb{F}_3}), \qquad \overline{c} \mapsto \begin{pmatrix} 1 & 1 & 1 \\ 0 & 1 & 1 \\ 0 & 0 & 1 \end{pmatrix}.$$

This defines an indecomposable but reducible representation of G^F.

We claim that there does not exist a rational representation of $G := \mathrm{SL}_2(\overline{\mathbb{F}_3})$ whose restriction to G^F is the above described representation. Suppose the contrary and let V be the underlying $\overline{\mathbb{F}_3}G$-module which affords the representation. Let $T \leq B \leq G$ be the subgroups of SL_2 defined in Example 6.7(2) and $W = N_G(T)/T$. For $1 \neq w \in W$ there exists $\dot{w} \in N_G(T)$ such that $\dot{w} \in N \leq G^F$. So W acts trivially on all weights in the representation. But this implies that the only weight occurring is the 0 weight. (See Exercise 20.18.) So V has an $\overline{\mathbb{F}}_3 G$-composition series $0 \leq M_1 \leq M_2 \leq V$ with one-dimensional quotients. But this implies that G is solvable, contradicting the simplicity of G.

The goal of this section is to discuss results that generalize Theorem 29.6. It is perhaps not surprising that one can obtain a similar result when $\mathrm{GL}(V)$ is replaced by one of the classical groups $\mathrm{SL}(V)$, $\mathrm{Sp}(V)$ or $\mathrm{SO}(V)$, making some appropriate assumptions on the action of H^{F_1} on the natural module for the classical group. In fact, one can also say something for representations inside groups of exceptional type. We just need to know that a semisimple group is "recovered" by its action on its Lie algebra, as shown in [69, 4.2]:

Theorem 29.9 (Seligman, Steinberg) *Let G be a semisimple linear algebraic group over k with Lie algebra $\mathfrak{g}$, and $\mathrm{Ad} : G \to \mathrm{GL}(\mathfrak{g})$ the adjoint representation. Then $\mathrm{Ad}\,(G)$ has finite index in $\mathrm{Aut}(\mathfrak{g})$, the group of Lie algebra automorphisms of $\mathfrak{g}$. In particular, $\mathrm{Ad}\,(G) = \mathrm{Aut}(\mathfrak{g})^\circ$.*

Returning now to the question of lifting homomorphisms between finite groups to morphisms of algebraic groups, let's first observe the following: if H is a simple algebraic group with a Steinberg endomorphism $F_1 : H \to H$, and $X := [H^{F_1}, H^{F_1}]$ is perfect, then $X \cong H^{F_1}_{\mathrm{sc}}/Z^{F_1}$ for some group H_{sc} of simply connected type and some $Z \leq Z(H_{\mathrm{sc}})$ (with F_1 lifted to H_{sc} by Proposition 22.7) by Proposition 24.21(b), so in particular any homomorphism $\varphi : X \to G^F$ can be regarded as a homomorphism $H^{F_1}_{\mathrm{sc}} \to G^F$. Thus it suffices to construct lifts of homomorphism in the case that H is simply connected and $X = H^{F_1}_{\mathrm{sc}}$ is perfect (note that the latter is not a serious restriction by Theorems 24.15 and 24.17). We start by treating a special case. Its proof is a good illustration of the interplay of many of the methods and concepts presented in this text.

Proposition 29.10 *Let X, H and F_1 be as above and let G be a simply connected simple algebraic group of exceptional type with Steinberg morphism $F : G \to G$. Let $\varphi : X \to G^F$ be a homomorphism such that $\varphi(X)$ lies in no F-stable, closed connected proper subgroup of G. Assume in addition that $p > 3 \dim G$. Then φ can be extended to a morphism $\overline{\varphi} : H \to G$ of algebraic groups with $\overline{\varphi}|_X = \varphi$.*

Sketch of proof Since $p > 3 \dim G$ we have in particular $p \geq 11$.

(1) We first construct a one-dimensional subtorus of the group $GL(\mathfrak{g})$, where $\mathfrak{g} = \mathrm{Lie}(G)$, which will play a key role in what follows. Let $\mathrm{Ad}_G \colon G \to GL(\mathfrak{g})$ denote the adjoint representation of G. By Exercise 30.24, there exists $J \leq X$ with $J \cong \mathrm{SL}_2(p)$. Now let S be the subgroup of J corresponding to the group of diagonal matrices in $\mathrm{SL}_2(p)$. So $S \leq J$ is isomorphic to $\mathbb{F}_p^\times$; let $\mathbb{F}_p^\times \to S$, $c \mapsto S(c)$, denote such an isomorphism. As S is cyclic, Proposition 26.6 shows that S lies in an F-stable maximal torus T of G.

Let Φ be the set of roots of G with respect to T. Fix a basis $\mathcal{C}_T$ of $\mathrm{Lie}(T)$ and for each $\alpha \in \Phi$, choose $v_\alpha \in \mathfrak{g}_\alpha \setminus \{0\}$, so that $\mathcal{C} = \mathcal{C}_T \cup \{v_\alpha \mid \alpha \in \Phi\}$ is a basis of $\mathfrak{g}$. Using this basis, identify $GL(\mathfrak{g})$ with the group of invertible $n \times n$-matrices (where $n = \dim \mathfrak{g}$); so we have $\mathrm{Ad}_G(T) \leq \mathrm{D}_n$, the group of diagonal matrices.

By Theorem 29.6 and Remark 29.7, the composition factors of J on $\mathfrak{g}$ are realized as restrictions of restricted irreducible representations of SL_2. As $p > 3 \dim G = 3 \dim \mathfrak{g}$, the J-composition factors on $\mathfrak{g}$ are of dimension strictly less than $\frac{p}{3}$. By Exercise 20.18, the action of S on any J-composition factor is diagonalizable with weights $l \in \mathbb{Z}$ satisfying $-\frac{p-1}{3} < l < \frac{p-1}{3}$, where the weights are defined by $S(c) \mapsto c^l$, for $c \in \mathbb{F}_p^\times$. So in the present situation we have, for $c \in \mathbb{F}_p^\times$, $\mathrm{Ad}_G(S(c))v = v$ for all $v \in \mathcal{C}_T$ and $\mathrm{Ad}_G(S(c))v_\alpha = c^{l_\alpha}v_\alpha$ for some integers l_α satisfying $-\frac{p-1}{3} < l_\alpha < \frac{p-1}{3}$.

We now define a cocharacter $\gamma \in Y(\mathrm{D}_n)$: for $a \in k^\times$, set

(i) $\gamma(a)v = v$ for all $v \in \mathcal{C}_T$.
(ii) $\gamma(a)v_\alpha = a^{l_\alpha}v_\alpha$, for $\alpha \in \Phi$.

Set $\overline{S} := \mathrm{im}(\gamma)$, a one-dimensional subtorus of $GL(\mathfrak{g})$; so $\gamma(c) = \mathrm{Ad}\,(S(c))$, for all $c \in \mathbb{F}_p^\times$.

(2) We claim that $\overline{S} \leq \mathrm{Ad}_G(G)$. Indeed, we will show that $\overline{S}$ acts as a group of Lie algebra automorphisms of $\mathfrak{g}$ and so $\overline{S} = \overline{S}^\circ \leq \mathrm{Aut}(\mathfrak{g})^\circ = \mathrm{Ad}_G(G)$, by Theorem 29.9.

Let $\alpha, \beta \in \Phi$ with $[v_\alpha, v_\beta] \neq 0$. Then considering the action of the torus T, we see that $[v_\alpha, v_\beta]$ is a scalar multiple of $v_{\alpha+\beta}$. Thus for all $c \in \mathbb{F}_p^\times$,

$$c^{l_\alpha} c^{l_\beta}[v_\alpha, v_\beta] = [\gamma(c)v_\alpha, \gamma(c)v_\beta] = \gamma(c)[v_\alpha, v_\beta] = c^{l_{\alpha+\beta}}[v_\alpha, v_\beta]$$

so $c^{l_\alpha + l_\beta} = c^{l_{\alpha+\beta}}$ for all $c \in \mathbb{F}_p^\times$. Using the fact that l_α, l_β lie in the interval $[-\frac{p-1}{3}, \frac{p-1}{3}]$ we see that $l_\alpha + l_\beta = l_{\alpha+\beta}$ and so $a^{l_\alpha + l_\beta} = a^{l_{\alpha+\beta}}$ for all $a \in k^\times$. One easily checks that the action on the remaining commutators $[v, v']$, for $v, v' \in \mathcal{C}$ is also preserved by $\gamma(a)$ and so $\overline{S}$ acts as a group of automorphisms of the Lie algebra $\mathfrak{g}$, as claimed.

(3) We next show that the closed subgroup $R = (\mathrm{Ad}_G^{-1}(\overline{S}))^\circ$ is F-stable. Write $F|_X = q\phi$, for some positive integral power q of p and for some

$\phi \in \mathrm{Aut}(X_{\mathbb{R}})$, as in Proposition 22.2 (recall $p \geq 11$). As $p \geq 11$, Theorem 15.20 shows that $\mathfrak{g}$ is an irreducible kG-module; moreover by Example 15.19 its highest weight is the highest root α_0, and hence invariant under all graph automorphisms of G by the defining property of the highest root. The highest weight of the representation $\mathrm{Ad}_G \circ F$ of G is then just q times the highest weight of the adjoint representation. Theorem 15.17(a) implies that the representations $\mathrm{Ad}_G \circ F$ and $F' \circ \mathrm{Ad}_G$ are equivalent, where F' is a q-power standard Frobenius morphism of $\mathrm{GL}(\mathfrak{g})$. Thus there exists $x \in \mathrm{GL}(\mathfrak{g})$ such that $x\mathrm{Ad}_G(F(g))x^{-1} = F'(\mathrm{Ad}_G(g))$ for all $g \in G$. Applying Theorem 21.7, we get that $x = y^{-1}F(y)$ for some $y \in \mathrm{GL}(\mathfrak{g})$, which then implies that

$$\mathrm{Ad}_G(F(g)) = \omega\mathrm{Ad}_G(g)\omega^{-1} \text{ for each } g \in G,$$

where ω is the invertible semilinear map $\sum c_j v_j \mapsto \sum c_j^q v_j$ on $\mathfrak{g}$ with respect to our fixed basis of $\mathfrak{g}$.

For $t \in T$, $\alpha \in \Phi$, and setting $g = F^{-1}(t)$ in the above, we have

$$\mathrm{Ad}_G(t)\omega v_\alpha = \omega\mathrm{Ad}_G(F^{-1}(t))v_\alpha = (\alpha(F^{-1}(t))^q)\omega v_\alpha.$$

But $\alpha(F^{-1}(t))^q = (F^{-1}(\alpha))(t)^q = (\phi^{-1}(\alpha))(t) = (\rho(\alpha)(t))$ (with ρ as in Proposition 22.2). So $\mathrm{Ad}_G(t)\omega v_\alpha = \rho(\alpha)(t)\omega v_\alpha$ for all $t \in T$ and we conclude that $\omega v_\alpha = bv_{\rho(\alpha)}$ for some $b \in k^\times$. Hence for $a \in k^\times$,

$$\omega\gamma(a)\omega^{-1}v_{\rho(\alpha)} = \omega\gamma(a)b^{-1/q}v_\alpha = \omega a^{l_\alpha}b^{-1/q}v_\alpha$$
$$= a^{ql_\alpha}b^{-1}\omega v_\alpha = a^{ql_\alpha}v_{\rho(\alpha)}.$$

Similarly, $\omega\gamma(a)\omega^{-1}v = v$, for $v \in \mathcal{C}_T$.

Taking a in the prime field we have $\gamma(a) \in \mathrm{Ad}_G(S)$, so ω centralizes $\gamma(a)$. But then the above equality and the restriction on the l_α show that $l_{\rho(\alpha)} = l_\alpha$ and then applying the same equality for arbitrary $a \in k^\times$, we see that $\omega\gamma(a)\omega^{-1} = \gamma(a)^q$. Finally, we have

$$\mathrm{Ad}_G(F(\mathrm{Ad}_G^{-1}(\gamma(a)))) = \omega\mathrm{Ad}_G(\mathrm{Ad}_G^{-1}(\gamma(a)))\omega^{-1} = \gamma(a)^q \leq \overline{S}$$

and so $\mathrm{Ad}_G^{-1}(\overline{S})$ is F-stable and hence $R = (\mathrm{Ad}_G^{-1}(\overline{S}))^\circ$ as well.

(4) We can now show that X acts irreducibly on $\mathfrak{g}$. For suppose that V is a proper X-invariant subspace of $\mathfrak{g}$, so V is spanned by weight vectors for S. We claim that weight vectors for S are in fact weight vectors for $\overline{S}$. For suppose two $\overline{S}$-weights on $\mathfrak{g}$ have equal restrictions to S. Then $c^{l_\alpha} = c^{l_\beta}$ for some $\alpha, \beta \in \Phi$ and for some generator $c \in \mathbb{F}_p^\times$; that is, $c^{l_\alpha - l_\beta} = 1$, so $(p-1)|(l_\alpha - l_\beta)$. But $-\frac{p-1}{3} \leq l_\alpha, l_\beta \leq \frac{p-1}{3}$, so $l_\alpha = l_\beta$ as claimed. Now set $D = \langle xRx^{-1} \mid x \in X \rangle$, a closed connected subgroup of G by Proposition 1.16, which is F-stable as R is. Moreover D stabilizes the subspace V and so by

Theorem 15.20, D is proper in G. But the group $[X, S]$ is normal in X and contains the subgroup J, so we have

$$X = [X, S] \leq [X, RZ(G)] = [X, R] \leq D,$$

which contradicts our standing assumption. Hence X acts irreducibly as claimed.

(5) A lengthy argument relying upon detailed considerations of the representation theory of X and G, in particular, Theorem 29.6 and Proposition 16.3, shows that the groups H and G have isomorphic root systems and the absolutely irreducible representation $\mathrm{Ad}_G \circ \varphi : X \to \mathrm{GL}(\mathfrak{g})$ is the restriction of a twist of the adjoint representation of H, that is, there exists a standard Frobenius endomorphism F' of H such that $\mathrm{Ad}_G \circ \varphi = (\mathrm{Ad}_H \circ F')|_X$. We refer the reader to [62, pp.565–566] for the details.

We now have that X stabilizes two Lie brackets on $\mathfrak{g}$: $[\ ,\ ]_G$ coming from $\mathfrak{g}$ and $[\ ,\ ]_H$ coming from $\mathrm{Lie}(H)$. Now the existence of a Lie bracket on $\mathfrak{g}$ shows that $\mathrm{Hom}_{kG}(\mathfrak{g} \wedge \mathfrak{g}, \mathfrak{g})$ is non-trivial. Moreover, using the theory of highest weights and the assumption on p, one can check that for each exceptional group G, $\mathfrak{g}$ occurs with multiplicity 1 as a composition factor of $\mathfrak{g} \wedge \mathfrak{g}$ and so $\mathrm{Hom}_{kG}(\mathfrak{g} \wedge \mathfrak{g}, \mathfrak{g})$ is a 1-space. (See Exercise 20.21 for the case $G = G_2$.) Hence the two Lie brackets are scalar multiples of each other and Theorem 29.9 implies that $\mathrm{Ad}_H(F'(H)) = \mathrm{Ad}_G(G)$.

Recall that H is simply connected. Moreover the isogeny $\mathrm{Ad}_G : G \to \mathrm{Ad}_G(G)$ satisfies $\ker d\mathrm{Ad} = \ker \mathrm{ad} = 0$, by the restriction on p and Theorem 15.20. Hence we may apply Proposition 9.18 to obtain a morphism $\psi : H \to G$ such that $\mathrm{Ad}_H = \mathrm{Ad}_G \circ \psi$. But then $\mathrm{Ad}_G(\varphi(x)) = \mathrm{Ad}_H(F'(x)) = \mathrm{Ad}_G(\psi(F'(x)))$ for all $x \in X$. By Theorem 24.15, X is generated by its unipotent elements and Ad_G is bijective on unipotent elements, so we have $\varphi = \psi \circ F'|_X$, whence $\psi \circ F'$ is the desired extension. $\qquad\square$

One of the first generalizations of Theorem 29.6 is the following result, where the condition "absolute irreducibility" is replaced by requiring that the subgroup stabilize no proper totally singular subspace of the natural module in the classical group case. (See Proposition 12.13.)

Theorem 29.11 *Let G be a semisimple algebraic group defined over $k = \overline{\mathbb{F}_p}$ with Steinberg endomorphism $F : G \to G$, and H, X as before. Let $\varphi : X \to G^F$ be a homomorphism such that $\varphi(X)$ lies in no proper F-stable parabolic subgroup of G. Then there exists an integer N, depending on the maximal dimension of a simple factor of G, such that if $p > N$, φ can be extended to a morphism $\overline{\varphi} : H \to G$ of algebraic groups with $\overline{\varphi}|_X = \varphi$. Moreover, if G is classical, no restriction on p is necessary.*

Proof We prove the result by induction, taking G as a counterexample of minimal dimension. Take $N = 3m$ where m is the maximal dimension of a simple factor of G; in particular, $p \geq 11$. As pointed out above we may assume that H is simply connected.

Step 1: There is no F-stable, closed connected proper subgroup $D < G$ with $\varphi(X) \leq D$.

Indeed, suppose $\varphi(X) \leq D < G$ as above. If D is not reductive, then $1 \neq R_u(D)$ is an F-stable unipotent subgroup. Hence as in the proof of Theorem 26.5, $\varphi(X) \leq D \leq N_G(R_u(D))$ lies in a proper F-stable parabolic subgroup, contradicting the assumption on $\varphi(X)$. Hence D is reductive and as X is perfect, $\varphi(X) \leq [D, D]$, a semisimple F-invariant subgroup of dimension less than $\dim G$. So by minimality of G, we have the desired extension of φ.

Step 2: We may suppose that G is of simply connected type.

Indeed, let $\pi : G_{sc} \to G$ be the natural isogeny from a group G_{sc} of simply connected type and $\hat{F} : G_{sc} \to G_{sc}$ a Steinberg morphism with $\pi \circ \hat{F} = F \circ \pi$ (see Proposition 22.7). Since H is simply connected, the Schur multiplier of X has order prime to $|\ker \pi|$ (see [44, Thm. 5.1.4]). Thus, by the defining property of the Schur multiplier (see [39, Cor. 11.20(a)]) the full preimage $\pi^{-1}(\varphi(X))$ is a direct product of the abelian group $\ker \pi$ with a perfect subgroup $\hat{X}$ isomorphic to $\varphi(X) = [\varphi(X), \varphi(X)]$. Since $\hat{X} = [\pi^{-1}(\varphi(X)), \pi^{-1}(\varphi(X))]$, $\hat{X}$ is characteristic and $\pi|_{\hat{X}}$ is an isomorphism. Hence, $\hat{X}$ is centralized by $\hat{F}$. So assuming the result holds for G simply-connected, there exists a homomorphism $\hat{\varphi} : X \to \hat{X} \leq G_{sc}$ such that $\varphi = \pi \circ \hat{\varphi}$. Now, if $\overline{\varphi} : H \to G_{sc}$ lifts $\hat{\varphi}$, then $\pi \circ \overline{\varphi}$ is the desired lift of φ.

Now let $M_1, \ldots, M_t$ be the simple factors of the simply connected group G.

Step 3: F acts transitively on the set $\{M_1, \ldots, M_t\}$.

Let A be the product of the elements of an F-orbit on $\{M_1, \ldots, M_t\}$. If γ is the projection of G onto A, then $\gamma \circ \varphi$ is a homomorphism from X to the F-invariant subgroup A of G. If F is not transitive, then $A < G$ is of smaller dimension, and φ can be lifted by Step 1.

Step 4: G is simple.

By Step 3 we may assume that $F^i(M_1) = M_{i+1}$ for $1 \leq i \leq t - 1$. Then

$$M = \{x \cdot F(x) \cdot F^2(x) \cdot \ldots \cdot F^{t-1}(x) \mid x \in M_1\} \cong M_1$$

is a closed connected F^t-stable simple subgroup of G. Moreover, as $\varphi(X) \leq G^F$, $\varphi(X) \leq M$. Note that $\varphi(X)$ does not lie in an F^t-stable proper parabolic of M, as such a parabolic would be diagonally embedded in a subgroup

$P \cdot F(P) \cdots F^{t-1}(P)$ for some F^t-stable proper parabolic subgroup of M_1, contradicting our hypothesis on X. So we now have $F^t : M \to M$ with $\varphi(X) \leq M^{F^t}$, not contained in any F^t-stable parabolic of M, and $p > 3 \dim G \geq \dim M$, so by Step 1 we necessarily have that $M = G$ is simple.

If G is simple of simply connected type, we may extend φ to $\overline{\varphi} : H \to G$ by Proposition 29.10, contradicting our choice of G. $\qquad\square$

With the same proof, but looking carefully at possible X-composition factors on $\mathrm{Lie}(G)$ and the corresponding weights for S, as in the proof of Proposition 29.10, one gets an improved result, where we may take $N = 13$, unless H is of type A_1, A_2, or B_2. In particular, the worst case occurs if H is of type A_1 and G has a factor of type E_8, with $N = 113$ (see [62, Thm. 2] for a precise statement).

Finally we conclude this discussion with the statement of the most general lifting result for exceptional groups established at this point in time. We first introduce one additional notion:

Let Φ be an irreducible root system; we call an element of the root lattice $\mathbb{Z}\Phi$ a *root difference* if it is of the form $\alpha - \beta$ for some $\alpha, \beta \in \Phi$. Given a sublattice L of $\mathbb{Z}\Phi$, we write $t(L)$ for the exponent of the torsion subgroup of the quotient $\mathbb{Z}\Phi/L$; we set

$$t(\Phi) = \max\{t(L) \mid L \text{ a sublattice of } \mathbb{Z}\Phi \text{ generated by root differences}\}.$$

Take H, F_1 and X as before the statement of Proposition 29.10 and let $F_1 = q\phi$ as in Proposition 22.2. Write $\Phi(H)$ for the root system of H with respect to a fixed maximal torus. Then we have the following result ([51, Thms. 1, 4 and 10]):

Theorem 29.12 (Liebeck–Seitz) *Let G be an exceptional algebraic group of adjoint type over $\overline{\mathbb{F}}_p$ with root system Φ and Steinberg endomorphism $F : G \to G$, such that $H^{F_1} = X \leq G^F$.*

(a) *Assume further that $X \neq \mathrm{SL}_3(16), \mathrm{SU}_3(16)$ and*

$$\begin{cases} q > t(\Phi(G)) \cdot (2, p-1) & \text{if } X = A_1(q), \ {}^2B_2(q^2), \text{ or } {}^2G_2(q^2), \\ q > 9 & \text{otherwise.} \end{cases}$$

Then there exists a closed connected F-stable subgroup $\overline{X}$ of G normalized by $N_G(X)$ with $X \leq \overline{X}$ and such that $\overline{X}$ stabilizes every X-invariant subspace of $\mathrm{Lie}(G)$. Moreover, if X is not of the same type as G, that is $\Phi(H)$ and Φ are not isomorphic, then $\overline{X}$ may be chosen to be a proper subgroup of G.

(b) *Assume in addition that $p > 7$. Then X lies in a closed connected semisimple F-stable subgroup $\overline{X}$ of G, where each simple factor of $\overline{X}$ is of type $\Phi(H)$.*

In view of this result it is interesting to know the value $t(\Phi)$ for each of the exceptional type root systems. This has been calculated by Lawther (see [46, Thm. 2]):

Proposition 29.13   *The values $t(\Phi)$ for Φ an indecomposable root system of exceptional type are as follows: $t(G_2) = 12$, $t(F_4) = 68$, $t(E_6) = 124$, $t(E_7) = 388$, $t(E_8) = 1312$.*

30

Exercises for Part III

Throughout all algebraic groups are considered over $k = \overline{\mathbb{F}_p}$, an algebraic closure of $\mathbb{F}_p$.

Exercise 30.1 Let G be a linear algebraic group and $\sigma : G \to G$ a surjective endomorphism. Show the following:

(a) $\ker(\sigma)$ is finite and lies in $Z(G)$.
(b) If G° is semisimple, then $\ker(\sigma) = 1$, so σ is an automorphism of abstract groups.
(c) If $H \leq G$ is connected and $\sigma(H) \leq H$, then $\sigma(H) = H$.

[**Hint:** Use that $\dim G = \dim \sigma(G)$ and Exercise 10.4 for (a). For (b), consider the powers σ^n of σ. For (c), use again Corollary 1.20.]

Exercise 30.2 Let $F : G \to G$ be a surjective endomorphism of a group G and define

$$F_1 : G^m \to G^m, \qquad (g_1, \ldots, g_m) \mapsto (g_2, \ldots, g_m, F(g_1))$$

and

$$F_2 : G^m \to G^m, \qquad (g_1, \ldots, g_m) \mapsto (F(g_2), \ldots, F(g_m), F(g_1)).$$

Show that $(G^m)^{F_1} \cong G^F$ and $(G^m)^{F_2} \cong G^{F^m}$.

Exercise 30.3 Let W be a finite group and $\sigma \in \mathrm{Aut}(W)$. Show that σ-classes in W are in natural bijection with W-conjugacy classes in the coset $W.\sigma$, and the size of such a conjugacy class is given by $|[w\sigma]| = |W|/|C_W(w\sigma)|$.

Exercise 30.4 Let $F : G \to G$ be a Steinberg endomorphism on the connected group G. Then in the semidirect product $G \rtimes \langle F \rangle$ we have $G.F = F^G$. In particular, G^{gF} is G-conjugate to G^F for all $g \in G$.

Exercise 30.5   Let G be a connected group with Steinberg endomorphism $F : G \to G$, acting transitively on a set V with compatible F-action. Let $v \in V^F$ with closed stabilizer G_v. Show that the map $g.v \mapsto g^{-1}F(g)$ induces a bijection between G^F-orbits on V^F and F-classes in G_v.
[**Hint:** Use the Lang–Steinberg Theorem for the connected group G_v°.]

Exercise 30.6   Let G be simple with Steinberg endomorphism $F : G \to G$, T, B, X, Φ, ρ, q_α as in Proposition 22.2. Assume that q_α is not constant on Φ. Show that G is of type B_2, G_2 or F_4 with $p = 2, 3, 2$ respectively, ρ interchanges long and short roots and $q_\alpha q_\beta = q^2$, $q_\beta/q_\alpha = p$ for all long roots α and all short roots β.
[**Hint:** Let $\alpha, \beta \in \Phi$ with α long, β short and $q_\alpha \neq q_\beta$. By Proposition 22.2 we have $F \circ s_{\rho(\alpha)} = s_\alpha \circ F$. Apply this to $\rho(\beta)$ to deduce that $n(\rho(\beta), \rho(\alpha))q_\alpha = n(\beta, \alpha)q_\beta$. By Table A.1 we have $n(\beta, \alpha) = 1$ and $n(\rho(\beta), \rho(\alpha)) \in \{1, 2, 3\}$. But q_β/q_α is a power of p.]

Exercise 30.7   Let G be a linear algebraic group with a Steinberg endomorphism $F : G \to G$, and H_1, H_2 two F-stable closed subgroups of G such that H_2 is connected and normal in H_1. Show that the natural map $H_1^F/H_2^F \to (H_1/H_2)^F$ is an isomorphism.

Exercise 30.8   (Tits)   Let G be a group with a BN-pair (B, N), let G_1 be a subgroup of G and $B_1 := B \cap G_1$, $N_1 := N \cap G_1$. Show the following:

(a) If $(B \cap N)B_1 = B$ then (B_1, N_1) is a BN-pair in G_1.
(b) If moreover G_1 is normal in G, then its Weyl group can be naturally identified with the one of G and we have $G = (B \cap N)G_1$.

Exercise 30.9   Let G be connected reductive with a Steinberg endomorphism $F : G \to G$, T an F-stable maximal torus contained in an F-stable Borel subgroup of G, with root system Φ and corresponding positive system Φ^+. Show that

$$|G^F| = Q\,|T^F| \sum_{w \in W^F} q_w$$

with

$$Q := \prod_{\alpha \in \Phi^+} q_\alpha \quad \text{and} \quad q_w := \prod_{\alpha \in \Phi^+,\, w(\alpha) \in \Phi^-} q_\alpha$$

with q_α as in Proposition 22.2.

Exercise 30.10   Let G be connected reductive with Steinberg endomorphism $F : G \to G$, $H \leq G$ a closed connected reductive F-stable subgroup, with complete root data $\mathbb{G}$, respectively $\mathbb{H}$. Then the polynomial $|\mathbb{H}|$ divides $|\mathbb{G}|$ in $\mathbb{Z}[x]$.

[**Hint**: You may want to use that both polynomials are monic and that the rational function $|\mathbb{G}|/|\mathbb{H}|$ takes integral values for infinitely many prime powers q.]

Exercise 30.11 Use Example 23.10(2) to show that $\mathrm{SU}_3(q)$ is perfect for $q > 2$

Exercise 30.12 (Steinberg) Let G be a group, $Z \le G$ a central subgroup, and $F : G \to G$ an automorphism normalizing Z. Then the exact sequence

$$1 \longrightarrow Z \longrightarrow G \xrightarrow{\pi} G/Z \longrightarrow 1$$

induces a long exact sequence

$$1 \longrightarrow Z^F \longrightarrow G^F \xrightarrow{\pi} (G/Z)^F \xrightarrow{\delta} (L(G) \cap Z)/L(Z) \longrightarrow 1,$$

where $L : G \to G$ is defined by $L(g) := g^{-1}F(g)$.

[**Hint**: Show that $\delta := \nu \circ L \circ \pi^{-1}$, with $\nu : Z \to Z/L(Z)$ the natural map, is a well-defined homomorphism. See for example [26, Prop. 4.2.3].]

Exercise 30.13 Let G be a simple algebraic group with Steinberg endomorphism $F : G \to G$. Show that $G^F = [G^F, G^F] T^F$ for any F-stable maximal torus T of G.

[**Hint**: Let $\pi : G_{\mathrm{sc}} \to G$ be an isogeny from a simply connected group with central kernel Z, let T_{sc} be the full preimage of T, a maximal torus of G_{sc}. Let $\tilde{G} := G_{\mathrm{sc}} \circ_Z T_{\mathrm{sc}}$ be the central product of G_{sc} with T_{sc} over Z and let F act componentwise on $\tilde{G}$. Then $\tilde{\pi} : \tilde{G} \to G$, $(g,t) \mapsto \pi(g)$, is an epimorphism with connected kernel T_{sc}. Apply Proposition 23.2 to see that $\tilde{\pi}|_{\tilde{G}^F}$ is surjective onto G^F and Proposition 24.21 to conclude.]

Exercise 30.14 (Maximal tori in finite classical groups)

(a) Classify the maximal tori in $\mathrm{GU}_n(q)$ and work out their orders.

(b) Classify the maximal tori in $\mathrm{Sp}_{2n}(q)$ and work out their orders.

Exercise 30.15 (Molien's formula) Let V be a finite-dimensional complex vector space, $W \le \mathrm{GL}(V)$ finite and $\phi \in N_{\mathrm{GL}(V)}(W)$. Denote by V^W the invariants of W in V.

(a) Show that $\mathrm{tr}(\phi, V^W) = \frac{1}{|W|} \sum_{w \in W} \mathrm{tr}(\phi w, V)$.

(b) For a graded complex W-module $\oplus_{i \ge 0} V_i$ and $g \in \mathrm{GL}(\oplus V_i)$ define the *graded trace* $\mathrm{grtr}(g, \oplus V_i) := \sum_{i \ge 0} \mathrm{tr}(g, V_i) x^n \in \mathbb{C}[[x]]$. Now the symmetric algebra $S(V)$ of V is a graded W-module. For $g \in \mathrm{GL}(V)$ show that $\mathrm{grtr}(g, S(V)) = \det_V (1 - gx)^{-1}$.

(c) Compute the graded trace of ϕ on the W-invariants $S(V)^W$ in $S(V)$ as

$$\mathrm{grtr}(\phi, S(V)^G) = \frac{1}{|W|} \sum_{w \in W} \mathrm{grtr}(\phi w, S(V)) = \frac{1}{|W|} \sum_{w \in W} \frac{1}{\det_V(1 - \phi wx)}.$$

(d) Let $f_1, \ldots, f_n$ be homogeneous generators of $S(V)^W$ which are eigenvectors for ϕ with eigenvalues $\epsilon_1, \ldots, \epsilon_n$. Conclude that

$$\prod_{i=1}^{n} \frac{1}{1 - \epsilon_i x^{d_i}} = \frac{1}{|W|} \sum_{w \in W} \frac{1}{\det_V(1 - \phi wx)}.$$

[**Hint**: For (a) use that $v \mapsto \frac{1}{|W|} \sum_{w \in W} w.v$ defines a projection from V to V^W. For (b) choose a basis of V with respect to which g is triangular and compute the trace on each homogeneous component. See [9, V, Lemma 5.3] for the case where $\phi = \mathrm{id}_V$.]

Exercise 30.16   Let (T, F) be a torus with complete root datum $\mathbb{T} = (X, \emptyset, Y, \emptyset, \phi)$. Then $\mathbb{S} = (X', \emptyset, Y', \emptyset, \phi)$, with $Y' = \ker_Y(\Phi_d(\phi))$ and $X' = X/\mathrm{Ann}(Y')$, is the complete root datum of the Sylow d-torus (S, F) of T.
[**Hint**: Show that $\ker_Y(\Phi_d(\phi))$ is a pure submodule of Y, then compare the generic orders of S, T.]

Exercise 30.17 (Generic Sylow Theorem for GL_n)   Let $W = \mathfrak{S}_n$ in its permutation representation in $\mathrm{GL}_n(\mathbb{Q})$, $d \in \mathbb{N}$, and $\zeta_d \in \mathbb{C}$ a primitive dth root of unity. Show the following:

(a) All ζ_d-eigenspaces of maximal possible dimension of elements $g \in \mathfrak{S}_n$ are conjugate.
(b) The ζ_d-eigenspace of any $w \in \mathfrak{S}_n$ is contained in a maximal such.
(c) The stabilizer in W of a ζ_d-eigenspace as in (a) is isomorphic to $Z_d \wr \mathfrak{S}_a \times \mathfrak{S}_b$, where $n = ad + b$ with $0 \le b < d$ and $Z_d \wr \mathfrak{S}_a$ denotes the wreath product of the cyclic group of order d with the symmetric group $\mathfrak{S}_a$.

Exercise 30.18   Let G be connected reductive with Steinberg endomorphism $F : G \to G$, and S a Sylow d-torus of G. Then $N_{G^F}(S)$ controls G^F-fusion in $C_{G^F}(S)$, that is, if $s, t \in C_{G^F}(S)$ are G^F-conjugate, they are conjugate by an element of $N_{G^F}(S)$.
[**Hint**: Apply the Sylow Theorem 25.11 to the connected component of $C/R_u(C)$, where C is the centralizer of an element in $C_G(S)$. See [56, Prop. 5.11].]

Exercise 30.19   Let G be connected reductive with Steinberg endomorphism $F : G \to G$. Show that the centralizers of 1-tori in G are precisely the F-stable Levi subgroups contained in F-stable parabolic subgroups of G.

Exercise 30.20 Show that the general unitary group $GU_n(q)$ has exactly two orbits on the lines in its natural representation on $\mathbb{F}_{q^2}^n$, and determine the stabilizers.

[**Hint**: Theorem 21.11 does *not* apply, since here the F-structures on GL_n and on $V = k^n$ are not compatible. You should find the groups constructed in Example 26.3.]

Exercise 30.21 Let G be connected reductive with Steinberg endomorphism F, $L \leq G$ an F-stable Levi subgroup of some parabolic subgroup of G, with Weyl group W_L.

(a) The G^F-classes of F-stable G-conjugates of L are in bijection with ϕ-classes of $N_W(W_L)/W_L$.

(b) Write L_w for a Levi subgroup corresponding to the ϕ-class of w under the map in (a). Then $N_{G^F}(L_w)/L_w^F \cong N_W(W_L w\phi)/W_L$.

Exercise 30.22 Let G be connected reductive with Steinberg endomorphism $F : G \to G$, $T \leq G$ an F-stable maximal torus of G, $Z := Z(G)^\circ$, $G' := [G,G]$ and $S := T \cap G'$, an F-stable maximal torus of G'. Show that $|T^F| = |Z^F| \cdot |S^F|$.

[**Hint**: We have $T = SZ$. Show that the intersection of the kernels of the (surjective) restriction morphisms from $X(T)$ to $X(S)$ and to $X(Z)$ is trivial. Conclude that $X(T) \otimes \mathbb{R} \cong (X(S) \otimes \mathbb{R}) \oplus (X(Z) \otimes \mathbb{R})$ as F-modules.]

Exercise 30.23 Assume $n \geq 2$, and let $\rho\colon Sp_{2n}(q) \to GL_d$ be a representation of $Sp_{2n}(q)$ with central kernel over a field k of characteristic $r \neq p$, where $q = p^f$ is odd. Then $d \geq q^{n-1}(q-1)/2$.

[**Hint**: Let $P \leq G = Sp_{2n}(q)$ be the maximal parabolic subgroup with Levi complement of type $CSp_{2n-2}(q)$, and consider its action on the faithful characters of the unipotent radical U of P.]

Exercise 30.24 Let $1 \neq G$ be semisimple of simply connected type, $F : G \to G$ a Steinberg endomorphism. Show that G^F contains a subgroup $SL_2(p)$, where $p = \mathrm{char}(k)$, unless for every simple component of G there is a power of F stabilizing it and which is very twisted. (The claim holds more generally unless G^F is a direct product of groups $^2B_2$.)

[**Hint**: Check that there is an F-stable completely disconnected subset of the Dynkin diagram of G, unless G is a product of groups of type A_2. In the first case apply Proposition 12.14 and Exercise 30.2. For type A_2, use the computations in Example 23.10.]

Appendix A
Root systems

The structure theory for linear algebraic groups shows that many questions on these groups can be translated into questions on the associated root system, which are then of a purely combinatorial nature.

In this appendix we collect some basic results on root systems as they are pertinent to our study of algebraic groups and finite groups of Lie type. Most of these are well-known (see for example [9, §VI] or [33, Chap. III]), except possibly for the discussion of maximal subsystems in Sections B.3 and B.4.

We do not present the proof of the classification of indecomposable root systems, since it is not used here and is very well-documented in the literature (see for example [9, VI, §4] or [33, §11]), nor the properties of invariant rings of finite groups generated by reflections (see [9, V, §5] or [47]).

A.1 Bases and positive systems

Recall that an element $s \in \mathrm{GL}(V)$, where V is a finite-dimensional vector space, is called a (linear) *reflection* along $\alpha \in V$, if α is an eigenvector of s with eigenvalue -1, and s fixes a hyperplane of V pointwise.

The action of reflections in a Euclidean space E can be described by an easy formula:

Proposition A.1 *Assume that $s \in \mathrm{GL}(E)$ is a reflection stabilizing the positive definite bilinear form $(\ ,\ )$ on E. Then for any eigenvector α for the non-trivial eigenvalue of s we have*

$$s.v = v - 2\frac{(v, \alpha)}{(\alpha, \alpha)}\alpha.$$

Proof Let $H \subset E$ denote the eigenspace for the eigenvalue 1, the fixed space of s. Then for $v \in H$ we have

$$(v, \alpha) = (s.v, s.\alpha) = (v, -\alpha),$$

so $(v, \alpha) = 0$ for all $v \in H$. Thus the linear map $E \to E$, $v \mapsto v - 2\frac{(v,\alpha)}{(\alpha,\alpha)}\alpha$, agrees with s on H as well as on α, so on $E = H \oplus \langle\alpha\rangle_{\mathbb{R}}$. □

Definition A.2 A subset Φ of a finite-dimensional real vector space E is called an *(abstract) root system* in E if the following properties are satisfied:

(R1) Φ is finite, $0 \notin \Phi$, $\langle\Phi\rangle_{\mathbb{R}} = E$;
(R2) if $c \in \mathbb{R}$ is such that $\alpha, c\alpha \in \Phi$, then $c = \pm 1$;
(R3) for each $\alpha \in \Phi$ there exists a reflection $s_\alpha \in \mathrm{GL}(E)$ along α stabilizing Φ;
(R4) *(crystallographic condition)* for $\alpha, \beta \in \Phi$, $s_\alpha.\beta - \beta$ is an integral multiple of α.

The group $W = W(\Phi) := \langle s_\alpha \mid \alpha \in \Phi \rangle$ is called the *Weyl group* of Φ. The dimension of E is called the *rank* of Φ.

Let Φ be a root system in E. Since Φ is finite, generates E and is stabilized by W, the Weyl group of an abstract root system is always finite. Thus, it stabilizes a positive definite W-invariant symmetric bilinear form $(\, , \,)$ on E, which is unique up to non-zero scalars on each irreducible W-submodule of E. We'll always assume such a form to have been chosen so that E is Euclidean and we can speak of lengths of vectors and angles between them.

Using the non-degenerate bilinear form $(\, , \,)$ we may identify E with its dual $E^* = \mathrm{Hom}(E, \mathbb{R})$ by $v \mapsto (E \to \mathbb{R}, \; u \mapsto v(u) := (u, v))$. For $\alpha \in \Phi$ define the corresponding *coroot* $\alpha^\vee := 2\alpha/(\alpha, \alpha)$. Then the formula in Proposition A.1 for an orthogonal reflection s along α can be written as

$$s.v = v - (v, \alpha^\vee)\alpha = v - (v, \alpha)\alpha^\vee = v - \alpha^\vee(v)\alpha.$$

Definition A.3 Let Φ be a root system. The set $\Phi^\vee := \{\alpha^\vee \mid \alpha \in \Phi\}$ is called the *dual root system* of Φ.

It is easily seen that $\Phi^\vee$ is indeed a root system in E, see Exercise A.1.

Lemma A.4 *Let Φ be a root system with Weyl group W. Then for all $\alpha \in \Phi$, $w \in W$ we have*

$$w s_\alpha w^{-1} = s_{w.\alpha}.$$

Proof By Proposition A.1 we have

$$w s_\alpha w^{-1}.v = w.\left(w^{-1}.v - 2\frac{(w^{-1}.v, \alpha)}{(\alpha, \alpha)}\alpha\right) = v - 2\frac{(v, w.\alpha)}{(w.\alpha, w.\alpha)}w.\alpha = s_{w.\alpha}.v$$

for all $v \in E$. □

Given a total ordering ">" on E compatible with addition and scalar multiplication by positive real numbers we say that $v \in E$ is *positive* if $v > 0$. Clearly such orderings exist: choose a basis B of E and take the lexicographical ordering on the coefficient vectors with respect to this basis.

Definition A.5 A subset $\Phi^+ \subseteq \Phi$ consisting of all positive roots in Φ with respect to some total ordering on E as above is called a *system of positive roots* or *positive system* of Φ. A subset $\Delta \subset \Phi$ is called a *base of* Φ if it is a vector space basis of E and any $\beta \in \Phi$ is a linear combination $\beta = \sum_{\alpha \in \Delta} c_\alpha \alpha$ with either all $c_\alpha \geq 0$ or all $c_\alpha \leq 0$.

Note that if $\Phi^+ \subseteq \Phi$ is a positive system, then $\Phi = \Phi^+ \sqcup -\Phi^+$ by (R2) and (R3). We'll then also write $\Phi^- := -\Phi^+$.

Lemma A.6 *Let $\Phi^+ \subseteq \Phi$ be a positive system and $\Delta \subseteq \Phi^+$ minimal with the property that any element of Φ^+ is a non-negative linear combination of elements in Δ. Then $(\alpha, \beta) \leq 0$ for all $\alpha, \beta \in \Delta$, $\alpha \neq \beta$.*

Proof Suppose that this fails for some α, β. By Proposition A.1 we then have $s_\alpha.\beta = \beta - c\alpha$ for some $c > 0$, and by (R3) this has to lie in Φ, so either $s_\alpha.\beta$ or $-s_\alpha.\beta$ lies in Φ^+. If $s_\alpha.\beta \in \Phi^+$ then we can write it as $s_\alpha.\beta = \sum_{\gamma \in \Delta} c_\gamma \gamma$ with $c_\gamma \geq 0$. Now, if $c_\beta \geq 1$ we get

$$0 = s_\alpha.\beta - (\beta - c\alpha) = c\alpha + (c_\beta - 1)\beta + \sum_{\gamma \neq \beta} c_\gamma \gamma$$

which is absurd since all summands on the right are non-negative, and the first is strictly positive. On the other hand, if $c_\beta < 1$ then the above equation implies that $(1 - c_\beta)\beta$ is a non-negative linear combination of $\Delta \setminus \{\beta\}$ in contradiction to the minimal choice of Δ. Thus $-s_\alpha.\beta \in \Phi^+$, but then a similar argument leads to a contradiction. □

Proposition A.7 *Let Φ be an abstract root system. Then every positive system contains a unique base. Conversely, any base is contained in a unique positive system.*

Proof Let Φ^+ be a positive system in Φ and choose $\Delta \subseteq \Phi^+$ minimal with the property that any element of Φ^+ is a non-negative linear combination of elements in Δ. (Clearly such a subset exists.) It suffices to show that Δ is linearly independent. For this let $\sum_{\alpha \in \Delta} c_\alpha \alpha = 0$. With $\Delta' := \{\alpha \in \Delta \mid c_\alpha \geq 0\} \subseteq \Delta$, we can rewrite this as $\sum_{\alpha \in \Delta'} c_\alpha \alpha = -\sum_{\beta \notin \Delta'} c_\beta \beta =: v$ with non-negative coefficients on both sides. Then

$$0 \leq (v, v) = -\sum_{\alpha \in \Delta'} \sum_{\beta \notin \Delta'} c_\alpha c_\beta (\alpha, \beta) \leq 0$$

by Lemma A.6, so $v = 0$ and hence, since $c_\alpha \alpha \geq 0$ for all α, all $c_\alpha = 0$, showing linear independence of Δ. Clearly Δ is characterized as the set of roots in Φ^+ which are not expressible as a non-negative linear combination of more than one element of Φ^+, so it is unique.

Conversely, given a base Δ, choose a total ordering on Δ and define a total ordering on E by letting $\sum_{\alpha \in \Delta} c_\alpha \, \alpha$ be positive if and only if the first non-zero coefficient c_α (with respect to the ordering on Δ) is positive. Then the set of positive elements in Φ is a positive system containing Δ. Again, uniqueness is obvious since any positive system containing the base Δ must consist precisely of those roots which are non-negative linear combinations over Δ. □

In particular this shows that bases of root systems exist, and we can speak of *the* base in a positive system, or *the* positive system containing a given base. The elements of Δ are then also called the *simple roots* with respect to Φ^+.

We next aim to show that the Weyl group acts transitively on the set of bases. For this we need:

Lemma A.8 *Let Δ be a base of the root system Φ, and $\alpha \in \Delta$. Then α is the only positive root made negative by s_α, that is, $s_\alpha(\Phi^+ \setminus \{\alpha\}) \subseteq \Phi^+$.*

Proof By (R3) we have $s_\alpha(\Phi^+) \subset \Phi$. Any $\beta \in \Phi^+$ can be written as $\beta = \sum_{\alpha \in \Delta} c_\alpha \, \alpha$ with $c_\alpha \geq 0$. Then all coefficients of $s_\alpha.\beta = \beta - (\beta, \alpha^\vee)\alpha$ are still non-negative, except possibly the coefficient at α. By the defining property of a base, if $s_\alpha.\beta \in \Phi^-$, then necessarily $c_\gamma = 0$ for $\gamma \neq \alpha$, that is, β is a (positive) multiple of α. By (R2) this implies that $\beta = \alpha$. □

Proposition A.9 *Any two bases (respectively positive systems) of a root system Φ are conjugate under the Weyl group W of Φ.*

Proof By the existence and uniqueness assertions in Proposition A.7 it is enough to show this for positive systems. So let Φ_1^+, Φ_2^+ be two positive systems of Φ. We induct on $n := |\Phi_1^+ \cap -\Phi_2^+|$. If $n = 0$ there is nothing to show. Now assume that $n > 0$. Then the base Δ of Φ_1^+ cannot be contained in Φ_2^+, say $\alpha \in \Delta \cap -\Phi_2^+$. Then Lemma A.8 shows that $|s_\alpha.\Phi_1^+ \cap -\Phi_2^+| = n - 1$. The result now follows from the inductive hypothesis applied to $s_\alpha.\Phi_1^+, \Phi_2^+$. □

Definition A.10 Let Δ be a base of the root system Φ. The *height* of a root $\beta = \sum_{\alpha \in \Delta} c_\alpha \alpha \in \Phi$ (with respect to Δ) is defined as $\mathrm{ht}(\beta) := \sum_{\alpha \in \Delta} c_\alpha$.

Proposition A.11 *If Δ is a base, then $W = \langle s_\alpha \mid \alpha \in \Delta \rangle$. Furthermore, for every $\alpha \in \Phi$ there is $w \in W$ such that $w.\alpha \in \Delta$.*

Proof Let $W' := \langle s_\alpha \mid \alpha \in \Delta \rangle$, a subgroup of W. For any $\beta \in \Phi^+$, $W'.\beta \cap \Phi^+$ is a non-empty set of positive roots. Let $\gamma \in W'.\beta \cap \Phi^+$ of smallest height, say $\gamma = \sum_{\alpha \in \Delta} c_\alpha \alpha$ with $c_\alpha \geq 0$. Then

$$0 < (\gamma, \gamma) = \sum_{\alpha \in \Delta} c_\alpha (\gamma, \alpha),$$

whence $(\gamma, \alpha) > 0$ for some α. We claim that $\gamma = \alpha$. Otherwise $s_\alpha.\gamma$ is positive by Lemma A.8, but on the other hand it is obtained from γ by subtracting a positive multiple of α, so it is of strictly smaller height than γ. Since $s_\alpha.\gamma \in W'.\beta \cap \Phi^+$ this contradicts the choice of γ. So we have shown that $\gamma = \alpha \in \Delta$.

In particular the W'-orbit of any $\beta \in \Phi^+$ contains a simple root, so $\Phi^+ \subseteq W'.\Delta$. Since $s_\alpha.\alpha = -\alpha$ we conclude that in fact also $-\Phi^+ \subseteq W'.(-\Delta) \subseteq W'.\Delta$. Now take any generator s_β of W, where $\beta \in \Phi$. By the above we have $\beta = w.\alpha$ for some $w \in W'$, $\alpha \in \Delta$, so $s_\beta = s_{w.\alpha} = ws_\alpha w^{-1} \in W'$ by Lemma A.4, whence $W = W'$. $\square$

The reflections in $S := \{s_\alpha \mid \alpha \in \Delta\}$ along simple roots of Φ are called the *simple reflections* of W (with respect to Δ). As a direct consequence, we obtain the following integrality result, which for the first time uses axiom (R4).

Corollary A.12 *Let Δ be a base of Φ. Then every $\alpha \in \Phi$ is an integral linear combination of the roots in Δ.*

Proof The result is trivially true for the roots in Δ. Now let $\beta \in \Phi$ and $\alpha \in \Delta$. If the claim holds for β, then so it holds as well for $s_\alpha.\beta = \beta - 2\frac{(\beta,\alpha)}{(\alpha,\alpha)}\alpha$ by (R4). Since any $\beta \in \Phi$ is of the form $w.\gamma$ for some $w \in W$ and $\gamma \in \Delta$, and w can be written as a product of reflections s_α for $\alpha \in \Delta$ by Proposition A.11, the claim follows by induction. $\square$

A.2 Decomposition of root systems

Definition A.13 A non-empty root system Φ with base Δ is called *decomposable* if there exists a non-trivial partition $\Delta = \Delta_1 \sqcup \Delta_2$ such that

$(\alpha_1, \alpha_2) = 0$ for all $\alpha_i \in \Delta_i$, $i = 1, 2$; if no such decomposition exists then Φ is said to be *indecomposable*.

Note that by Proposition A.1 this notion does not depend on the choice of a W-invariant scalar product on E.

Proposition A.14 *Let Φ be decomposable with corresponding partition of a base $\Delta = \Delta_1 \sqcup \Delta_2$, and set $E_i := \mathbb{R}\Delta_i$, $\Phi_i := \Phi \cap E_i$, $i = 1, 2$. Then we have:*

(a) Φ_i *is a root system in E_i with base Δ_i, for $i = 1, 2$,*
(b) $\Phi = \Phi_1 \sqcup \Phi_2$, *and $(\alpha_1, \alpha_2) = 0$ for all $\alpha_1 \in \Phi_1, \alpha_2 \in \Phi_2$,*
(c) $W(\Phi) = W(\Phi_1) \times W(\Phi_2)$ *by letting $W(\Phi_i)$ act trivial on E_{3-i}.*

Proof By Proposition A.1 for $\alpha \in \Delta_1$, $\beta \in \Delta_2$ we have $s_\beta.\alpha = \alpha$, so s_α, s_β commute by Lemma A.4. Since W is generated by the s_α, $\alpha \in \Delta$, by Proposition A.11, any $w \in W$ can be written as a commuting product $w = w_1 w_2$ with $w_i \in W_i$, where $W_i := \langle s_\alpha \mid \alpha \in \Delta_i \rangle$, $i = 1, 2$. Now any $w \in W_i$ fixes all elements of Δ_{3-i}, and again by Proposition A.1 also $W_i.\Delta_i \subset E_i$. The intersection $W_1 \cap W_2$ thus fixes the basis Δ of E, hence is trivial, so $W = W_1 \times W_2$.

Furthermore, by Proposition A.11 we have

$$\Phi = W.\Delta = (W_1 \times W_2).(\Delta_1 \cup \Delta_2) = W_1.\Delta_1 \cup W_2.\Delta_2 \subset E_1 \cup E_2.$$

Since Δ is a basis of E, $E = E_1 \oplus E_2$, so $\Phi_i = \Phi \cap E_i = W_i.\Delta_i$, whence (b). Clearly axioms (R1)–(R4) are satisfied for $\Phi_i \subset E_i$, Δ_i is a base of Φ_i, and thus $W_i = W(\Phi_i)$ by Proposition A.11, showing (a) and (c). $\square$

Induction then gives:

Corollary A.15 *Any root system Φ can be decomposed uniquely (up to reordering) into a disjoint orthogonal union $\Phi_1 \sqcup \ldots \sqcup \Phi_r$ of indecomposable root systems Φ_i, and then $W(\Phi) \cong W(\Phi_1) \times \cdots \times W(\Phi_r)$.*

The Φ_i in Corollary A.15 are called the *indecomposable components* of the root system Φ.

Proposition A.16 *A root system $\Phi \subset E$ is indecomposable if and only if its Weyl group W acts irreducibly on E.*

Proof If Φ is decomposable then W is reducible by Proposition A.14(c). Conversely, if $E_1 < E$ is a non-trivial W-invariant subspace then we obtain a W-invariant orthogonal decomposition $E = E_1 \oplus E_2$, with $E_2 = E_1^\perp$. We claim that any $\alpha \in \Phi$ lies in either E_1 or E_2. Indeed, as s_α acts semisimply on

E there is a basis of E_1 consisting of eigenvectors of s_α. If the eigenvalue -1 occurs then, since the -1-eigenspace of s_α is spanned by α, we have $\alpha \in E_1$. Otherwise, $E_1 \le \ker(s_\alpha - 1)$. The same reasoning applies to E_2. Since clearly we can't have both $E_1, E_2 \le \ker(s_\alpha - 1)$, $\alpha \in E_i$ for some i. So $\Delta \subset E_1 \cup E_2$ gives an orthogonal (and non-trivial) decomposition of Δ. $\qquad\square$

Irreducible linear groups can preserve at most one scalar product up to scalars, hence for indecomposable root systems the W-invariant scalar product on E is unique up to scalars.

For roots $\alpha, \beta \in \Phi$ let's write

$$n(\alpha, \beta) := (\alpha, \beta^\vee) = 2\frac{(\alpha, \beta)}{(\beta, \beta)}.$$

By the crystallographic condition (R4) and our Proposition A.1 we have $n(\alpha, \beta) \in \mathbb{Z}$, and by elementary geometry,

$$n(\alpha, \beta)n(\beta, \alpha) = \frac{4(\alpha, \beta)^2}{(\alpha, \alpha)(\beta, \beta)} = 4\cos^2 \angle(\alpha, \beta) \le 4,$$

where $\angle(\alpha, \beta)$ denotes the angle between α and β. Here, $n(\alpha, \beta)n(\beta, \alpha) = 4$ can only occur if $\cos \angle(\alpha, \beta) = \pm 1$, that is, when α, β are proportional. But in that case $\beta = \pm\alpha$ by (R2), so $n(\alpha, \beta) = n(\beta, \alpha) = \pm 2$. Thus, for $\beta \ne \pm\alpha$ we always have $|n(\alpha, \beta)| \le 3$.

The various possibilities can now easily be enumerated; they are given in Table A.1, up to interchanging α, β and up to replacing β by $-\beta$. We have also listed the order $o(s_\alpha s_\beta)$ of the product $s_\alpha s_\beta$.

Table A.1 *Relation between two roots*

$n(\alpha, \beta)$	$n(\beta, \alpha)$	$\angle(\alpha, \beta)$	lengths	$o(s_\alpha s_\beta)$
0	0	$\pi/2$	any	2
1	1	$\pi/3$	$(\alpha, \alpha) = (\beta, \beta)$	3
1	2	$\pi/4$	$2(\alpha, \alpha) = (\beta, \beta)$	4
1	3	$\pi/6$	$3(\alpha, \alpha) = (\beta, \beta)$	6
2	2	0	$\alpha = \beta$	1

From this, the various two-dimensional root systems can be classified. If $\Delta = \{\alpha, \beta\}$ denotes a base, then $(\alpha, \beta) \le 0$ by Lemma A.6 and thus either α, β are perpendicular, or $n(\alpha, \beta) = -1$ with $n(\beta, \alpha) \in \{-1, -2, -3\}$ and $\angle(\alpha, \beta)$ as in Table A.1. Application of the corresponding reflections gives the possibilities depicted in Table A.2.

Proposition A.17 *Any two-dimensional root system is of one of the four types in Table A.2, except that the two root lengths in type $A_1 \times A_1$ may be different.*

Table A.2 *The two-dimensional root systems*

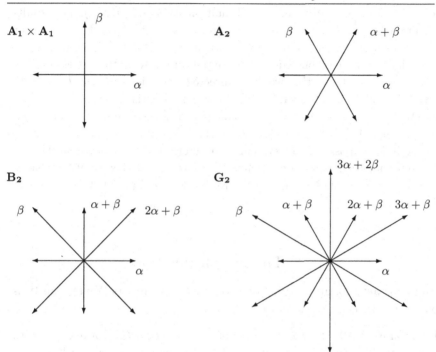

Corollary A.18 *Let Φ be indecomposable. Then:*

(a) *There are at most two different root lengths in Φ.*

(b) *All roots of Φ of the same length are conjugate under W.*

Proof As W acts irreducibly on E by Proposition A.16, the W-orbit of any $\alpha \in \Phi$ generates E. Thus, for any $\alpha, \beta \in \Phi$ there exists a W-conjugate β' of β (hence of the same length) with $(\alpha, \beta') \neq 0$. Now by the above considerations $(\beta', \beta')/(\alpha, \alpha) \in \{1, 2, 3, 1/2, 1/3\}$, so we may rescale the scalar product such that $(\alpha, \alpha) = 1$ for some $\alpha \in \Phi$, and all squared root lengths are among $1, 2, 3$. By what we saw before, neither $2/3$ nor $3/2$ is a quotient of squared root lengths, so not both squared root lengths 2, 3 may occur.

For the second assertion, see Exercise A.3. $\qquad\qquad\square$

The roots $\alpha \in \Phi$ of maximal length are called *long roots*. If there are two distinct root lengths, then the shorter ones are called *short roots*.

We now introduce a convenient description of root systems. By Proposition A.11 any root system can be recovered from a base. The base is encoded in terms of a directed graph which we now describe. Its nodes are in bijection with the elements of a base Δ, and two different nodes corresponding to $\alpha, \beta \in \Delta$ are joined by an edge of multiplicity $n(\alpha, \beta)n(\beta, \alpha)$. The resulting graph is called the *Coxeter diagram* associated to Φ or W. It does not determine Φ uniquely. So in addition, whenever α, β are of different lengths and joined by at least one edge, then we put an arrow on this edge, pointing towards the shorter of the two. This directed graph is called the *Dynkin diagram* of Φ. It can be shown that the base can essentially be recovered from its Dynkin diagram (that is, up to changing lengths in different connected components of the diagram), see [33, 11.1].

The indecomposable root systems can be classified, see Theorem 9.6 and Table 9.1 for their Dynkin diagrams. We'll not repeat this proof here, since it is well documented in the literature, see for example [9, VI, §4] or [33, Thm. 11.4].

A.3 The length function

An important tool in the study of a root system Φ and its Weyl group W is played by the length function.

Definition A.19 Let Δ be a base of Φ. The *length* $\ell(w)$ of $w \in W$ (with respect to Δ) is the minimal integer l such that $w = s_{\alpha_1} \cdots s_{\alpha_l}$ with $\alpha_i \in \Delta$. Any such expression of w of minimal length is called *reduced*.

It is clear that $\ell(ws_\alpha) = \ell(w) \pm 1$ for $w \in W$, $\alpha \in \Delta$, and also $\ell(w) = \ell(w^{-1})$ since all s_α are involutions.

We define a second function on W by

$$n(w) := |\{\beta \in \Phi^+ \mid w.\beta \in \Phi^-\}|,$$

the number of positive roots made negative by $w \in W$. It will turn out that this coincides with the length $\ell(w)$ introduced above. For this we need:

Lemma A.20 *Let $\alpha \in \Delta$ and $w \in W$. Then*

$$n(ws_\alpha) = \begin{cases} n(w) + 1 & \text{if } w.\alpha \in \Phi^+, \\ n(w) - 1 & \text{if } w.\alpha \in \Phi^-. \end{cases}$$

Proof Write $N(w) := (w.\Phi^+) \cap \Phi^-$, so that $n(w) = |N(w)|$. Then by Lemma A.8 we have

$$N(ws_\alpha) = (ws_\alpha.\Phi^+) \cap \Phi^- = w.\left(\Phi^+ \setminus \{\alpha\} \sqcup \{-\alpha\}\right) \cap \Phi^-.$$

Thus, if $w.\alpha \in \Phi^+$ then

$$N(ws_\alpha) = (w.(\Phi^+ \setminus \{\alpha\}) \sqcup \{-w.\alpha\}) \cap \Phi^- = N(w) \sqcup \{-w.\alpha\},$$

whence the result. On the other hand, if $w.\alpha \in \Phi^-$ then this shows that $N(ws_\alpha) = N(w) \setminus \{w.\alpha\}$. □

Proposition A.21 *We have $\ell(w) = n(w)$ for all $w \in W$.*

Proof Let $w \in W$. An easy induction, starting with $w = 1$ and using Lemma A.20 shows that $n(w) \le \ell(w)$ for all $w \in W$. Assume that $n(w) < \ell(w)$ and choose a reduced expression $w = s_1 \cdots s_l$ with $s_i = s_{\alpha_i}$ for $\alpha_i \in \Delta$ and $l = \ell(w)$. According to Lemma A.20 there is $j < l$ with $n(s_1 \cdots s_{j+1}) < n(s_1 \cdots s_j)$ and $s_1 \cdots s_j.\alpha_{j+1} \in \Phi^-$. Since $\alpha_{j+1} \in \Phi^+$ there is $i \le j$ such that $s_i \cdots s_j.\alpha_{j+1} \in \Phi^-$ and $s_{i+1} \cdots s_j.\alpha_{j+1} \in \Phi^+$, so that $s_{i+1} \cdots s_j.\alpha_{j+1} = \alpha_i$ by Lemma A.8. With $u := s_{i+1} \cdots s_j$ this gives

$$s_{i+1} \cdots s_j \cdot s_{j+1} \cdot s_j \cdots s_{i+1} = us_{j+1}u^{-1} = s_{u.\alpha_{j+1}} = s_{\alpha_i} = s_i$$

by Lemma A.4, so $s_{i+1} \cdots s_{j+1} = s_i \cdots s_j$. Substituting the right-hand side for the left in w and using $s_i^2 = 1$ we find a shorter expression for w, a contradiction. Thus $n(w) = \ell(w)$ as claimed. □

Theorem A.22 *W acts simply transitively on the set of bases (respectively positive systems) in Φ, that is, the only element of W fixing a given base setwise is the identity.*

Proof The transitivity was already shown in Proposition A.9. Now if $w \in W$ stabilizes the base Δ, then clearly it fixes Φ^+, whence $\ell(w) = n(w) = 0$ by Proposition A.21. □

Corollary A.23 *Let Δ be a base. Then there is a unique element w_0 of W with $w_0(\Delta) = -\Delta$. It is the element of maximal length in W, and $w_0^2 = 1$.*

Proof If Δ is a base of Φ, then clearly so is $-\Delta$, so by Theorem A.22 there exists a unique $w_0 \in W$ with $w_0(\Delta) = -\Delta$. Its length $\ell(w_0) = n(w_0) = |\Phi^+|$ is maximal by Proposition A.21. Since $w_0^2(\Delta) = \Delta$, the last claim follows. □

Definition A.24 The element w_0 in Corollary A.23 is called the *longest element* of W (with respect to Δ).

A.4 Parabolic subgroups

We next introduce a natural collection of subsystems of a root system Φ and corresponding subgroups of its Weyl group W, where a subset $\Psi \subset \Phi$ is called a *subsystem* if it is a root system in $\mathbb{R}\Psi$.

For this, let Δ denote a base and $S = \{s_\alpha \mid \alpha \in \Delta\}$ the set of simple reflections of W. For a subset $I \subseteq S$, $W_I := \langle s \in I \rangle$ is called a *standard parabolic subgroup* of W. A *parabolic subgroup* of W is any conjugate of a standard parabolic subgroup. We let $\Delta_I := \{\alpha \in \Delta \mid s_\alpha \in I\}$ and

$$\Phi_I := \Phi \cap \mathbb{Z}\Delta_I$$

be the corresponding *parabolic subsystem* of roots.

Proposition A.25 *Let $I \subseteq S$. Then Φ_I is a root system in $\mathbb{R}\Phi_I$ with base Δ_I and Weyl group W_I.*

Proof Conditions (R1)–(R4) are trivially satisfied, as Φ_I is a subset of Φ with $\Phi_I = -\Phi_I$ (for (R3) use the explicit formula for s_α in Proposition A.1). Now any root in Φ_I is a $\mathbb{Z}$-linear combination of roots in Δ_I. As Δ_I is contained in a base, the coefficients are either all positive or all negative. Thus Δ_I is a base of Φ_I. By Proposition A.11, the Weyl group of Φ_I is generated by the corresponding simple reflections, that is, by I, as claimed. $\square$

We have the following generalization of Lemma A.8 (which is the case $|I| = 1$):

Corollary A.26 *Let $I \subseteq S$. Then we have:*

(a) *The length function ℓ_I (with respect to Δ_i) on W_I is the restriction of the length function ℓ (with respect to Δ) on W.*
(b) *Let $w_I \in W_I$ be the longest element of W_I. Then*

$$w_I(\Phi_I^+) = \Phi_I^- \qquad \text{and} \qquad w_I(\Phi^+ \setminus \Phi_I^+) = \Phi^+ \setminus \Phi_I^+.$$

Proof Let $w \in W_I$. Then certainly $\ell(w) \leq \ell_I(w)$ since $I \subseteq S$. On the other hand we have $\ell_I(w) = n_I(w) \leq n(w) = \ell(w)$ by Proposition A.21.

The first part of (b) holds by the definition of the longest element and Proposition A.25. Since $n_I(w_I) = n(w_I)$ by (a), w_I cannot make any further roots negative, so $w_I(\Phi^+ \setminus \Phi_I^+) \subseteq \Phi^+$. The claim now follows as $w_I^2 = 1$. $\square$

We next study the action of W on E in some more detail. The set

$$C := \{v \in E \mid (v, \alpha) > 0 \text{ for all } \alpha \in \Delta\}$$

is called the *fundamental chamber* of W (with respect to Δ). We write

$$\bar{C} = \{v \in E \mid (v,\alpha) \geq 0 \text{ for all } \alpha \in \Delta\}$$

for the closure of C in the natural topology of E.

Theorem A.27 *The closure $\bar{C}$ of C is a fundamental domain for the action of W on E.*

Proof Let $v \in E$ and choose $w \in W$ such that $|\{\beta \in \Phi^+ \mid (w.v, \beta) < 0\}|$ is minimal. We claim that then $w.v \in \bar{C}$. Otherwise, $(w.v, \alpha) < 0$ for some $\alpha \in \Delta$, but then $(s_\alpha w.v, \alpha) = (w.v, s_\alpha.\alpha) = -(w.v, \alpha) > 0$, and by Lemma A.8 this contradicts our choice of w. Hence the W-orbit of any $v \in E$ has a representative in $\bar{C}$.

To finish the proof it is enough to show that when $v_1, v_2 \in \bar{C}$ with $w.v_1 = v_2$ for some $w \in W$ then $v_1 = v_2$. We argue by induction on $\ell(w)$, the case $\ell(w) = 0$ being trivial. If $\ell(w) > 0$ then by Proposition A.21 there is some simple root $\alpha \in \Delta$ made negative by w. By Lemma A.20 this means that $n(ws_\alpha) = n(w) - 1$. Now $v_1, v_2 \in \bar{C}$, so

$$0 \leq (v_1, \alpha) = (w.v_1, w.\alpha) = (v_2, w.\alpha) \leq 0,$$

whence $(v_1, \alpha) = 0$ and so $s_\alpha.v_1 = v_1$ by Proposition A.1. Thus $ws_\alpha.v_1 = v_2$. Since $\ell(ws_\alpha) = n(ws_\alpha) < \ell(w)$ we conclude that $v_1 = v_2$ by the induction hypothesis. $\qquad\square$

The above proof actually showed a bit more: choosing $v = v_1 = v_2$ we saw that any $w \in W$ fixing $v \in \bar{C}$ is a product of simple reflections which fix v:

Corollary A.28 *Let $v \in \bar{C}$. Then $C_W(v) = W_I$ with $I = \{s \in S \mid s.v = v\}$, a standard parabolic subgroup of W.*

More generally this shows the following characterization of parabolic subgroups:

Corollary A.29 *Let $M \subseteq E$. Then the pointwise stabilizer $C_W(M)$ of M in W is a parabolic subgroup. Conversely, any parabolic subgroup $H \leq W$ is the centralizer of its fixed space, that is, $H = C_W(V^H)$.*

Proof Clearly, the pointwise stabilizer of M is the same as that of its linear span, so we may replace M by a basis of the latter. Now let $v \in M$. We may assume that $v \in \bar{C}$ by Theorem A.27 and then $H = C_W(v)$ is a parabolic subgroup by the above observation. Now, by induction on $|M|$, the centralizer in H of $M \setminus \{v\}$ is a parabolic subgroup of H, hence of W, as claimed.

For the converse, note that for $I \subseteq S$ the fixed space $M := V^{W_I}$ of the corresponding parabolic subgroup W_I satisfies $\dim M = \dim V - |I|$, since W is generated by $|S| = \dim V$ reflections. Then $W_I \le C_W(M) = W_J$ for some $I \subseteq J \subseteq S$ by the first part, and comparison of dimensions shows that $I = J$, so $C_W(M) = W_I$. $\square$

Recall the longest element w_0 of W from Definition A.24.

Proposition A.30   *Let W be the Weyl group of a root system Φ of rank l. Then $-\mathrm{id} \in W$ if and only if Φ contains a subset of l mutually orthogonal roots.*

Proof   First note that if $-\mathrm{id} \in W$ then necessarily $-\mathrm{id} = w_0$ by Corollary A.23. Now if $\alpha_1, \dots, \alpha_l \in \Phi$ are mutually orthogonal, then the corresponding reflections $s_1, \dots, s_l$ commute. Thus by Proposition A.1 with $w := s_1 \cdots s_l$ we have $w.\alpha_i = -\alpha_i$ for $i = 1, \dots, l$, whence $w = -\mathrm{id} \in W$.

Conversely, if $w_0 = -\mathrm{id}$ let $\alpha \in \Phi$ with corresponding reflection s_α. The stabilizer W_1 of α in W is a parabolic subgroup of W by Corollary A.29. Any reflection in W_1 fixes α, so the root system Ψ of W_1 (see Proposition A.25) lies in $E_1 = \langle \alpha \rangle^\perp$, and $s_\alpha w_0 \in W_1$ acts as $-\mathrm{id}$ on E_1. The claim now follows by induction on the rank applied to $\Psi \subset E_1$. $\square$

Let $I \subseteq S$; for $\beta \in \Phi^+$ with

$$\beta = \sum_{\alpha \in \Delta_I} c_\alpha \alpha + \sum_{\gamma \in \Delta \setminus \Delta_I} d_\gamma \gamma$$

we call $(d_\gamma)_{\gamma \in \Delta \setminus \Delta_I}$ the *shape* of β. Note that W_I preserves the set of roots of a given shape. Let's note the following property:

Lemma A.31   *Let Φ be a root system with base Δ, and $I \subseteq S$. Then:*

(a) *Any two roots in $\Phi^+ \setminus \Phi_I$ of the same shape and the same length are conjugate under W_I.*

(b) *There is a unique root of minimal height of a given shape.*

Proof   Let $\beta, \gamma \in \Phi^+ \setminus \Phi_I$ have the same shape. We may assume that $\beta \ne \gamma$ are of minimal height in their respective W_I-orbits. Then $(\beta, \alpha) \le 0$, $(\gamma, \alpha) \le 0$ for all $\alpha \in \Delta_I$ by Proposition A.1. As $\beta \in \Phi^+ \setminus \Phi_I^+$ the set $\{\beta\} \cup \Delta_I$ consists of linearly independent positive roots. Since β, γ have the same shape, γ lies in the span of this set.

If $(\beta, \gamma) \le 0$, then $\{\beta, \gamma\} \cup \Delta_I$ would be a set of positive roots all of whose pairwise angles are obtuse, which would force them to be linearly independent. Thus $(\beta, \gamma) > 0$. Now first assume that β, γ are of the same length. Then $n(\beta, \gamma) = 1$ (see Table A.1) and so $s_\gamma(\beta) = \beta - \gamma =: \delta \in \Phi_I$, a root of the same length as β and γ. But then $s_\delta(\beta) = \beta - n(\beta, \delta)\delta = \beta - \delta = \gamma$, with $s_\delta \in W_I$, proving (a). In general, either $n(\beta, \gamma) = 1$ or $n(\gamma, \beta) = 1$, whence $\beta - \gamma \in \Phi$, a contradiction if β, γ have the same (minimal) height. $\quad\square$

Exercises

A.1 Let Φ be a root system with base Δ. Show that $\Phi^\vee$ is a root system with base $\Delta^\vee := \{\alpha^\vee \mid \alpha \in \Delta\}$.

A.2 Let Φ be a root system with dual root system $\Phi^\vee$. Then the Weyl groups of Φ and of $\Phi^\vee$ are isomorphic via $s_\alpha \mapsto s_{\alpha^\vee}$.

A.3 Let $\Phi \subset E$ be an indecomposable root system with Weyl group W and fix a W-invariant scalar product on E. Show the following: There are at most two W-orbits of roots in Φ, and any two roots of the same length are conjugate.

[**Hint:** First argue that given $\alpha, \beta \in \Phi$ we can always find $w \in W$ such that $\alpha, w.\beta$ are not orthogonal, then apply Proposition A.17 and Corollary A.18(a).]

A.4 Let Φ be an indecomposable root system with two different root lengths and let $p = (\alpha, \alpha)/(\beta, \beta)$ for $\alpha, \beta \in \Phi$ with α long and β short.

(a) The map $\alpha \mapsto \begin{cases} p\alpha^\vee & \alpha \text{ long,} \\ \alpha^\vee & \alpha \text{ short,} \end{cases}$ extends to a homothety.

(b) A root $\gamma = \sum_{\alpha \in \Delta} c_\alpha \alpha$ is long if and only if $p | c_\alpha$ for all short $\alpha \in \Delta$.

A.5 Let Φ be a root system, $\alpha, \beta \in \Phi$ with $\beta \neq -\alpha$. Show that there exists a base Δ of Φ with $\alpha \in \Delta$ such that $\beta \in \Phi^+$ (with respect to Δ).

A.6 Let Φ be a root system in E with Weyl group W. Show that the only reflections in W are of the form s_α for $\alpha \in \Phi$.

[**Hint:** You may want to use Corollary A.29.]

A.7 Let Φ be a root system with base Δ, $I \subsetneq S$ a subset of the corresponding set of simple reflections, and $\mathcal{S}$ a shape with respect to the subset I. Then there exists a unique root of maximal height of shape $\mathcal{S}$.

[**Hint:** Use Lemma A.31, and show that if β is of maximal height and of shape $\mathcal{S}$, then $w_I(\beta)$ is of minimal height and of shape $\mathcal{S}$, where w_I is the longest element of the parabolic subgroup W_I.]

Appendix B
Subsystems

In this appendix we collect various results on subsystems of an indecomposable root system Φ, including a classification of maximal subsystems. This turns out to be closely related to the so-called highest root of Φ which we introduce first. We follow mainly [74, §1] and [9, Ex. VI.4.4].

B.1 The highest root

Throughout $\Phi \subset E$ denotes an indecomposable root system with Weyl group W, $\Delta \subset \Phi$ a base with positive system Φ^+, and $(\ ,\ )$ a fixed W-invariant scalar product on E.

Recall from Section A.2 that for $\alpha, \beta \in \Phi$,

$$n(\alpha, \beta) = (\alpha, \beta^\vee) = 2\frac{(\alpha, \beta)}{(\beta, \beta)}$$

can only take values $0, \pm 1, \pm 2, \pm 3$.

Lemma B.1 *Let $\alpha, \beta \in \Phi$ with $\alpha \neq \pm\beta$.*

(a) *If $(\alpha, \beta) > 0$ then $\alpha - \beta \in \Phi$;*
(b) *if $(\alpha, \beta) < 0$ then $\alpha + \beta \in \Phi$.*

Proof By assumption α, β are neither perpendicular nor proportional, hence at least one of $n(\alpha, \beta), n(\beta, \alpha)$ is equal to ± 1 by Table A.1

Now if $(\alpha, \beta) > 0$ then moreover both terms are positive. If $n(\alpha, \beta) = 1$ then Proposition A.1 shows that $s_\beta.\alpha = \alpha - \beta \in \Phi$, while $s_\alpha.\beta = \beta - \alpha \in \Phi$ when $n(\beta, \alpha) = 1$. By (R2) this proves (a). The second part is an immediate consequence by replacing β by $-\beta$ in (a). $\square$

Corollary B.2 *Let* $\beta = \sum_{\alpha \in \Delta} c_\alpha \alpha \in \Phi^+$ *be a positive root. Then for all* $\alpha \in \Delta$ *and all* $1 \leq j \leq c_\alpha$ *there exists a root* $\gamma \in \Phi^+$ *whose coefficient at* α *equals* j.

Proof We have $0 < (\beta, \beta) = \sum_{\alpha \in \Delta} c_\alpha(\beta, \alpha)$, so there is $\alpha \in \Delta$ with $c_\alpha > 0$ and $(\beta, \alpha) > 0$. If $\beta \in \Delta$ there is nothing to prove. Else, $\beta \neq \alpha$, so $\beta_1 := \beta - \alpha \in \Phi^+$ by the preceding result. The coefficient vector of β_1 differs from that of β only at α, by 1. A straightforward induction now completes the proof. □

Proposition B.3 *Let* $\alpha, \beta \in \Phi$ *with* $\alpha \neq \pm\beta$. *Then:*

(a) *The set of integers* $I := \{i \mid \alpha + i\beta \in \Phi\}$ *forms an interval* $[-q, p]$ *containing 0.*
(b) *We have* $p - q = -n(\beta, \alpha)$ *and thus* $s_\alpha.(\beta + p\alpha) = \beta - q\alpha$.

Proof Let $q := -\min I$, $p := \max I$. Since $0 \in I$ we have $p, q \geq 0$. Assume that I is not the interval $[-q, p]$. Then there exist $r, t \in I$ with $t > r + 1$ and $r + i \notin I$ for $1 \leq i \leq t - r - 1$. As $(\beta + r\alpha) + \alpha, (\beta + t\alpha) - \alpha \notin \Phi$, Lemma B.1 gives

$$0 \leq (\alpha, \beta + r\alpha) < (\alpha, \beta + r\alpha) + (\alpha, (t - r)\alpha) = (\alpha, \beta + t\alpha) \leq 0,$$

which is not possible. This proves (a).

We have $s_\alpha.(\beta + i\alpha) = \beta - n(\beta, \alpha)\alpha - i\alpha = \beta + (-i - n(\beta, \alpha))\alpha$, so $-i - n(\beta, \alpha) \in I$, and $i \mapsto -i - n(\beta, \alpha)$ is an order reversing involution on I. Thus, for $i = p$ we get $-p - n(\beta, \alpha) = -q$, showing (b). □

The choice of a base Δ of Φ induces a partial order relation on E and hence on Φ as follows: we say that μ is smaller than λ if $\lambda - \mu$ is a non-negative linear combination of positive roots.

Lemma B.4 *Let* $v \in E$. *Then there exists a unique maximal element* $v' \in W.v$ *with respect to the partial order introduced before, and* $v' \in \bar{C}$.

Proof Let v' be a maximal element in the orbit $W.v$. We claim that $v' \in \bar{C}$. Indeed, if $(v', \alpha) < 0$ for some $\alpha \in \Delta$ then $s_\alpha.v' = v' - 2(v', \alpha)/(\alpha, \alpha)\alpha > v'$, contradicting the maximality of v'. By Theorem A.27 this implies the uniqueness of v'. □

The maximal element of Φ with respect to this order plays an important role:

Proposition B.5 *Let* Φ *be an indecomposable root system with base* Δ. *Then there exists a unique maximal root* $\alpha_0 \in \Phi^+$ *in the above partial order,* *that is, with the following property. Writing $\alpha_0 = \sum_{\alpha \in \Delta} n_\alpha \, \alpha$, then for every root $\beta = \sum_{\alpha \in \Delta} c_\alpha \, \alpha \in \Phi^+$ we have $c_\alpha \leq n_\alpha$ for all $\alpha \in \Delta$.*

Proof Let $\alpha_0 = \sum_{\alpha \in \Delta} n_\alpha \, \alpha \in \Phi^+$ be maximal with respect to the partial order. Since $\alpha_0 \in \Phi^+$, all n_α are non-negative. Let $\Delta' \subseteq \Delta$ be the set of simple roots α for which $n_\alpha > 0$. Clearly $\Delta' \neq \emptyset$. If $\Delta' \neq \Delta$ then by Lemma A.6 the indecomposability of Φ gives the existence of $\alpha_1 \in \Delta'$ and $\alpha_2 \in \Delta \setminus \Delta'$ with $(\alpha_1, \alpha_2) < 0$. Then

$$(\alpha_0, \alpha_2) = \sum_{\alpha \in \Delta} n_\alpha(\alpha, \alpha_2) = \sum_{\alpha \in \Delta'} n_\alpha(\alpha, \alpha_2) < 0$$

by Lemma A.6, so $\alpha_0 + \alpha_2 \in \Phi$ by Lemma B.1, which is not possible. Thus, $n_\alpha > 0$ for all $\alpha \in \Delta$.

Now let $\beta = \sum_{\alpha \in \Delta} c_\alpha \, \alpha \in \Phi^+$ be another maximal root. Then Lemma B.1(b) shows that $(\beta, \alpha) \geq 0$ for all $\alpha \in \Delta$, and at least one of them is strictly positive. We conclude that $(\beta, \alpha_0) = \sum_{\alpha \in \Delta} n_\alpha(\beta, \alpha) > 0$. If $\alpha_0 - \beta \in \Phi$, then it's either positive or negative, both of which contradict the maximality of α_0, β. But then we must have $\beta = \alpha_0$ by Lemma B.1(a). $\square$

Definition B.6 The root α_0 above is called the *highest root* of Φ with respect to Δ (since, clearly it has the largest height (see Definition A.10) of all roots in Φ).

Corollary B.7 *Let Φ be indecomposable with highest root α_0. Then we have:*

(a) *$(\alpha_0, \alpha_0) \geq (\alpha, \alpha)$ for all $\alpha \in \Phi$, that is, α_0 is a long root.*

(b) *If $\alpha \in \Phi^+$ is different from α_0 then $n(\alpha, \alpha_0) \in \{0, 1\}$.*

Proof For (a) let $\alpha \in \Phi$. Since $\bar{C}$ is a fundamental domain for the action of W by Theorem A.27 we may assume that $\alpha \in \bar{C}$. Now $\alpha_0 - \alpha \geq 0$ by Proposition B.5, so $(\alpha_0 - \alpha, v) \geq 0$ for all $v \in \bar{C}$. In particular $(\alpha_0 - \alpha, \alpha_0) \geq 0$ and $(\alpha_0 - \alpha, \alpha) \geq 0$, which gives $(\alpha_0, \alpha_0) \geq (\alpha_0, \alpha) \geq (\alpha, \alpha)$.

In (b) we have $n(\alpha, \alpha_0) \in \{0, \pm 1\}$ by Table A.1. But since $\alpha_0 \in \bar{C}$ this integer is non-negative. $\square$

The coefficients of the highest root in the indecomposable root systems are given in Table B.1 (see for example [9, pp. 250–275]).

The coefficients of the highest root will come up again in the investigation of bad primes and torsion primes, see Section B.5.

Table B.1 *Highest roots of indecomposable root systems*

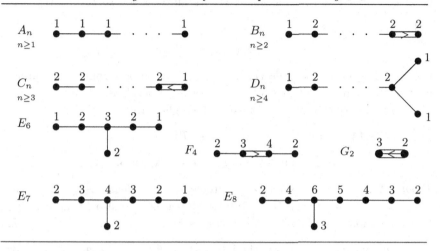

B.2 The affine Weyl group

We now consider the extension of the Weyl group of a root system by the group of affine translations in the coroots.

For $v \in E$ let t_v denote the corresponding affine translation

$$t_v : E \to E, \qquad u \mapsto u + v.$$

For $\alpha \in \Phi$ and $j \in \mathbb{Z}$ we define the *affine reflection*

$$s_{\alpha,j} : E \to E, \qquad s_{\alpha,j}(v) := v - ((v, \alpha) - j)\,\alpha^\vee.$$

The following is easy to see (see Exercise B.1):

Lemma B.8 *The transformation $s_{\alpha,j}$ satisfies $s_{\alpha,j} = t_{j\alpha^\vee}\, s_\alpha$, i.e., it is a reflection in the hyperplane $H_{\alpha,j} := \{v \in E \mid (v, \alpha) = j\}$ along $\alpha^\vee$.*

Definition B.9 The *affine Weyl group* W_{a} of a root system Φ is the group of affine transformations of E generated by all $s_{\alpha,j}$, for $\alpha \in \Phi$, $j \in \mathbb{Z}$.

Let $L := \mathbb{Z}\Phi^\vee$, the lattice generated by the coroots. Then we have the following structure result for W_{a}:

Proposition B.10 *The affine Weyl group is the semidirect product of the translation subgroup corresponding to L with W.*

Proof The Weyl group W normalizes L, and their intersection is trivial, so their product is a semidirect product. By Lemma B.8 all generators of W_{a} lie in $L.W$, but also $t_{\alpha^\vee} = s_{\alpha,1}s_\alpha \in W_{\mathrm{a}}$ for all $\alpha^\vee \in \Phi^\vee$, so $L.W = W_{\mathrm{a}}$. □

The set

$$A := \{v \in E \mid 0 \le (v,\alpha) \le 1 \text{ for all } \alpha \in \Phi^+\}$$

is called the *fundamental alcove*. It is easily seen to be a connected component of the complement in E of the reflecting hyperplanes $\cup_{\alpha,j}H_{\alpha,j}$. There is another description of this set using the highest root α_0 of Φ:

Lemma B.11   *Let Φ be indecomposable. Then*

$$A = \{v \in E \mid 0 \le (v,\alpha) \text{ for all } \alpha \in \Delta, \ (v,\alpha_0) \le 1\}.$$

Proof Write B for the set in the statement. Then clearly $A \subseteq B$. On the other hand, if $v \in B$ then $(v,\alpha) \ge 0$ for all $\alpha \in \Phi^+$, so also $(v,\alpha_0 - \alpha) \ge 0$ by Proposition B.5. Thus $(v,\alpha) \le (v,\alpha_0) \le 1$ and so $v \in A$.   □

The lemma shows that A is the intersection of the half-spaces defined by $0 \le (v,\alpha)$, $\alpha \in \Delta$, and $(v,\alpha_0) \le 1$. The corresponding hyperplanes are called the *walls* of the fundamental alcove A.

Proposition B.12   *Let Φ be indecomposable. Every W_{a}-orbit on E contains a point of A.*

Proof Let $v \in E$ and u be in the interior of A. The orbit of v under the translation subgroup L of W_{a} is clearly discrete in E. Since W_{a} is a finite extension of L by Proposition B.10, the W_{a}-orbit of v is also discrete. So there exists $w \in W_{\mathrm{a}}$ such that $w.v$ has minimal distance to u.

If $w.v \notin A$ there is a wall of A separating u and $w.v$. Let $s \in W_{\mathrm{a}}$ denote the reflection in that hyperplane. Then $w.v, sw.v, u, s.u$ form a trapezoid. Now in any trapezoid the length of any diagonal is greater than the lengths of the two non-parallel sides, so $(u - w.v, u - w.v) > (u - sw.v, u - sw.v)$ and $sw.v$ has smaller distance to u than $w.v$. But this contradicts the choice of $w \in W_{\mathrm{a}}$. Thus $w.v \in A$ and the claim is proved.   □

In fact, A is even a fundamental domain for the action of W_{a} on E, see for example [34, Thm. 4.8].

B.3 Closed subsystems

Definition B.13   Let Φ be a root system. A subset $\Psi \subseteq \Phi$ is said to be *closed* if

(C1) for all $\alpha, \beta \in \Psi$ we have $s_\alpha.\beta \in \Psi$, and

(C2) for $\alpha, \beta \in \Psi$ with $\alpha + \beta \in \Phi$, we have $\alpha + \beta \in \Psi$.

Proposition B.14 *A subset Ψ of a root system Φ is closed if and only if it satisfies (C2) and it is* symmetric, *that is, for all $\alpha \in \Psi$ we have $\mathbb{R}\alpha \cap \Psi = \{\pm\alpha\}$.*

Proof Let $\Psi \subseteq \Phi$ be symmetric with property (C2). Now for $\alpha, \beta \in \Psi$, $\alpha \neq \pm\beta$, we have $s_\alpha.\beta = \beta - n(\beta, \alpha)\alpha$. If $n(\beta, \alpha)$ is negative, then by Proposition B.3(a) all $\beta + i\alpha$ for $1 \leq i \leq -n(\beta, \alpha)$ are contained in Φ, hence in Ψ by (C2), which shows (C1). The same argument with β replaced by $-\beta$ applies when $n(\beta, \alpha) > 0$. The converse is clear. □

The proposition shows that closed subsets satisfy (R1)–(R4), and therefore are automatically subsystems.

Example B.15 Let Φ be a root system.

(1) If I is a subset of the set of simple reflections S of Φ, then Φ_I is a closed subset.

(2) More generally, if $\Psi \subseteq \Phi$ is any subset then $\Phi \cap \mathbb{Z}\Psi$ is closed in Φ.

(3) Let Φ be indecomposable. Any subsystem consisting only of long roots is closed. In particular, the subset Φ_l consisting of all long roots is a closed subsystem of Φ (see Exercise B.2).

(4) On the other hand, the subset Φ_s consisting of all short roots in Φ is, in general, not a closed subset although it is a subsystem (by the classification of two-dimensional root systems).

Proposition B.16 *The closed subsystems of Φ are precisely the sets $\Phi \cap H$ for H a subgroup of $\mathbb{Z}\Phi$ of finite index.*

Proof Let $H \leq \mathbb{Z}\Phi$ be a subgroup. Then clearly $\Phi \cap H$ is symmetric and satisfies (C2). For the converse we claim that for any finite subset M of a finitely generated abelian group A not containing 0 there is a subgroup of finite index having empty intersection with M. Indeed, let $v \in M$. By the structure theorem for finitely generated abelian groups there exists a generating system $e_1, \ldots, e_l$ of A such that $v = me_1$ for some $m \in \mathbb{N}$. Let n denote the order of v, respectively $n = 2$ when v has infinite order. Clearly, the subgroup $\langle mne_1, e_2, \ldots, e_l \rangle$ has finite index in A and does not contain v. The claim now follows by induction on $|M|$.

Now, for Ψ a closed subset, choosing $A = \mathbb{Z}\Phi/\mathbb{Z}\Psi$ and M the image of $\Phi \setminus \Psi$ in A we obtain a subgroup H as required. □

Proposition B.17   *Let Φ be indecomposable, $\Psi \subset \Phi$ be a maximal closed subsystem. Then either Ψ lies in a proper subspace of E, or there exists a W-conjugate Ψ' of Ψ and $\gamma \in \Delta$ such that $\Delta \setminus \{\gamma\} \cup \{\alpha_0\} \subseteq \Psi'$.*

Proof Let $H \le \mathbb{Z}\Phi$ of finite index such that $\Psi = \Phi \cap H$ as in Proposition B.16. Since Ψ is supposed to be maximal we may assume, replacing H by an overgroup if necessary, that $\mathbb{Z}\Phi/H$ is cyclic. Choosing adapted bases of $\mathbb{Z}\Phi$ and H as in the previous proof we see that there is $u \in E$ such that Ψ is the set of roots α for which $(u, \alpha) \in \mathbb{Z}$. By (R4) this condition is not affected by adding to u integral multiples of $\beta^\vee$, $\beta \in \Phi$. Application of elements of W sends Ψ to a conjugate maximal subsystem. So by Proposition B.12 we may assume that u lies in the alcove A, that is, $(u, \alpha) \ge 0$ for $\alpha \in \Delta$ and $(u, \alpha_0) \le 1$ by Lemma B.11.

Let $\Delta' := \{\alpha \in \Delta \mid (u, \alpha) \ne 0\}$. Since Ψ is proper, $\Delta' \ne \emptyset$. Now

$$1 \ge (u, \alpha_0) = \sum_{\alpha \in \Delta} n_\alpha(u, \alpha) = \sum_{\alpha \in \Delta'} n_\alpha(u, \alpha) > 0.$$

Thus, if $\beta = \sum_\alpha c_\alpha \alpha \in \Psi$, then

$$(u, \beta) = \sum_{\alpha \in \Delta'} c_\alpha(u, \alpha) \in \{0, 1\}$$

by Proposition B.5, and this is non-zero only if

$$\sum_{\alpha \in \Delta'} c_\alpha(u, \alpha) = \sum_{\alpha \in \Delta'} n_\alpha(u, \alpha) = 1,$$

that is, $c_\alpha = n_\alpha$ for $\alpha \in \Delta'$. So $\Psi \subseteq \mathbb{Z}(\Delta \setminus \Delta') + \mathbb{Z}(\sum_{\Delta'} n_\alpha \alpha)$ lies in a proper subspace unless $|\Delta'| = 1$.

If $\Delta' = \{\gamma\}$, then for $\beta = \sum_\alpha c_\alpha \alpha \in \Psi$ with $c_\gamma \ne 0$ we have

$$0 < (u, \beta) = c_\gamma(u, \gamma) \le 1,$$

so in fact $c_\gamma(u, \gamma) = 1$. In particular, $1 \le n_\gamma(u, \gamma) \le (u, \alpha_0) \le 1$, so $\alpha_0 \in \Psi$. $\qquad\square$

Theorem B.18 (Borel–de Siebenthal)   *Let Φ be an indecomposable root system with base $\Delta = \{\alpha_1, \dots, \alpha_l\}$ and highest root $\alpha_0 = \sum_{i=1}^l n_i \alpha_i$ with respect to Δ. Then the maximal closed subsystems of Φ up to conjugation by W are those with bases:*

(1) $\Delta \setminus \{\alpha_i\}$ *for $1 \le i \le l$ with $n_i = 1$, and*
(2) $\Delta \setminus \{\alpha_i\} \cup \{-\alpha_0\}$ *for $1 \le i \le l$ with n_i a prime.*

Proof For $1 \leq i \leq l$ let Φ_i be the subset of Φ consisting of the roots which are linear combinations of $\Delta \setminus \{\alpha_i\}$. This is a closed subsystem by Example B.15 with base $\Delta \setminus \{\alpha_i\}$ (see Proposition A.25). If $n_i = 1$ then by Proposition B.5 the coefficient at α_i of all roots $\beta \in \Phi \setminus \Phi_i$ equals ± 1. Thus any subgroup $H \leq \mathbb{Z}\Phi$ containing Φ_i and one of these roots must also contain α_i, and thus be equal to $\mathbb{Z}\Phi$. Hence Φ_i is maximal by the characterization of closed subsystems in Proposition B.16.

Next, for $1 \leq i \leq l$ with $n_i > 1$ consider the closed subset Ψ_i of Φ consisting of all $\mathbb{Z}$-linear combinations of $\Delta_i^0 := \Delta \setminus \{\alpha_i\} \cup \{-\alpha_0\}$ in Φ. Note that the coefficient at α_i for all roots in Ψ_i is divisible by n_i, hence equal to $0, \pm n_i$ by Proposition B.5. So Ψ_i is a subsystem of Φ lying properly between Φ_i and Φ, whence Φ_i is not maximal in this case. We claim that Δ_i^0 is a base of Ψ_i. Indeed, if $\beta = \sum_{j=1}^{l} c_j \alpha_j \in \Psi_i$ with $c_i = 0$, then it is a non-negative or non-positive linear combination of $\Delta \setminus \{\alpha_i\}$. If $c_i < 0$, then $c_i = -n_i$ and again $\beta = -\alpha_0 + \sum_{j \neq i}(n_j + c_j)\alpha_j$ is a non-negative linear combination of Δ_i^0 since $n_j + c_j \geq 0$ by Proposition B.5.

To investigate maximality of Ψ_i, first assume that $n_i = cd$ with $c, d > 1$. Then the set of all roots whose coefficient at α_i is an integral multiple of c is closed, strictly smaller than Φ (since it does not contain α_i), and contains Ψ_i. By Corollary B.2 there exist positive roots whose coefficient at α_i equals c, so the last containment is also strict and Ψ_i is not maximal in this case. On the other hand, if n_i is prime then obviously $\mathbb{Z}\Phi/\mathbb{Z}\Psi_i$ is cyclic of prime order n_i, so Ψ_i is maximal.

It remains to show that any maximal closed subsystem Ψ is of one of the above two forms. By the previous proposition either Ψ is contained in a proper subspace, and then it is contained in (and hence equal to) a parabolic subsystem by Corollary A.29, or it contains α_0 and all but one of the α_i. $\square$

Example B.19 Let Φ be a root system of type E_8. The corresponding Dynkin diagram is as shown in Figure B.1. We have also given the coefficients n_α of the highest root (see Table B.1).

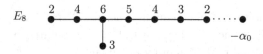

Figure B.1 Coefficients of the highest root of E_8.

Since $n_\alpha > 1$ for all $\alpha \in \Delta$ there are no maximal closed subsystems as in

Theorem B.18(1). The maximal closed subsystems as in Theorem B.18(2) are of types D_8, A_8, $A_4 \times A_4$, $E_6 \times A_2$ and $E_7 \times A_1$. (See also Exercise B.3.)

B.4 Other subsystems

We investigate the subsystems of root systems which are not necessarily closed.

Lemma B.20　*Let Φ be an indecomposable root system containing roots of two different lengths. Then $\Phi = \Phi \cap \mathbb{Z}\Phi_s$, that is, Φ is generated by its short roots.*

Proof　By Proposition A.11 any base Δ of Φ contains short and long roots. Since Φ is indecomposable, there is a short root α and a long root β in Δ with $(\alpha, \beta) \neq 0$. Then by Table A.2, $\Phi \cap (\mathbb{Z}\alpha + \mathbb{Z}\beta)$ is of type B_2 or G_2, and in both cases, all long roots are integral linear combinations of the short ones. Since W acts transitively on the set of long roots (see Corollary A.18(b)) this completes the proof.　　　　　　　　　　　　　　　　　　□

Proposition B.21　*Let Φ be an indecomposable root system, $\Psi \subset \Phi$ a maximal subsystem which is not closed. Then one of the following occurs:*

(1) *there is an indecomposable parabolic subsystem $\Psi_1 \subset \Phi$ containing short roots such that $\Psi = \Psi_1 \sqcup (\Phi \cap \Psi_1^{\perp})$ is decomposable; or*
(2) *the subsystem Ψ_s of Ψ consisting of its short roots is of maximal rank in Φ, $\Psi_s^{\vee}$ is closed in $\Psi^{\vee}$, and $\Phi \subset \mathbb{Z}\Psi_s$.*

Proof　First note that Ψ is of the same rank as Φ, since otherwise by Corollary A.29 it is contained in the proper (closed) parabolic subsystem of the parabolic subgroup pointwise fixing $\Psi^{\perp}$. Since Ψ is not closed, it contains some short root by Example B.15(3). So there is an indecomposable component Ψ_1 of Ψ containing short roots. If Ψ_1 is proper in Ψ, then

$$\Psi = \Psi_1 \sqcup (\Psi \cap \Psi_1^{\perp}) \subseteq (\Phi \cap \mathbb{Z}\Psi_1) \sqcup (\Phi \cap \Psi_1^{\perp}),$$

and by maximality we have equality, as in (1). Now suppose $\Psi = \Psi_1$ is indecomposable. Hence, by Lemma B.20 it is generated by its short roots, so Ψ_s is of maximal rank. Again by maximality, we have $\Phi \cap \mathbb{Z}\Psi_s = \Phi$ since Ψ is not closed. Since $\Psi_s^{\vee}$ is a subsystem of long roots in $\Psi^{\vee}$, it is closed by Example B.15(3).　　　　　　　　　　　　　　　　　　□

Note that the conditions in Proposition B.21 are necessary, but not always sufficient for a subsystem to be maximal. Still, in conjunction with Theorem B.18 it can be used to determine the maximal non-closed subsystems of each indecomposable root system in turn.

In Table B.2 we describe the subroot systems Φ_l, Φ_s of long, respectively short roots in the various indecomposable root systems.

Table B.2 *Long and short roots in indecomposable root systems*

Φ	A_n	B_n	C_n	D_n	E_6	E_7	E_8	F_4	G_2
Φ_l	A_n	D_n	A_1^n	D_n	E_6	E_7	E_8	D_4	A_2
Φ_s	$-$	A_1^n	D_n	$-$	$-$	$-$	$-$	D_4	A_2
$|\Phi|$	$n(n+1)$	$2n^2$	$2n^2$	$2n(n-1)$	72	126	240	48	12

Since duality of root systems (see Definition A.3) interchanges long and short roots, we see from the classification that type C_n is dual to type B_n, and all other indecomposable root systems are self-dual.

Example B.22 (Maximal non-closed subsystems)

(1) If Φ is indecomposable with only one root length, then any subsystem is automatically closed, by Example B.15(3).

(2) Let Φ be of type B_n. The indecomposable proper parabolic subsystems Ψ_1 containing short roots are of types B_r, $1 \leq r < n$, and then $\Psi_1 \sqcup (\Phi \cap \Psi_1^\perp)$ are maximal subsystems of type $B_r B_{n-r}$. As for the situation in Proposition B.21(2), Φ_s is of type A_1^n by Table B.2, without proper subsystems of maximal rank. The only indecomposable root system of rank n having fewer roots than B_n is A_n, but this does not contain a maximal rank subsystem of type A_1^n. Thus, the only maximal non-closed subsystems of B_n are of type $B_r B_{n-r}$, $1 \leq r < n$.

(3) Let Φ be of type C_n. This root system is dual to that of type B_n, so its maximal subsystems can be obtained as duals of those in B_n. The duals of parabolic subsystems are parabolic, hence closed. According to Theorem B.18 the closed maximal rank subsystems in type B_n are $D_r B_{n-r}$ with $2 \leq r \leq n$, with duals $D_r C_{n-r}$ in C_n. Finally the dual of the maximal non-closed subsystem $B_r B_{n-r}$ from part (2) is $C_r C_{n-r}$, which is closed. So here the maximal non-closed subsystems are of type $D_r C_{n-r}$, $2 \leq r \leq n$.

(4) Let Φ be of type F_4. The indecomposable proper parabolic subsystems containing short roots are of types A_1, A_2, C_3, which by Proposition B.21(1) lead to subsystems $A_1 B_3$, $A_2 A_2$, $C_3 A_1$, only the first of

which is not closed. Further, Φ_s is of type D_4, with only proper maximal rank subsystem A_1^4. The only indecomposable rank 4 system (apart from F_4) containing a maximal rank subsystem of type D_4 or A_1^4 consisting of long roots is B_4 by Theorem B.18, so dualizing we obtain a subsystem C_4.

(5) Let Φ be of type G_2. The only indecomposable proper parabolic subsystem containing short roots is of type A_1, with corresponding maximal subsystem of type $A_1 A_1$ which is closed. Further, Φ_s is of type A_2. This has no proper maximal rank subsystems and it is obvious that any larger subsystem equals Φ, so the only non-closed maximal subsystems of G_2 are of type A_2 (the short roots).

Corollary B.23 *Let Φ be an indecomposable root system. Then $-\mathrm{id} \in W$ if and only if Φ is of type*

$$A_1, \ B_n, \ C_n, \ D_n \ (n \ \mathrm{even}), \ E_7, \ E_8, \ F_4, \ \mathrm{or} \ G_2.$$

Proof This follows from the above description of subsystems with Proposition A.30.   □

In the remaining cases A_n ($n \geq 2$), D_n (n odd) and E_6, the longest element necessarily acts as the (unique) non-trivial graph automorphism of order 2 on the Dynkin diagram.

B.5 Bad primes and torsion primes

Here we investigate two types of primes which play special roles for a given indecomposable root system.

Definition B.24 Let Φ be a root system. A prime r is called *bad for* Φ if $\mathbb{Z}\Phi/\mathbb{Z}\Psi$ has r-torsion for some closed subsystem $\Psi \subseteq \Phi$.

Proposition B.25 *Let $\Psi \subseteq \Phi$ be closed. Then the bad primes of Ψ are among those for Φ.*

Proof If $\Psi_1 \subseteq \Psi$ is closed in Ψ, then it is also closed in Φ, and if $\mathbb{Z}\Psi/\mathbb{Z}\Psi_1$ has r-torsion then so has $\mathbb{Z}\Phi/\mathbb{Z}\Psi_1$.   □

Let Φ be indecomposable with base $\Delta = \{\alpha_1, \ldots, \alpha_l\}$ and highest root $\alpha_0 = \sum_{i=1}^{l} n_i \alpha_i$. We set $n_0 := 1$. Let

$$n(\Phi) = \max\{n_i \mid 1 \leq i \leq l\}$$

denote the largest coefficient of the highest root of Φ. Clearly, this is independent of the chosen base.

Proposition B.26 *Let Φ be indecomposable, $\Psi \subseteq \Phi$ a closed indecomposable subsystem. Then $n(\Psi) \leq n(\Phi)$.*

Proof Let Δ_1 be a base of Ψ. Extend this to a basis of E and choose an ordering on E with respect to which this basis is positive. Let Φ^+ be the positive system of Φ with respect to this ordering, with base Δ. Then all roots in Δ_1 are positive in this ordering, hence all elements of Δ_1 are non-negative integral combinations over Δ. Substituting these combinations into the expression for the highest root of Ψ shows the claim. $\square$

Lemma B.27 *Let Φ be indecomposable with $\Delta = \{\alpha_1, \ldots, \alpha_l\}$ labeled such that $\alpha_1, \ldots, \alpha_r$ is a sequence in Δ of minimal length $r \geq 0$ subject to*

(1) $n_r = n(\Phi)$ is maximal among the n_i, $1 \leq i \leq l$, and
(2) $(\alpha_i, \alpha_{i+1}) \neq 0$ for $i = 0, \ldots, r-1$.

Then for $i = 1, \ldots, r$ we have $(\alpha_i, \alpha_i) = (\alpha_0, \alpha_0)$ and $n_i = i + 1$.

Proof If $n_r = 1$ then $r = 0$ and there is nothing to prove. Now assume that $n_r > 1$ and let $0 \leq j < r$. Taking the scalar product of $\alpha_j^\vee$ with $\alpha_0 = \sum_{i=1}^l n_i \alpha_i$ we obtain

$$2 = (\alpha_0, \alpha_0^\vee) = \sum_{i=1}^l n_i (\alpha_i, \alpha_0^\vee) \geq n_1 (\alpha_1, \alpha_0^\vee) \qquad \text{and}$$

$$0 \leq (\alpha_0, \alpha_j^\vee) = \sum_{i=1}^l n_i (\alpha_i, \alpha_j^\vee) \leq \sum_{i=j-1}^{j+1} n_i (\alpha_i, \alpha_j^\vee) \qquad \text{for } j = 1, \ldots, r-1$$

since $(\alpha_0, \alpha_j^\vee) \geq 0$ by Corollary B.7(b) and $n_i (\alpha_i, \alpha_j^\vee) \leq 0$ for $i, j \geq 1$, $i \neq j$ by Lemma A.6. Using that $(\alpha_j, \alpha_{j+1}^\vee) \leq -1$ by (2) this yields

$$2 - n_1 \geq 0 \qquad \text{and} \qquad -n_{j-1} + 2n_j - n_{j+1} \geq 0 \qquad \text{for } j = 1, \ldots, r-1.$$

Adding up these equations gives $1 + n_{r-1} - n_r \geq 0$, hence $n_r = n_{r-1} + 1$ by the minimality of the sequence. Thus equality holds here, and hence in fact in all of the inequalities above. By induction this shows that $n_j = j + 1$. From $(\alpha_i, \alpha_{i\pm1}^\vee) = -1$ for $1 \leq i \leq r-1$, respectively $(\alpha_1, \alpha_0^\vee) = 1$, we get $(\alpha_{i-1}, \alpha_{i-1}) = (\alpha_i, \alpha_i)$ as claimed. $\square$

Corollary B.28 *Let Φ be an indecomposable root system with highest root $\alpha_0 = \sum_{\alpha \in \Delta} n_\alpha \alpha$ and r a prime. The following are equivalent:*

(i) r is bad for Φ.

(ii) r *equals some* n_α.

(iii) r *divides some* n_α.

(iv) $r \leq n_\alpha$ *for some* $\alpha \in \Delta$.

Proof The equivalence of (ii), (iii), and (iv) is a direct consequence of Lemma B.27. Now assume that r is bad for Φ. So there is a closed subsystem $\Psi \subseteq \Phi$ such that $\mathbb{Z}\Phi/\mathbb{Z}\Psi$ has r-torsion. Let $I \subseteq \Delta$ be minimal such that $\Psi \subseteq \Phi_I$ up to conjugation. Then by Theorem B.18 we have that $r = n_\alpha^I$ for some $\alpha \in I$, where n_α^I denotes the coefficient of α in the highest root of Φ_I. But $n_\alpha^I \leq n(\Psi) \leq n(\Phi)$ by Proposition B.26 so we get (iv).

Conversely, if $r = n_\alpha$ for some $\alpha \in \Delta$ then the maximal subsystem Ψ_α in Theorem B.18(2) gives rise to r-torsion. $\qquad\square$

This characterization of bad primes could also easily be verified "by inspection" of the various indecomposable root systems, see Table B.1. Steinberg [74, §1] has given the above proof relying on Lemma B.27 which does not use the classification.

We now come to the second type of primes:

Definition B.29 Let Φ be a root system with dual root system $\Phi^\vee$. A prime r is said to be a *torsion prime for* Φ if $\mathbb{Z}\Phi^\vee/\mathbb{Z}\Psi^\vee$ has r-torsion for some closed subsystem $\Psi \subseteq \Phi$.

Note that here we do not require that $\Psi^\vee$ be closed in $\Phi^\vee$.

Lemma B.30 *Let* Φ *be indecomposable with highest root* $\alpha_0 = \sum_\alpha n_\alpha \alpha$. *Then* $\alpha_0^\vee = \sum_\alpha n_\alpha^\vee \alpha^\vee$ *with* $n_\alpha^\vee = n_\alpha(\alpha, \alpha)/(\alpha_0, \alpha_0)$. *Moreover, if* $\beta \in \Phi^+$ *is long, and* $\beta^\vee = \sum_\alpha c_\alpha^\vee \alpha^\vee$, *then* $c_\alpha^\vee \leq n_\alpha^\vee$ *for all* $\alpha \in \Delta$.

Proof By definition, for $\beta = \sum_\alpha c_\alpha \alpha \in \Phi$ we have

$$\beta^\vee = 2 \sum_\alpha \frac{c_\alpha}{(\beta, \beta)} \alpha = \sum_\alpha c_\alpha \frac{(\alpha, \alpha)}{(\beta, \beta)} \alpha^\vee =: \sum_\alpha c_\alpha^\vee \alpha^\vee.$$

Since $c_\alpha \leq n_\alpha$ for all $\alpha \in \Delta$ by Proposition B.5 this shows that $c_\alpha^\vee \leq n_\alpha^\vee$ whenever $(\beta, \beta) = (\alpha_0, \alpha_0)$. $\qquad\square$

In particular, if Φ only contains roots of one fixed length, then $\alpha_0^\vee$ is the highest root of $\Phi^\vee$. Else, by definition $\beta \in \Phi$ is long if and only if $\beta^\vee \in \Phi^\vee$ is short, so the preceding result can be paraphrased as saying that $\alpha_0^\vee$ is the *highest short root* of $\Phi^\vee$. A list of highest short roots in the case where two different root lengths occur can readily be derived from Table B.1, see Table B.3.

We let $n^\vee(\Phi) := \max\{n_i^\vee \mid 1 \leq i \leq l\}$, the largest coefficient of the highest

Table B.3 *Highest short roots of indecomposable root systems*

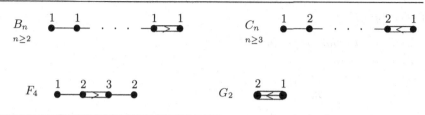

short root in the dual root system. Then we have the following analogue of Proposition B.26:

Proposition B.31 *Let Φ be indecomposable, $\Psi \subseteq \Phi$ a closed indecomposable subsystem. Then $n^{\vee}(\Psi) \leq n^{\vee}(\Phi)$.*

Proof As in the proof of Proposition B.26 we may choose compatible bases in Ψ and Φ. Then, by induction we may assume that Ψ is a maximal subsystem of Φ. Now let $\tilde{\alpha}_0$ denote the highest root of Ψ and write $\tilde{\alpha}_0 = \sum_i \tilde{n}_i \tilde{\alpha}_i$, where $\{\tilde{\alpha}_1, \ldots, \tilde{\alpha}_r\}$ is the chosen base of Ψ. Expressing $\tilde{\alpha}_i = \sum_j m_{ij}\alpha_j$ in the base $\{\alpha_1, \ldots, \alpha_l\}$ of Φ with non-negative integral coefficients, we have

$$\tilde{\alpha}_0 = \sum_j \Big(\sum_i \tilde{n}_i m_{ij} \Big) \alpha_j$$

and thus $\sum_i \tilde{n}_i m_{ij} \leq n_j$ for all j by Proposition B.5. Now sums of long roots are long, by Example B.15(3), so for any short root $\tilde{\alpha}_i$ there is some j with α_j short and $m_{ij} \neq 0$. Thus,

$$\max\{\tilde{n}_i \mid \tilde{\alpha}_i \text{ short}\} \leq \max\{n_j \mid \alpha_j \text{ short}\},$$

and similarly for long roots. Now if $\tilde{\alpha}_0$ has the same length as α_0 (hence is long) then by Lemma B.30 we find

$$n^{\vee}(\Psi) = \max\{\tilde{n}_i^{\vee}\} \leq \max\{n_j^{\vee}\} = n^{\vee}(\Phi).$$

Thus we may assume $\tilde{\alpha}_0$ is short, whence all roots in Ψ are short, and among the possible maximal closed subsystems described in Theorem B.18 this can only happen when $\Psi = \mathbb{Z}(\Delta \setminus \{\alpha_i\}) \cap \Phi$ for some i, with α_i long and all other simple roots short. Let s denote the reflection in $\tilde{\alpha}_0$. Then $\beta := s.\alpha_i = \alpha_i - (\alpha_i, \tilde{\alpha}_0^{\vee})\tilde{\alpha}_0$ is long, and since $\tilde{\alpha}_0$ is a positive linear combination in $\Delta \setminus \{\alpha_i\}$, $(\alpha_i, \tilde{\alpha}_0^{\vee}) < 0$ by Lemma A.6. So the coefficients of $\tilde{\alpha}_0^{\vee}$ are dominated by those of $\beta^{\vee}$, whose coefficients in turn are dominated by those of $\alpha_0^{\vee}$, because both β, α_0 are long. $\qquad\square$

Proposition B.32 *Let Φ be an indecomposable root system with highest short root $\alpha_0^\vee = \sum_\alpha n_\alpha^\vee \alpha^\vee$ of the dual root system and r a prime. The following are equivalent:*

(i) *r is a torsion prime for Φ.*
(ii) *r equals some $n_\alpha^\vee$.*
(iii) *r divides some $n_\alpha^\vee$.*
(iv) *$r \le n_\alpha^\vee$ for some $\alpha \in \Delta$.*

Proof The proof is completely analogous to that of Corollary B.28, using Proposition B.31 in place of Proposition B.26. ☐

The lists of bad primes and torsion primes for the various indecomposable root systems can now be extracted from the highest roots in Tables B.1 and B.3. They are given in Table 14.1.

Exercises

B.1 Let Φ be a root system and $\alpha \in \Phi$. Show that the affine transformation $s_{\alpha,j}$ is a reflection in the hyperplane $H_{\alpha,j} := \{v \in E \mid (v, \alpha) = j\}$ along $\alpha^\vee$ and that $s_{\alpha,j} = t_{j\alpha^\vee} s_\alpha$.

B.2 Let Φ be an indecomposable root system, Ψ a subsystem consisting only of long roots. Then Ψ is closed in Φ.
[**Hint**: Use the classification of two-dimensional root systems in Proposition A.17.]

B.3 Consider the root system Φ of type E_8 (see Example B.19). By Theorem B.18, the subsystems Ψ_i generated by $\Delta \setminus \{\alpha_i\} \cup \{-\alpha_0\}$ are not maximal when n_i is not prime. In each such case find a (base of a) maximal subsystem containing Ψ_i.

B.4 Let Φ be a root system, S the set of simple reflections corresponding to some base of Φ.

(a) Show that $\mathbb{Z}\Phi/\mathbb{Z}\Phi_I$ has no torsion, for any $I \subseteq S$.
(b) Conclude that Φ of type A_n has no bad primes nor torsion primes.
(c) Determine the torsion primes for Φ of type C_n.

Appendix C
Automorphisms of root systems

In this appendix we consider the following situation: $\Phi \subset E$ is a root system in the Euclidean space E with positive system Φ^+, and $F \in \mathrm{GL}(E)$ is a linear transformation which permutes Φ^+ up to positive multiples, that is, F stabilizes the set of half-lines $\{\mathbb{R}^+\alpha \mid \alpha \in \Phi^+\}$. Such automorphisms arise naturally in the study of finite groups of Lie type viewed as groups of fixed points under Steinberg endomorphisms in Part III.

Given these data there exists a permutation ρ of Φ stabilizing Φ^+ such that $F(\rho(\alpha)) = q_\alpha\alpha$ for all $\alpha \in \Phi$ and suitable $q_\alpha > 0$. By Proposition A.1, F permutes the reflections $\{s_\alpha \mid \alpha \in \Phi\}$, so normalizes the Weyl group W of Φ. Note that ρ also stabilizes the simple system Δ. Indeed, for all $\beta = \sum_{\alpha \in \Delta} c_\alpha\alpha \in \Phi^+$ we have that

$$\rho(\beta) = q_\beta F^{-1}(\beta) = q_\beta \sum_{\alpha \in \Delta} c_\alpha F^{-1}(\alpha) = q_\beta \sum_{\alpha \in \Delta} c_\alpha q_\alpha^{-1} \rho(\alpha) \in \Phi^+$$

is a non-negative linear combination of $\rho(\Delta)$, so $\rho(\Delta)$ is another base in Φ^+, hence equal to Δ by Proposition A.7. Thus, F also permutes the set $S = \{s_\alpha \mid \alpha \in \Delta\}$ of simple reflections of W and hence induces a graph automorphism of the corresponding Coxeter diagram.

We first investigate the centralizer of F in W. For $I \subseteq S$ we let $\Delta_I \subseteq \Delta$ denote the corresponding set of simple roots and $W_I = \langle I \rangle$ the associated parabolic subgroup of W. We write W^F for the centralizer of F in W.

Lemma C.1 *Let $F \in \mathrm{GL}(E)$ be as above.*

(a) *For each F-orbit $I \subseteq S$ we have $W_I^F = \langle w_I \rangle$, where w_I is the longest element of the parabolic subgroup W_I.*

(b) *The group of fixed points W^F is generated by $\{w_I \mid I \subseteq S$ an F-orbit$\}$.*

Proof Let $w_I \in W_I$ denote the longest element of W_I. Since I is an F-orbit,

F normalizes W_I and we have $F w_I F^{-1} \in W_I$. Since ρ also preserves Φ^+, $F w_I F^{-1}$ sends Φ_I^+ to Φ_I^-. But then by the uniqueness of the longest element (Corollary A.23) we must have $F w_I F^{-1} = w_I$, that is, $w_I \in W_I^F$.

Now let $w \in W^F$. If $w \neq 1$ then there exists $\alpha \in \Delta$ such that $w.\alpha \in \Phi^-$. But then also $w.\beta \in \Phi^-$ for all β in the ρ-orbit Δ_I of α. By Corollary A.26, w_I also sends Φ_I^+ to Φ_I^- but leaves all other positive roots positive, so

$$\ell(w w_I) = n(w w_I) = n(w) - n(w_I) = \ell(w) - \ell(w_I).$$

Thus $w w_I \in W^F$ has smaller length than w, and we conclude by induction that W^F is generated by the w_I. Applying this to W_I we obtain that $W_I^F = \langle w_I \rangle$. $\qquad \square$

The preceding result implies in particular that the Coxeter diagram of W^F can be obtained by a suitable folding process from the one of W, according to the graph automorphism induced by F. The various possibilities for W^F when W is irreducible and F is non-trivial are displayed in Table 23.1.

We now study the action of F on the root system. If δ denotes the order of ρ, then F^δ fixes all roots up to positive constants. To simplify exposition we make the further assumption that all these constants agree (which is the case for example if Φ is indecomposable). Then $F = q\phi$ for some $q > 0$ and an automorphism $\phi \in \mathrm{GL}(E)$ of order δ. Since $\langle W, \phi \rangle$ is finite, we may and will assume that the scalar product on E is $\langle W, \phi \rangle$-invariant. Consider the homomorphism

$$\pi : E \longrightarrow E^\phi, \qquad \chi \mapsto \frac{1}{\delta} \sum_{i=0}^{\delta-1} \phi^i(\chi);$$

this is clearly surjective and the identity on E^ϕ, hence a projection.

Lemma C.2 *Let* $\Omega := \{w(\Phi_I^+) \mid I \subseteq S \text{ an } F\text{-orbit}, \ w \in W^F\}$. *Then:*

(a) Ω *defines a partition of* Φ. *If* $\sim$ *denotes the associated equivalence relation, then* $\alpha \sim \beta$ *if and only if* $\pi(\alpha), \pi(\beta)$ *are positive multiples of each other.*

(b) *Each* $\omega \in \Omega$ *is contained in either* Φ^+ *or in* Φ^-.

Proof Let w_0 be the longest element of W. As F preserves the set of positive roots up to multiples, $F w_0 F^{-1}$ has the same property as w_0, so equals w_0. Hence $w_0 \in W^F$.

Now let $\alpha \in \Phi^+$. Then $w_0.\alpha \in \Phi^-$. Write $w_0 = w_1 \cdots w_r$, where $w_j = w_{I_j} \in W^F$ for suitable F-orbits $I_j \subseteq S$, by Lemma C.1(b). Then there exists $1 \leq j \leq r$ such that

$$w_{j+1} \cdots w_r.\alpha \in \Phi^+ \qquad \text{and} \qquad w_j w_{j+1} \cdots w_r.\alpha \in \Phi^-.$$

By Corollary A.26 all positive roots made negative by $w_j \in W_I$ lie in Φ_I^+, where $I := I_j$. Thus $w_{j+1} \cdots w_r.\alpha \in \Phi_I^+$, whence α is contained in $\omega :=$ $(w_{j+1} \cdots w_r)^{-1}(\Phi_I^+) \in \Omega$. As $w_0 \in W^F$, the same applies to negative roots.

Now suppose that $\alpha, \beta \in \Phi$ with $\pi(\beta) = a\pi(\alpha)$ for some $a > 0$. By the previous argument, $w.\alpha \in \Phi_I^+$ for some $w \in W^F$ and some F-orbit $I \subseteq S$. So $\pi(w.\alpha)$ is a non-negative linear combination of roots in Δ_I. However

$$\pi(w.\beta) = w.\pi(\beta) = w.(a\pi(\alpha)) = a\pi(w.\alpha),$$

so $\pi(w.\beta)$ also is a non-negative linear combination of roots in Δ_I. By the definition of π, this must also hold for $w.\beta$, so $w.\beta \in \Phi_I^+$ as well, whence α, β lie in the same equivalence class $w^{-1}(\Phi_I^+)$.

Finally, π is clearly constant on Δ_I for any F-orbit $I \subseteq S$. Now any root $\gamma \in \Phi_I^+$ is a non-negative linear combination of roots in Δ_I, so $\pi(\gamma)$ is a positive multiple of $\pi(\alpha)$, for $\alpha \in \Delta_I$. This completes the proof of (a).

For (b) let $I \subseteq S$ be an F-orbit and $\alpha \in \Delta_I$. If $w.\alpha \in \Phi^+$ for some $w \in W^F$, then also $w.\rho^i(\alpha) = \rho^i(w.\alpha) \in \Phi^+$ for all i since ρ preserves Φ^+ and commutes with W^F. Thus $w(\Delta_I) \subseteq \Phi^+$ and then also $w(\Phi_I^+) \subseteq \Phi^+$. $\qquad\square$

Lemma C.3 *Let $I \subseteq S$ be an F-orbit. Then the restriction to E^ϕ of the longest element $w_I \in W^F$ is a reflection in $\pi(\alpha)$ for all $\alpha \in \Delta_I$.*

Proof Let $\alpha \in \Delta_I$. Since $w_I(\Delta_I) = -\Delta_I$ we have

$$w_I.\pi(\alpha) = \frac{1}{\delta} \sum_{i=0}^{\delta-1} w_I.\phi^i(\alpha) = \frac{1}{\delta} \sum_{i=0}^{\delta-1} \phi^i(w_I.\alpha) = -\pi(\alpha).$$

On the other hand, if $v \in E^\phi$ satisfies $(\pi(\alpha), v) = 0$, then

$$0 = \sum_{i=0}^{\delta-1} (\phi^i(\alpha), v) = \sum_{i=0}^{\delta-1} (\alpha, v),$$

so $(\alpha, v) = (\phi^i(\alpha), v) = 0$ for all i. Since w_I is a product of reflections s_β, with $\beta \in \Delta_I$, this forces $w_I.v = v$. $\qquad\square$

Definition C.4 For each equivalence class ω for the equivalence relation $\sim$ defined in Lemma C.2 we let α_ω denote the vector of maximal length among the $\{\pi(\alpha) \mid \alpha \in \omega\}$. We set $\Phi_F := \{\alpha_\omega \mid \omega \in \Omega\}$ and $\Delta_F := \{\alpha_\omega \mid \omega \subseteq \Delta\} \subseteq \Phi_F$.

Theorem C.5 *The set Φ_F satisfies axioms (R1)–(R3) of an abstract root system in E^ϕ, with Weyl group W^F and base Δ_F.*

Proof Since Φ spans E, it is immediate that Φ_F spans E^ϕ. By construction, Φ_F is finite and none of the α_ω is zero. This shows (R1). By Lemma C.2(a) and the definition of Φ_F, if $\mu, \nu \in \Phi_F$ are positive multiples of each other, they are equal, proving (R2). Next, if $\mu \in \Delta_F$ then by Lemma C.3 there is $w \in W^F$ whose restriction to E^ϕ is the orthogonal reflection in μ. By Lemma A.4 we obtain reflections for arbitrary $\mu \in \Phi_F$. Then for $\beta \in \Phi$, $w \in W^F$, we have $w.\pi(\beta) = \pi(w.\beta) \in \Phi_F$, which proves (R3).

Each element of Φ is a non-negative or non-positive linear combination in Δ, so by construction every element of Φ_F is a non-negative or non-positive linear combination in Δ_F. Since Δ is linearly independent, so is Δ_F. Finally, the Weyl group of Φ_F is generated by the reflections in the $\alpha \in \Delta_F$ by Lemma C.1. □

For Φ indecomposable it can be shown case by case that Φ_F also satisfies the integrality axiom (R4) unless Φ is of type F_4 and ρ is non-trivial. See also Exercise C.2.

Exercises

C.1 (Automorphisms of root systems) Let Φ be a root system in the Euclidean space V and $\sigma \in \mathrm{GL}(V)$ stabilizing a positive system $\Phi^+ \subset \Phi$. Then σ induces a graph automorphism of the Dynkin diagram of Φ. [**Hint**: First check that σ stabilizes the base Δ in Φ^+. Then use the classification of root systems of rank 2 to see that both the number of edges between two roots $\alpha, \beta \in \Delta$, as well as the direction of the arrow, is already determined by the set $(\mathbb{Z}\alpha + \mathbb{Z}\beta) \cap \Phi$.]

C.2 Verify the structure of W^F and Φ_F for the cases occurring in Table 23.1.

References

[1] M. Aschbacher, On the maximal subgroups of the finite classical groups. *Invent. Math.* **76** (1984), 469–514.

[2] M. Aschbacher, *Finite Group Theory*. Second edition. Cambridge Studies in Advanced Mathematics, 10. Cambridge University Press, Cambridge, 2000.

[3] H. B. Azad, M. Barry, G. M. Seitz, On the structure of parabolic subgroups. *Comm. Algebra* **18** (1990), 551–562.

[4] A. Borel, *Linear Algebraic Groups*. Second edition. Graduate Texts in Mathematics, 126. Springer-Verlag, New York, 1991.

[5] A. Borel, J. de Siebenthal, Les sous-groupes fermés de rang maximum des groupes de Lie clos. *Comment. Math. Helv.* **23** (1949), 200–221.

[6] A. Borel, J. Tits, Groupes réductifs. *Inst. Hautes Études Sci. Publ. Math.* **27** (1965), 55–150.

[7] A. Borel, J. Tits, Eléments unipotents et sous-groupes paraboliques de groupes réductifs. I. *Invent. Math.* **12** (1971), 95–104.

[8] A. V. Borovik, The structure of finite subgroups of simple algebraic groups. (Russian) *Algebra i Logika* **28** (1989), 249–279, 366; translation in *Algebra and Logic* **28** (1989), 163–182 (1990).

[9] N. Bourbaki, *Groupes et Algèbres de Lie*. IV, V, VI, Hermann, Paris, 1968.

[10] M. Broué, G. Malle, Théorèmes de Sylow génériques pour les groupes réductifs sur les corps finis. *Math. Ann.* **292** (1992), 241–262.

[11] J. Brundan, Double coset density in classical algebraic groups. *Trans. Amer. Math. Soc.* **352** (2000), 1405–1436.

[12] M. Cabanes, Unicité du sous-groupe abélien distingué maximal dans certains sous-groupes de Sylow. *C. R. Acad. Sci. Paris Sér. I Math.* **318** (1994), 889–894.

[13] R. W. Carter, *Simple Groups of Lie Type*. John Wiley & Sons, London, 1972.

[14] R. W. Carter, *Finite Groups of Lie Type—Conjugacy Classes and Complex Characters*. John Wiley & Sons, New York, 1985.

[15] C. Chevalley, *Classification des Groupes Algébriques Semi-simples*. Collected Works. Vol. 3. Springer-Verlag, Berlin, 2005.

[16] E. Cline, B. Parshall, L. Scott, On the tensor product theorem for algebraic groups. *J. Algebra* **63** (1980), 264–267.

[17] A. M. Cohen, M. W. Liebeck, J. Saxl, G. M. Seitz, The local maximal subgroups of exceptional groups of Lie type, finite and algebraic. *Proc. London Math. Soc. (3)* **64** (1992), 21–48.

[18] B. N. Cooperstein, Maximal subgroups of $G_2(2^n)$. *J. Algebra* **70** (1981), 23–36.

[19] C. W. Curtis, I. Reiner, *Representation Theory of Finite Groups and Associative Algebras.* John Wiley & Sons, New York, 1962.

[20] D. I. Deriziotis, Centralizers of semisimple elements in a Chevalley group. *Comm. Algebra* **9** (1981), 1997–2014.

[21] J. A. Dieudonné, *La Géométrie des Groupes Classiques.* Third edition. Springer-Verlag, Berlin, 1971.

[22] F. Digne, J. Michel, *Fonctions L des Variétés de Deligne–Lusztig et Descente de Shintani.* Mém. Soc. Math. France (N.S.) **20** (1985).

[23] E. B. Dynkin, Semisimple subalgebras of semisimple Lie algebras. *Amer. Math. Soc., Transl., II. Ser.* **6** (1957), 111–243.

[24] E. B. Dynkin, Maximal subgroups of the classical groups. *Amer. Math. Soc., Transl., II. Ser.* **6** (1957), 245–378.

[25] A. Fröhlich, M. J. Taylor, *Algebraic Number Theory.* Cambridge Studies in Advanced Mathematics, 27. Cambridge University Press, Cambridge, 1993.

[26] M. Geck, *An Introduction to Algebraic Geometry and Algebraic Groups.* Oxford Graduate Texts in Mathematics, 10. Oxford University Press, Oxford, 2003.

[27] R. Goodman, N. Wallach, *Representations and Invariants of the Classical Groups.* Encyclopedia of Mathematics and its Applications, 68. Cambridge University Press, Cambridge, 1998.

[28] D. Gorenstein, *Finite Groups.* Chelsea Publishing Company, New York, 1980.

[29] D. Gorenstein, R. Lyons, R. Solomon, *The Classification of Finite Simple Groups, Number 3.* Mathematical Surveys and Monographs, 40. American Mathematical Society, Providence, RI, 1998.

[30] L. Grove, *Classical Groups and Geometric Algebra.* Graduate Studies in Mathematics, 39. American Mathematical Society, Providence, RI, 2002.

[31] G. Hiss, Die adjungierten Darstellungen der Chevalley-Gruppen. *Arch. Math. (Basel)* **42** (1984), 408–416.

[32] J. Humphreys, *Linear Algebraic Groups.* Graduate Texts in Mathematics, 21. Springer-Verlag, New York, 1975.

[33] J. Humphreys, *Introduction to Lie Algebras and Representation Theory.* Graduate Texts in Mathematics, 9. Springer-Verlag, New York, Second printing, 1980.

[34] J. Humphreys, *Reflection Groups and Coxeter Groups.* Cambridge University Press, Cambridge, 1992.

[35] J. Humphreys, *Conjugacy Classes in Semisimple Algebraic Groups.* Mathematical Surveys and Monographs, 43. American Mathematical Society, Providence, RI, 1995.

[36] J. Humphreys, *Modular Representations of Finite Groups of Lie Type.* LMS Lecture Notes Series, 326. Cambridge University Press, Cambridge, 2006.

[37] B. Huppert, *Endliche Gruppen. I.* Grundlehren der Mathematischen Wissenschaften, 134. Springer-Verlag, Berlin, 1967.

[38] B. Huppert, N. Blackburn, *Finite Groups. II.* Grundlehren der Mathematischen Wissenschaften, 242. Springer-Verlag, Berlin, 1982.

[39] I. M. Isaacs, *Character Theory of Finite Groups.* Dover, New York, 1994.

[40] J.C. Jantzen, Darstellungen halbeinfacher algebraischer Gruppen und zugeordnete kontravariante Formen. *Bonner math. Schr.* **67** (1973).

[41] J.C. Jantzen, *Representations of Algebraic Groups.* Second edition. Mathematical Surveys and Monographs, 107. American Mathematical Society, Providence, RI, 2003.

[42] P. Kleidman, The maximal subgroups of the Steinberg triality groups $^3D_4(q)$ and of their automorphism groups. *J. Algebra* **115** (1988), 182–199.

[43] P. Kleidman, The maximal subgroups of the Chevalley groups $G_2(q)$ with q odd, the Ree groups $^2G_2(q)$, and their automorphism groups. *J. Algebra* **117** (1988), 30–71.

[44] P. Kleidman, M. W. Liebeck, *The Subgroup Structure of the Finite Classical Groups.* London Mathematical Society Lecture Note Series, **129**. Cambridge University Press, Cambridge, 1990.

[45] V. Landazuri, G. M. Seitz, On the minimal degrees of projective representations of the finite Chevalley groups. *J. Algebra* **32** (1974), 418–443.

[46] R. Lawther, Sublattices generated by root differences, preprint.

[47] G. I. Lehrer, D. E. Taylor, *Unitary Reflection Groups.* Australian Mathematical Society Lecture Series, 20. Cambridge University Press, Cambridge, 2009.

[48] M. W. Liebeck, G. M. Seitz, Maximal subgroups of exceptional groups of Lie type, finite and algebraic. *Geom. Dedicata* **35** (1990), 353–387.

[49] M. W. Liebeck, G. M. Seitz, *Reductive Subgroups of Exceptional Algebraic Groups.* Memoirs Amer. Math. Soc., **121** (1996).

[50] M. W. Liebeck, G. M. Seitz, On the subgroup structure of classical groups. *Invent. Math.* **134** (1998), 427–453.

[51] M. W. Liebeck, G. M. Seitz, On the subgroup structure of exceptional groups of Lie type. *Trans. Amer. Math. Soc.* **350** (1998), 3409–3482.

[52] M.W. Liebeck, G.M. Seitz, A survey of maximal subgroups of exceptional groups of Lie type. *Groups, Combinatorics & Geometry* (Durham, 2001), World Sci. Publ., River Edge, NJ, 2003, pp. 139–146.

[53] M.W. Liebeck, G.M. Seitz, *The Maximal Subgroups of Positive Dimension in Exceptional Algebraic Groups.* Memoirs Amer. Math. Soc., **802** (2004).

[54] F. Lübeck, Small degree representations of finite Chevalley groups in defining characteristic. *LMS J. Comput. Math.* **4** (2001), 135–169.

[55] G. Malle, The maximal subgroups of $^2F_4(q^2)$. *J. Algebra* **139** (1991), 52–69.

[56] G. Malle, Height 0 characters of finite groups of Lie type. *Represent. Theory* **11** (2007), 192–220.

[57] R. Ree, A family of simple groups associated with the simple Lie algebra of type (F_4). *Amer. J. Math.* **83** (1961) 401–420.

[58] R. Ree, A family of simple groups associated with the simple Lie algebra of type (G_2). *Amer. J. Math.* **83** (1961), 432–462.

[59] R. W. Richardson, Finiteness theorems for orbits of algebraic groups. *Nederl. Akad. Wetensch. Indag. Math.* **47** (1985), 337–344.

[60] G. M. Seitz, *The Maximal Subgroups of Classical Algebraic Groups.* Memoirs Amer. Math. Soc., **67** (1987).

[61] G. M. Seitz, *Maximal Subgroups of Exceptional Algebraic Groups.* Memoirs Amer. Math. Soc., **90** (1991).

[62] G. M. Seitz, D. M. Testerman, Extending morphisms from finite to algebraic groups. *J. Algebra* **131** (1990), 559–574.

[63] S. D. Smith, Irreducible modules and parabolic subgroups. *J. Algebra* **75** (1982), 286–289.

[64] N. Spaltenstein, *Classes Unipotentes et Sous-Groupes de Borel.* Lecture Notes in Mathematics, 946. Springer-Verlag, Berlin, 1982.

[65] T. A. Springer, Regular elements of finite reflection groups. *Invent. Math.* **25** (1974), 159–198.

[66] T. A. Springer, *Linear Algebraic Groups.* Second edition. Progress in Mathematics, 9. Birkhäuser, Boston, 1998.

[67] T. A. Springer, R. Steinberg, Conjugacy classes. In: *Seminar on Algebraic Groups and Related Finite Groups.* Lecture Notes in Mathematics, 131. Springer-Verlag, Berlin, 1970, pp. 167–266.

[68] R. Steinberg, Variations on a theme of Chevalley. *Pacific J. Math.* **9** (1959), 875–891.

[69] R. Steinberg, Automorphisms of classical Lie algebras. *Pacific J. Math.* **11** (1961), 1119–1129.

[70] R. Steinberg, Representations of algebraic groups. *Nagoya Math. J.* **22** (1963), 33–56.

[71] R. Steinberg, Regular elements of semisimple algebraic groups. *Inst. Hautes Études Sci. Publ. Math.* **25** (1965), 49–80.

[72] R. Steinberg, *Endomorphisms of Linear Algebraic Groups.* Memoirs Amer. Math. Soc., **80** (1968).

[73] R. Steinberg, *Lectures on Chevalley Groups.* Notes prepared by J. Faulkner and R. Wilson. Yale University, New Haven, Conn., 1968.

[74] R. Steinberg, Torsion in reductive groups. *Advances in Math.* **15** (1975), 63–92.

[75] R. Steinberg, *Conjugacy Classes in Algebraic Groups.* Lecture Notes in Mathematics, Vol. 366. Springer-Verlag, Berlin, 1974.

[76] R. Steinberg, On theorems of Lie–Kolchin, Borel, and Lang. In: *Contributions to Algebra (Collection of Papers Dedicated to Ellis Kolchin).* Academic Press, New York, 1977, pp. 349–354.

[77] I. Suprunenko, Conditions on the irreducibility of restrictions of irreducible representations of the group $SL(n, K)$ to connected algebraic subgroups. Preprint #13, (222), Ins. Mat. Akad. Nauk BSSR (1985) (in Russian).

[78] M. Suzuki, On a class of doubly transitive groups. *Ann. of Math.* (2) **75** (1962), 105–145.

[79] D. E. Taylor, *The Geometry of the Classical Groups.* Heldermann Verlag, Berlin, 1992.

[80] D. Testerman, *Irreducible Subgroups of Exceptional Algebraic Groups.* Memoirs Amer. Math. Soc., **75** (1988).

[81] D. Testerman, A construction of certain maximal subgroups of the algebraic groups E_6 and F_4. *J. Algebra* **122** (1989), 299–322.

[82] J. Tits, Algebraic and abstract simple groups. *Ann. of Math.* (2) **80** (1964), 313–329.

Index

Printed in the United States
By Bookmasters